# Cases on Research-Based Teaching Methods in Science Education

Eugene de Silva
*Virginia Research Institute, USA & MRAS - Walters State Community College, USA*

A volume in the Advances in Educational Technologies and Instructional Design (AETID) Book Series

Information Science REFERENCE
An Imprint of IGI Global

Managing Director: Lindsay Johnston
Production Editor: Christina Henning
Development Editor: Erin O'Dea
Acquisitions Editor: Kayla Wolfe
Typesetter: Michael Brehm
Cover Design: Jason Mull

Published in the United States of America by
Information Science Reference (an imprint of IGI Global)
701 E. Chocolate Avenue
Hershey PA 17033
Tel: 717-533-8845
Fax: 717-533-8661
E-mail: cust@igi-global.com
Web site: http://www.igi-global.com



Library of Congress Cataloging-in-Publication Data

Cases on research-based teaching methods in science education / Eugene de Silva, editor.
pages cm
Includes bibliographical references and index.
ISBN 978-1-4666-6375-6 (hardcover) -- ISBN 978-1-4666-6376-3 (ebook) -- ISBN 978-1-4666-6378-7 (print & perpetual access) 1. Science--Study and teaching--United States. 2. Technology--Study and teaching--United States. 3. Science--Study and teaching--Great Britain. 4. Technology--Study and teaching--Great Britain. 5. Effective teaching. I. De Silva, Eugene, editor.
Q183.3.A1C387 2015
507.1--dc23

2014021075

This book is published in the IGI Global book series Advances in Educational Technologies and Instructional Design (AETID) (ISSN: 2326-8905; eISSN: 2326-8913)

British Cataloguing in Publication Data
A Cataloguing in Publication record for this book is available from the British Library.

# Advances in Educational Technologies and Instructional Design (AETID) Book Series

ISSN: 2326-8905
EISSN: 2326-8913

## Mission

Education has undergone, and continues to undergo, immense changes in the way it is enacted and distributed to both child and adult learners. From distance education, Massive-Open-Online-Courses (MOOCs), and electronic tablets in the classroom, technology is now an integral part of the educational experience and is also affecting the way educators communicate information to students.

The **Advances in Educational Technologies & Instructional Design (AETID) Book Series** is a resource where researchers, students, administrators, and educators alike can find the most updated research and theories regarding technology's integration within education and its effect on teaching as a practice.

## Coverage

- Virtual School Environments
- K-12 Educational Technologies
- Instructional Design
- E-Learning
- Social Media Effects on Education
- Hybrid Learning
- Classroom Response Systems
- Online Media in Classrooms
- Higher Education Technologies
- Curriculum Development

IGI Global is currently accepting manuscripts for publication within this series. To submit a proposal for a volume in this series, please contact our Acquisition Editors at Acquisitions@igi-global.com or visit: http://www.igi-global.com/publish/.

The Advances in Educational Technologies and Instructional Design (AETID) Book Series (ISSN 2326-8905) is published by IGI Global, 701 E. Chocolate Avenue, Hershey, PA 17033-1240, USA, www.igi-global.com. This series is composed of titles available for purchase individually; each title is edited to be contextually exclusive from any other title within the series. For pricing and ordering information please visit http://www.igi-global.com/book-series/advances-educational-technologies-instructional-design/73678. Postmaster: Send all address changes to above address. 

## Titles in this Series

*For a list of additional titles in this series, please visit: www.igi-global.com*

*Promoting Global Literacy Skills through Technology-Infused Teaching and Learning*
Jared Keengwe (University of North Dakota, USA) Justus G. Mbae (Catholic University of Eastern Africa, Kenya) and Simon K. Ngigi (Catholic University of Eastern Africa, Kenya)
Information Science Reference • copyright 2015 • 337pp • H/C (ISBN: 9781466663473) • US $185.00 (our price)

*Learning in Metaverses Co-Existing in Real Virtuality*
Eliane Schlemmer (ICAR, France) and Luciana Backes (ICAR, France)
Information Science Reference • copyright 2015 • 351pp • H/C (ISBN: 9781466663510) • US $180.00 (our price)

*Cases on Research-Based Teaching Methods in Science Education*
Eugene de Silva (Virginia Research Institute, USA & MRAS - Walters State Community College, USA)
Information Science Reference • copyright 2015 • 332pp • H/C (ISBN: 9781466663756) • US $195.00 (our price)

*Tablets in K-12 Education Integrated Experiences and Implications*
Heejung An (William Paterson University, USA) Sandra Alon (William Paterson University, USA) and David Fuentes (William Paterson University, USA)
Information Science Reference • copyright 2015 • 346pp • H/C (ISBN: 9781466663008) • US $175.00 (our price)

*E-Learning as a Socio-Cultural System A Multidimensional Analysis*
Vaiva Zuzevičiūtė (Mykolas Romeris University, Lithuania) Edita Butrimė (Lithuanian University of Health Sciences, Lithuania) Daiva Vitkutė-Adžgauskienė (Vytautas Magnus University, Lithuania) Vladislav Vladimirovich Fomin (Vytautas Magnus University, Lithuania) and Kathy Kikis-Papadakis (Foundation for Research and Technology, Greece)
Information Science Reference • copyright 2014 • 349pp • H/C (ISBN: 9781466661547) • US $195.00 (our price)

*Effects of Information Capitalism and Globalization on Teaching and Learning*
Blessing F. Adeoye (University of Lagos, Nigeria) and Lawrence Tomei (Robert Morris University, USA)
Information Science Reference • copyright 2014 • 300pp • H/C (ISBN: 9781466661622) • US $195.00 (our price)

# Table of Contents

# Detailed Table of Contents



## Section 1
## Curricula Changes



This review outlines a research-informed teaching case study based on adult learners of chemistry within the Scottish Further Education (FE) sector. It provides some strategies for success in supporting non-major chemistry students and provides some practical ways forward for improving attitudes to learning chemistry and in studying the subject further. An overview of why science, and in particular chemistry, may be perceived to be difficult to learn is discussed, as well as links to the evidence base that outlines the facilitation of meaningful and relevant learning. Through a chemistry curriculum redesign, discussions on the shift of attitudes, perceptions of learning, difficulties, and preferred topics in lessons are discussed. In addition, a general overview of the science education scene in Further Education is presented. Based on research evidence and educational neuroscience, there are suggested implications for educators in supporting adult students learning non-major science courses.

# Foreword

Education is a pivotal cornerstone of a successful society. Numerous books have been written on education and the need for change in the system, delivery, and management of education. The outcry for improvement, innovative teaching, and more interaction has been in the domain of school improvement systems over the past few decades. However, progress has been painfully slow with major resistance to change both by the policy makers and traditional teachers.

Over the last 30 years, I have observed that good teachers and professors disappear upon entering the administration. For a comfortable life, many educators compromise and become tick box specialists due to the nature of the university and school systems where the policies are driven by the number of students enrolled, number of students graduated, and the grants/funds received.

Whilst these are good economical and financial decisions, the education and teaching should be free of such throes and allowed to evolve and improve to be more effective for imparting knowledge and training to future generations. New methods and policies in education should be focused on the development of new teaching techniques and practices, funding more teacher training and recruitment of teachers, development of student friendly learning resources, effective student advice, and opportunities for students to shadow the expert professionals.

Teaching, in schools or colleges/universities, needs to change from the traditional teacher-centered system to student-centered approaches. In theory, this is recommended and to a certain extent practiced all over the world; however, there is still no real freedom for students and teachers to explore new ways of learning. Teaching and learning are severely restricted through well-placed policies by the administration.

The students, too, have become frustrated with the traditional teaching system but also find it daunting to experience such new teaching methods. I have seen excellent innovative teachers leave the profession due to rigid rules and regulations that restricted them from practicing new styles and methods of teaching. Students complained about the effectiveness of the new teaching methods since those methods challenged their established norms. However, if our educational system does not challenge the minds of students, if students are not compelled to seek new knowledge beyond classroom settings, and if they are not tested on new grounds, how could we ever expect the students to be truly effective and independent in their chosen professions?

This is why I was excited to read this book written by like-minded professors who not only believe in change to the educational system, but also have done something constructive to prove and push the policy decisions towards a novel, effective form of teaching. Science and mathematics are considered as the most difficult subjects by the majority of students due to the way they are taught in classroom settings. Yet, results of scientific discoveries and technological developments are refreshing and provided convenience and comfort enabling learning to take place at time and place of choice for students. An integrated link between the two is an ideal path to getting students to be more involved in seeking true knowledge in science and mathematics. Even science and mathematics alone will not survive in an ever-changing society. They need to be synergistically combined with arts, commerce, and businesses to be effective in the world arena. Hence, I am a strong advocate of multidisciplinary studies combined with latest innovative research. This compilation of cases on research-based teaching is a timely addition to the scholarly work that must be read by academics and administrators alike. Research-based teaching methods implemented and results revealed in this book have established a strong reason for change to the curricula, teaching styles, teacher training, and in the use of IT. I strongly encourage the administration to embrace and implement the methods described in this book at least to provide students with the opportunity to experience a new way of learning science and mathematics.

This book, which is a collective work of individual experts in the fields of science and mathematics education and edited by Prof. de Silva who himself has pioneered new teaching methods of those I have been very familiar over the past 20 years provide practical, easy to follow methods for teachers. This book also demands policy changes by the administrators. I highly recommend this book to those ready to spring in to action to change the current educational system. If the methods and curricula changes suggested in this book are implemented in schools and colleges/ universities, we can eventually expect a generational change in the attitudes and perceptions amongst students and parents towards science and mathematics.

*Waqar Ahmed*
*University of Central Lancashire, UK*

**Waqar Ahmed** *is the Director of the Institute of Nanotechnology and Bioengineering at the University of Central Lancashire, UK and Head of Medical Sciences Research in the School of Medicine and Dentistry. He is a Fellow of the Royal Society of Chemistry and Fellow of Institute Materials, Minerals, and Mining. He was the founding editor-in-chief of the International Journal of Nanomanufacturing, International Journal of Bio and Nanomaterials, and International Journal of Nanoparticles. He has also been on editorial boards and has been a peer reviewer of numerous peer journals. He has been a keynote speaker at numerous international conferences. Since 2010, he has led the UCLan Institute of Nanotechnology and Bioengineering. He has authored/co-authored over 500 publications and has authored/edited 18 books with leading publishers such as Elsevier, Springer, Wiley, etc. He has supervised over 50 PhD students. One of his PhD students won the Bunshah Prize for the best research paper presented at the International Conference on Thin Films and Metallurgical Coatings in 2002, and many are in highly prestigious positions including deans of faculty and heads of departments and institutes throughout the world. His work is wide ranging including thins films, medical and dental materials solar cells, and nanocarriers for medical applications and educational development.*

# Preface

Research is the food for an inquiring mind. Research opens up new avenues in the world of curiosity and has paved the way for novel information and findings. Research provides the investigator with the freedom to delve into untested territory. Therefore, it is logical to accept the importance of the introduction of research in schools and colleges/universities in the early years. Current difficulties to introduce research in schools is mainly due to the challenges in finding sufficient numbers of qualified faculty, time, and syllabi that accommodate research as a part and parcel of curricula. Rectification of this situation would need a policy change at a national/state level that will make it compulsory for research to be taught in schools as a subject with other core school subjects. The core subjects are considered as the main language of the country (e.g. English language), social studies, mathematics, and science. The solving of this problem in colleges and universities is not as challenging as schools, although community colleges and the first two years of universities in the USA tend to focus the least on research in their attempt to foster students to learn the core subjects, which is a main requirement in the USA style of education. It is my personal opinion that the situation needs to be changed by holding high schools more responsible in teaching the core subjects thoroughly and then let students at community colleges and the first two years of universities spend more time strengthening the knowledge in their respective chosen fields. This will deliver a group of students who are better equipped to take on real-life challenges with advanced knowledge, since more time could be then spent at colleges/universities in mastering the skills necessary in their chosen careers. If schools, colleges, and universities can introduce multidisciplinary research in combination with core subjects, then the students would benefit more in experiencing the real-life challenges rather than learning the core subjects mainly from textbook-based classroom activities and limited laboratory scale settings. In the British educational system, high school students, after their General Certificate in Education at Ordinary Level (G.C. E. O/L), select only the subjects relevant to their chosen fields. Yet, the lack of an in-depth knowledge in core subjects due to time constraints in high schools makes it difficult for science students to make an easy transition to the work force after graduation from universities since they lack the skills and knowledge required by employers. The introduction of multidisciplinary research should provide the answer; yet this would be possible only if a strong policy change is made to accommodate the inclusion of multidisciplinary research that would expose students to real challenges in their chosen and allied fields while strengthening their knowledge on core subjects. In addition to students not possessing the necessary knowledge or skills when they leave education and enter the workforce, in science education the lack of interest and understanding of the importance of science among students further exacerbate the situation.

Science has been the foundation of the progress of mankind. The 21st century technology, which we have proudly embraced, the distances we have travelled in space, the software development that has made the "world in my pocket" possible, and the longevity that has been promised in humans and animals have been the result of the realization of capabilities of science and its allied fields. Most of the predictions made by great futurists such as Jules Verne, H. G. Wells, Hugo Gernsback, and also my mentor for a brief period, the late Sir Arthur C. Clarke, have now become realities because of the consensus of the great writers and great scientists. Science teaches a systematic approach to solving problems. It is a compass that leads us to wisdom. It is a tool that verifies concepts. It is a subject that empowers the learners and strengthens the practitioners. The scientific method, the singularly utmost and unique concept, conveyed to learners through scientific experiments, laboratory work, and research projects, is an unrestricted application in any field of study. Whilst in the study of science and its allied subjects, one hones the art of the application of scientific method; young learners do not realize the potential of this method in any non-scientific fields and also in their own future lives. As educators, our failure to enlighten our students of the efficacy of the scientific method perhaps is mostly because of our own failure to recognize its uses in our daily activities. Socrates, Plato, Aristotle, Galileo, Newton, Maxwell, and Einstein, to name a few of the giants of scientific discovery upon whose shoulders we attempt to stand, always combined science with other fields of research.

The method of elenchus introduced to us by Socrates is used as a form of pedagogy and in discussions in subjects in science, arts, and commerce. Socrates was a soldier in the Athenian army. His teaching practice of pedagogy where a teacher questions a student that ultimately leads to the correct answer is a unique contribution to the world of philosophy.

Plato was instrumental in establishing the Academy of Athens dedicated to higher learning, which was the first academy of this nature in the Western world. He was a philosopher who was also a competent mathematician. To date, his dialogues are used to teach subjects such as logic, ethics, religion, philosophy, and mathematics.

Aristotle was a philosopher who wrote on subjects such as physics, poetry, politics, music, logic, biology, etc. His writing was not limited purely to one field of study. He was also responsible for classifying knowledge into different subjects, which is important for in-depth studying at higher levels. Yet, at lower levels such as high schools and the first two years of college studies, science should embrace other disciplines and be delivered as an integrated, multidisciplinary subject.

Galileo combined physics, mathematics, and astronomy in his philosophical thoughts and was inspired by the renaissance artists such as Cigoli. Galileo combined his science knowledge with fine art as evident in his designing of instruments that are useful in the scientific field. His work also inspires us to understand the importance of supporting and fostering analytical, critical thought processes in science by combining and connecting with other subjects.

Newton, with his famous discoveries in science and mathematics that changed our lives tremendously, was also interested in religion, politics, and business ventures. The application of Newton's laws is not limited to physics principles; they can be seen in everyday activities. In sports, martial arts, and dancing, one can easily display the instances where the three laws of motion are practiced.

Maxwell, who made the second great unification in physics after Newton, was also talented in English and poetry. His curiosity from a very young age helped in delving into many different subjects. His explorations during school years went beyond the syllabi, and he was not concerned by examination performances. The true test of his abilities was the publication of scientific papers at the age of 14.

Einstein was not just interested in physics and mathematics. He was also an avid player of violin and was interested in classical music. Yet, in his early years as a student, he found it difficult to follow the standard, rigid teaching model in school and was considered a failure by some in the academic and professional fields, until his theories became acceptable, and now his name is synonymous with genius.

It is evident that most of the great scientists of the past were competent in more than one subject, struggled or disliked the standardized teaching and assessments in education, and were always the thinkers who went beyond the limitations and boundaries set up by traditional system. Therefore, the teaching of science and mathematics should always display relevance to the world around us, to our daily activities, and should link to other subjects. The standard examinations alone would not always display the abilities or the limitations of every student. The vast, varying nature of learning styles and the range of abilities among students make it difficult to truly measure the limitations or the promising future expansion of young students. Therefore, our approach to teaching needs to be changed to accommodate the needs of the students and to recognize the varying abilities of learners.

Nevertheless, the teaching of science and mathematics in our classrooms in schools and lecture halls in colleges and universities mostly tend to promote science and mathematics subjects as separate learning from other academia. Science and mathematics educators tend to stay away from business studies, marketing, arts, philosophy, literature, music, theater, films, etc. Over the years, there have been attempts to get students interested in science and mathematics through different methods based on pedagogical connotations and theories. The effectiveness of these methods has been little to nothing as evident from the scores released from the 2012 Program for International Student Assessment (PISA). The US results for mathematics were below the average and for science and reading the results just met the average standards lagging behind Shanghai-China, Hong Kong, Singapore, Japan, Korea, Finland, Canada, Poland, Netherlands, and Switzerland. These results were not much different from the 2009 PISA results. This comparison clearly indicates that our attempts to improve standards have not been as effective as we had expected.

In spite of sincere efforts by the science and mathematics educators, educational policymakers, and the administrators, the students still tend to look at science as a disconnected subject delivering abstract knowledge. This belief adds to the alienation of science from other popular subjects. As science educators, our beliefs in the efficacy of science need to be reflected through our approaches to solving problems, our assessment of worldly occurrences, and in general our philosophy about life and nature. We need to deliver our material and impart our knowledge with commitment, conviction, and a clear set of objectives that go beyond classrooms.

According to the National Math and Science Initiative (n.d.) (NMS), the latest data show that only 36% of the high schools students were ready for science in 2013. Further analysis showed that 38% of students who took STEM did not graduate with one. The NMS initiative also stated that in 2007, about a third of middle school science teachers either did not major in the subject in college and/or were certified to teach it. Another two important aspects to note are that in 2009 only 29% of research papers were published in the most influential journals, when compared to the 40% published in 1981, and that over half of the US patents were awarded to non-US companies. These are the results of failing standards in the US science education and its policies. If we attempt to correct the mistakes and improve the effectiveness, in about two decades we should be able to reap the benefits, since we need a cultural change within our society to produce a generation of science-loving, scientifically analyzing, and problem-solving men and women.

This book, written by experts in the field of science and mathematics education, brings together a plethora of information, methods, ideas, and activities that have produced results in improving the standards of science and mathematics education. The cases on research-based teaching in science education deliver a collection of successful results that can be used by any school, college, or university to improve the standards of science and mathematics education. This book also echoes the call for a policy change at a national/state level.

This book has been divided into four sections for better understanding of the efforts of the experts, the effectiveness of the work, and the extension of the research in the future of science education.

The science curricula written for students should focus on actual, measurable learning. A mere delivery of material to satisfy the institutional objectives will not be an effective path to delivering the knowledge.

The first chapter, "Making Sense of Science: A Review in Scottish Further Education," presents the readers with strategies for teaching chemistry to non-majors. This chapter delves into the ways in which attitudes to learning chemistry can be improved. The chapter further attempts to answer why chemistry is perceived as a difficult subject. This chapter provides a good overview of the problems faced in schools and colleges not only in chemistry but also in science in general. Changing the perception of science among students is paramount to tackling the difficulties faced in science education. The author provides a variety of recommended principles in the designing and delivery of curriculum, assessment formats, peer and material interaction, and that support project-based learning.

The second chapter, "Developing Scientific Literacy: Introducing Primary-Aged Children to Atomic-Molecular Theory," proposes the introduction of a spiral curriculum to teach macroscopic and microscopic properties of matter. Supported by successful research data and strong evidence, the authors of this chapter challenge the schools to introduce atomic-molecular structure in the years 3 and 4. The lack of progressive and persuasive teaching of science to students in elementary schools and middle schools later creates difficulties in learning science and appreciating it as a useful subject far beyond the classroom. The confidence in students to continue to study science is also tested during the years in elementary and primary schools. This chapter provides the solution to teaching atomic-molecular theory but could easily be adapted to teaching other concepts within science.

In developing an effective curriculum in any field, it is important to consider the possibility to deliver the material effectively to students. It is also imperative that teachers thoroughly understand the objectives stated in the syllabus and be competent and qualified in their chosen fields. The third chapter, "Implementing the Understanding by Design Framework in Higher Education," provides an insight to an organizational initiative undertaken to develop and implement curriculum-planning framework. The developed curriculum has focused on knowledge, skills, and dispositions related to science teaching. The chapter places emphasis on the importance of effective design of curricula by science teachers to proficiently deliver the subject to students. The failure of professionally competent science teachers in schools/colleges to design and deliver an effective curriculum causes science students to become confused, which then leads to disaffection. Therefore, the points given in this chapter can be easily extended to produce a learner-centered and objectives-focused curriculum.

The effectiveness of hands-on experience in learning is a valid method established in the arena of education. One of the ways to achieve success through this approach is to introduce popular, novel, and engaging activities. The fourth chapter, "Martial Arts and Physics: A Multidisciplinary Approach to Increase Student Engagement and Interest in the Sciences," delivers a new syllabus to teach physics through martial arts and presents results over several years of students' attitudes after taking part in annual physics day programs. The delivery of physics syllabi through martial arts and hands-on experience in understanding physics topics mainly using a qualitative, conceptual approach is described with supporting evidence.

In order to direct students from non-science majors to science studies, it is necessary to make science appealing to a wider audience. This could be achieved through highlighting the importance of science in multidisciplinary fields. The fifth chapter, "The Inclusion of Multidisciplinary Research in Science Teaching: A Novel Teaching Method," introduces a new model to be used in science teaching that reaches to other disciplines. The chapter, which is built up on previously conducted research leading up to the model, supports the efficacy of the model through past results and places emphasis on the importance of providing students with opportunities to understand the use of science in other non-science fields. This model is further presented as a way to generate interest among students who are least likely to select science as their majors.

The STEM (Science, Technology, Education, and Mathematics) initiatives in education have focused largely on improving the standards in schools and colleges. The collective work of these initiatives have paved the way for collaborative projects to aid in developing and delivering tested, effective modules in science subjects that have been perfected through research-based knowledge. The sixth chapter, "Developing a Research-Informed Teaching Module for Learning about Electrical Circuits at Lower Secondary School Level: Supporting Personal Learning about Science and the Nature of Science," describes the importance of social mediation of learning and the inclusion of learning through dialogue. This project, implemented as a part of "Effecting Principled Improvement in STEM Education" is informed by the constructivist perspective and also connects previous research conducted on student learning and thinking of science subjects, their relevance in the developed module, and the factors affecting learning.

Our delivery of subjects must improve the creativity among students and encourage their curiosity. The curiosity to investigate, research, and learn more on concepts allows the students to develop a set of important skills in their future careers as scientists. The seventh chapter, "Presenting Physics Content and Fostering Creativity in Physics among Less-Academically Inclined Students through a Simple Design-Based Toy Project," offers a pathway to teaching physics through a hands-on approach. The authors explain how the content was taught while enhancing the creativity among technical students, thereby providing an opportunity for students to generate a desire for learning. The chapter relates directly to kinesthetic learners and the advantage of introducing design-based toy projects to such learners. An elaboration of the principles of physics taught through the project provides an interesting insight to how a simple project can be utilized to deliver an effective lesson.

The focus on multidisciplinary research should improve the number of learners of science and mathematics and support non-science majors and weak science majors to improve their learning and application of knowledge. The eighth chapter, "Using Multidisciplinary Research Experiences to Enhance STEM Learning through Undergraduate, Team-Based, Summer Research Projects for At-Risk Students," describes a project focused on at-risk undergraduate students with weak mathematics backgrounds. This chapter delivers the results of Summer Immersion Projects from 2011 – 2013. The success of this program is further ascertained through survey responses and Depth of Science Experience (DOSE) results. The importance in the integration of multiple disciplines in research is further stressed in this project.

As previously stated, STEM initiatives have opened doors for sharing information and knowledge. Therefore, it is unwise to not use collaboration as a tool to improve the delivery of material. The ninth chapter, "Collaborative Teams as a Means of Constructing Knowledge in the Life Sciences: Theory and Practice," details the importance of collaborative learning in STEM classes in post-secondary settings. Combined with previous research work in this context, the chapter describes the advantages of team-based work even in classes with high numbers of students. The work also includes a discussion of the application of the system in a large-enrollment, non-major biology classroom. Enhanced peer-learning, improved communication, increased student retention, and higher-order thinking are just some of the advantages offered in this proposed collaborative learning.

When combined with other disciplines, the learning can provide the learner with several opportunities to find answers and improve the understanding of the subject. The tenth chapter, "Interdisciplinary Problem-Based Learning Practices in Higher Education," delivers the results of a research project conducted to bring together two groups of students from different disciplines. Using IPBL as a way to eliminate ad hoc learning of skills and study students' actual learning, this project takes a case study approach. The end results of the research display students' satisfaction in working on interdisciplinary tasks.

Similar to the previous chapter, the eleventh chapter, "Transdisciplinary Research in Sustainable Scientific Education in the Field of Urbanism and Architecture," highlights the importance of transdisciplinary research in higher education. The importance of sustainable scientific education in answering complex issues is stressed through the practices in the fields of architecture and urbanism. This work further elaborates how transdisciplinary studies can improve the educational process at universities.

The educators who deliver science subjects need to be trained in the most effective teaching methods using the latest available technology. It is also imperative that they deliver the material with greater confidence and understanding using appropriate methodology. The twelfth chapter, "Pre-Service Teachers' Self-Efficacy and Attitudes toward Learning and Teaching Science in a Content Course," elaborates on the introduction of active and interactive teaching techniques in the preparation of teachers of pre K-8 level. This chapter brings in the perspective from the point of view of science teachers. When teachers continue to follow the now outdated methods introduced to them during their training in teaching children and when teachers are not exposed to novel teaching techniques or are not provided with an opportunity to learn new ways to improve their teaching, the educational system fails. Based on this concept, this research work introduces interactive teaching methods to improve the attitudes and self-efficacy of pre-service teachers.

The advent of online teaching has provided learners with ample opportunities to embark on programs with the least interruptions to their daily commitments. Yet, it has also brought new challenges in the delivery of material, which is limited as a result of the online environment. The thirteenth chapter, "Traditional Teaching and its Effect on Research-Based Teaching: Science via Online Instruction," delivers the types of on-ground educational methods that can be used to teach online science subjects. The chapter further discusses various technologies that can be used to produce classroom setting that are conducive to effective learning.

There are obvious challenges faced by those who deliver research through online teaching. The solutions to these challenges are provided through advanced technology. The fourteenth chapter, "Research Institutions: Research-Based Teaching through Technology," addresses the challenges faced in delivering online instructions and some barriers to teaching some skills to students via online institutions. The chapter further provides solutions to managing these challenges and suggestions to improve the systems, including the preparation necessary for students when they embark on research-based studies.

Finally, the fifteenth chapter, "Application of Information and Communication Technology to Create E-Learning Environments for Mathematics Knowledge Learning to Prepare for Engineering Education," discusses the effective ways to include ICT applications and create e-learning environments to prepare students in engineering education. This chapter further explains how the use of mathematical models supports education in interdisciplinary fields and the ways to include mathematical models in classrooms through the use of technology.

Most of the work presented here is a collection of continuous research work moving and improving from previous established methods. If the trend to improve science and mathematics among high schools, colleges, and universities continues with this momentum, then it is possible that in about two generations the world would witness a sea change in attitudes, perceptions, and applications not only among future school children but also among the parents. If the mind-set of a child can be changed at a very young age to appreciate and enjoy the power of science, then this is possible not only by the efforts of teachers, but also because of the parents. It is important to understand the ways to generate this interest. It should be a fun and exciting process. If a child who yearns for a bicycle is first given a lesson on circular and rotational motions, center of mass, and velocity, would that ever create interest in the riding of the bicycle? Should we deliver a lesson of software programming to a kid who asks parents for an iPad? If in the process of cycling and using the iPad, the child is guided through the principles behind the use of the bicycle, iPad, or for that matter any toy, the child is more likely to remember the information. This knowledge could be coupled to show how the child can reap the maximum benefits by possessing that scientific and technological knowledge. Can calculus be used to understand the stock market or the economy? If so, why shouldn't we teach these subjects using more real-life applications where students can actually relate to them, instead of seeking examples purely related to science? Why should physics and chemistry be taught as abstract knowledge when ample opportunities are available to produce easy-to-understand curricula using most popular subjects such as sports, martial arts, magic, and children's toys?

Why couldn't we use real-life problems in the classrooms to find solutions with students to improve their critical thinking, logical approach, and understanding problems? We can use a multi-faceted, multi-pronged approach to find answers to day-to-day challenges through science and mathematics and make that the starting point of learning. We do not have to stick to the same style of teaching in which we have been conditioned. Our objectives should not be to generate a group of students who think, apply, and solve problems in the exact way we do. We need to promote multi-dimensional approaches to solving problems. We need to encourage students to challenge us, test us, disagree with us, and ultimately find themselves under our guidance if their pathways to finding solutions are the most effective ways or not. We need to not resist the diversion from norm of our students as long as their goal also is to successfully learn the subjects. We need to provide our students with all the opportunities for them to build their own success. We can guide, advise, and walk with them, but we must give them the freedom to explore and research. This can come only through the introduction of research-based teaching in multidisciplinary fields. We need to collaborate with non-science disciplines to promote science and mathematics among our future students. The best minds in businesses, creative arts, law, etc. could have been great scientists, too! Similarly, great scientists can be great businessmen, lawyers, and politicians! In promoting education, we should not separate science from arts, arts from business, or business from science. Our answers to problems should cover all aspects, and this is only possible through a multidisciplinary approach.

I hope that the work presented here and other similar work all over the world will continue to grow and generate students who are qualified in science and mathematics whilst simultaneously appreciating other subjects and using that appreciation and understanding as the supporting force to do well in the fields of science and mathematics. I am confident that the current trend in science and mathematics studies can be changed through the use of multidisciplinary research approach in future education.

*Eugene de Silva*
*Virginia Research Institute, USA & MRAS – Walters State Community College, USA*
*March 30, 2014*

## REFERENCE

National Math and Science Initiative (NMS). (n.d.). Retrieved from https://www.nms.org/Programs/ResearchResults.aspx

# Acknowledgment

It is imperative that I extend my thanks to faculty and staff at Walters State Community College, including Dr. Jeffrey Horner, the Dean of Natural Science, Dr. Lori Campbell, Vice President for Academic Affairs, and Dr. Wade McCamey, the President of the College, for their continuous support in my endeavors to improve science education and research. My thanks to the Governing Board of Virginia Research Institute and National Accreditation Commission for Martial Arts, including Prof. Joseph Connolly and Mr. Troy Glen, Tennessee Board of Regents Drs. Tristan Denley and Pamela Knox, Tennessee Department of Education Dr. Gary Nixon, and the team at the Institute of Physics UK and the USA branch including Drs. Katherine Grzywacz-Jones and Jacqueline Johnson for collaborative support in my efforts to introduce science education through multidisciplinary research. My gratitude also goes to the faculty and staff at IISER (now a branch of Virginia Research Institute, USA/UK), including the Director Mark Holloren and faculty members Terry Frost and William Davidson. My thanks also to Prof. Waqar Ahmed who agreed to write a foreword for this book. His countless papers published in the field of scientific research speak volumes of his credibility to present a foreword for this book. I greatly appreciate his commitment to this task in spite of his very busy schedule as the head of several international initiatives. I finally want to thank my daughter, Eugenie, for being my staunch supporter in all my research, both as a research subject and also as an unpaid living assistant. I also want to thank the reviewers of chapters published here who took time from their daily, busy schedules. Invariably, thanks are due to the contributors of chapters, who are a group of like-minded researchers genuinely contriving to develop methods conducive to teaching science and allied fields. Their efforts and published results speak for their effectiveness and adaptability. It is hoped that this collaborative effort will pave the way for necessary policy and curricula changes in science education to produce a future generation of great scientists.

*Eugene de Silva*
*Virginia Research Institute, USA & MRAS – Walters State Community College, USA*

# Section 1
# Curricula Changes

# Chapter 1
# Making Sense of Science: A Review in Scottish Further Education

**Nancy El-Farargy**
*UK National Health Service, UK*

## EXECUTIVE SUMMARY

*This review outlines a research-informed teaching case study based on adult learners of chemistry within the Scottish Further Education (FE) sector. It provides some strategies for success in supporting non-major chemistry students and provides some practical ways forward for improving attitudes to learning chemistry and in studying the subject further. An overview of why science, and in particular chemistry, may be perceived to be difficult to learn is discussed, as well as links to the evidence base that outlines the facilitation of meaningful and relevant learning. Through a chemistry curriculum redesign, discussions on the shift of attitudes, perceptions of learning, difficulties, and preferred topics in lessons are discussed. In addition, a general overview of the science education scene in Further Education is presented. Based on research evidence and educational neuroscience, there are suggested implications for educators in supporting adult students learning non-major science courses.*

DOI: 10.4018/978-1-4666-6375-6.ch001

# ORGANIZATIONAL BACKGROUND

## Scottish Further Education

The Scottish educational landscape is distinct from the rest of the UK (Humes & Bryce, 2003), with Further Education (FE) having a long and distinguished history (Paterson, 2003). FE colleges are highly diverse in the nature and range of courses they offer; and subsequently, the students they attract are also diverse in nature. Some students enter FE to embark on courses with strong vocational emphases. Others may study to fill gaps left from school education, whilst some enter FE as adults after a gap of many years of formal learning. FE can also be used as a stepping stone for those wishing entry into formal Higher Education (HE). The emphasis is largely placed on the notion of lifelong learning that is accessible to everyone (Scottish Executive, 2003).

## Political and Strategic Background

The Scottish Government considers lifelong learning to be one of its top priorities, and its vision is that all individuals will be able to access education that is flexible and relevant to their needs and aspirations (Canning, 1999; Scottish Executive, 2003; Scottish Government, 2007, 2011). It is widely understood that the role of FE is to provide post-16 education and vocational training, which offers learners the opportunity to gain more skills and greater employment options (Scottish Government, 2011).

The Scottish Further and Higher Education Funding Council (The Scottish Funding Council, SFC) is a Non-Departmental Public Body (NDPB), with the sponsor department being the Scottish Government Employability, Skills and Lifelong Learning Directorate (SGESLLD). The SFC is responsible for funding teaching, learning provision and research over Scotland's colleges and universities and each year, announces the allocation of public funds to support these areas. This funding also includes costs related to staff, infrastructure, buildings, and equipment. Over the academic year 2011-2012 for instance, a total of £577.6m was administered to the college sector (Scottish Funding Council, 2012). In addition to the allocation of funds, the SFC, under the Further and Higher Education (Scotland) Act 2005 (Crown Copyright, 2005), needs to ensure that each of the 'fundable bodies' have accountable officers, arrangements for student support needs, and other systematic provisions. Furthermore, the SFC promotes a 'widening participation' agenda to Further and Higher Education via its national annual conference 'Learning for All' (Scottish Funding Council, 2013c).

Until 1993, publicly funded colleges were run by Local Authorities, and it was the implementation of the Further and Higher Education (Scotland) Act 1992 (Crown Copyright, 1992) that removed power from Local Authorities and established colleges as registered charities with Boards of Management. The result of this Act was that Ministers had the power to close, merge, or establish colleges (the legal requirement for non-Ministerial direction for registered charities did not and does not apply here). Nevertheless, alongside some governance requirements by statute, each college was largely autonomous in creating its own future direction (Griggs, 2012). It is only in recent times that the governance, operational ways of working and future directions of each college is being revisited (Scottish Government & Scottish Funding Council, 2011).

Previously (e.g. during 2004), there were 46 Colleges of Further and Higher Education, where provision, scale, and scope of activity varied. Given the competition between colleges and duplication of effort (within a few miles of each other) (Scottish Government, 2011), Scotland's colleges are currently being rationalised, giving rise to 26 'super' colleges, over 13 regional areas, (Scottish Funding Council, 2013a, 2013d). The Post-16 Education (Scotland) Bill proposes two college structures: "single college regions (with college boards) and multi-college regions (with regional strategic bodies and assigned colleges working within this structure)" (Liddell & Macpherson, 2013, p. 4). Given the new college structures, the potential implication on the status of the SFC as the governing body and funder of assigned colleges, is under question (Liddell & Macpherson, 2013).

There are two main organisations that support the Scottish college FE sector: (1) Colleges Scotland (Colleges Scotland, 2013) and (2) College Development Network (College Development Network, 2013). Colleges Scotland acts as a voice for the sector, campaigning and shaping public policy and debate. The Scottish Education Further Education Unit (SFEU) is the legal entity, but trades under the brand name of the College Development Network. This network delivers CPD activities for college staff, and provides advice and guidance on the curriculum, learning, and teaching and assessment practices. It also provides a medium for sharing learning tools and services, as well as delivering consultancy services. Both are companies limited by guarantee and registered charities.

The Scottish Qualifications Authority (SQA) is the national accreditation and awarding body in Scotland, and is sponsored by the Scottish Government's Learning Directorate (Scottish Qualifications Authority, 2013a). It accredits Scottish qualifications other than degrees, and approves and quality assures prospective and current establishments which offer SQA qualifications. Other roles include: (1) the development, validation, and review of qualifications; (2) overseeing the assessment of qualifications; and (3) the issuing of certificates to candidates (Scottish Qualifications Authority, 2013c). It also promotes the Scottish Credit and Qualifications Framework (SCQF), (Scottish Qualifications Authority, 2013b) which aims to support learners, employers and educational institutions to compare achievement levels (via a 'points' classification), and to support navigation through the Scottish qualifications system. By way of example, approximately 3600 Higher National (HN) students enter 2nd or 3rd year of university each year in Scotland, demonstrating articulation, academic linkages, and widening participation to post-16 Further and Higher Education (Scottish Government, 2011). According to the Higher Education (Scotland) Act 2005, Higher Education is defined at HN levels (SCQF Level 7) or above. However, FE colleges typically offer qualifications ranging from points one to ten, and universities typically offer qualifications ranging from eight to 12. Additionally, in aligning with the requirements of the workplace, colleges and educational institutions will be cognizant with the 'National Occupational Standards' (UK Commission for Employment and Skills, 2013); further supporting qualifications and the graduate transition into the workforce.

A number of changes in the Scottish educational arena are underway. A new qualifications structure – the 'Curriculum for Excellence' – has been recently introduced, and this aims to tailor young people across four main capacities: (1) Confident individuals; (2) Responsible citizens; (3) Effective contributors; and (4) Successful learners (Education Scotland, 2013). Furthermore, following a major review ('Teaching Scotland's Future') of the teaching profession in Scotland (Donaldson, 2011), 50 recommendations – covering the full spectrum of teacher education – are currently under discussion for implementation. These notions were implicated within the Scottish Government's overall vision for science and engineering (Scottish Government, 2008, 2010, 2012).

In particular FE colleges were actioned to (Scottish Government, 2008):

- Support implementation of the science elements of Curriculum for Excellence.
- Enhance the match of skills, competencies and qualifications to the needs of science industries, through flexible learning opportunities, and other forms of learning.
- Increase capacity further to work collaboratively with businesses to support science skills and knowledge programmes.
- Work collaboratively with universities to meet the needs of learners and employers through integrated provision and knowledge transfer.
- Enhance the range of work-based vocational learning, assessment and accreditation opportunities.
- Promote increasing participation in science, engineering and technology.

In addition, the action plan for science and engineering highlighted the following five work-streams and aspirations (Scottish Government, 2010):

Workstreams (p. 2):

1. "Building capacity and expertise of teachers.
2. Practical support for teachers and learners.
3. Increasing children and young people's engagement with, and understanding of real life science, engineering and technology.
4. Further learning, training and employment.
5. Improving the public knowledge, understanding and perception of science."

Aspirations included:

- Increases in the participation of science as measured through National Qualifications.
- Increases in success rates in science & engineering related school and college qualifications.
- Greater motivation for young people to enter further studies and employment in science, engineering, and technology.
- Positive media articles about science in education.

Whilst the case study presented here occurred in a slightly different political backdrop and qualifications infrastructure, it is understood that the principles and course components of the case study are still valid.

## SETTING THE STAGE

### Learning in Further Education

The role of FE in the development of qualified scientists and technicians is widely understood (Scottish Executive, 2001), and many FE courses align and articulate to Higher Education Science courses. Other courses are of a vocational nature and many (e.g. Sports Science and Nursing) will have a compulsory science component, with an assumption that student entrants will have an underlying awareness of basic chemistry and science. Nevertheless, given general life experience and ambitions of learning, the majority of students will be welcome to enrol, including those students with varying or no prior levels of chemical knowledge (e.g. Ang & van Reyk, 2013; El-Farargy, 2009a). Despite the notion that prior knowledge of chemistry proffers success in the future learning of chemistry (Childs & Sheehan, 2009), other evidence of learning intervention has demonstrated that success could be achieved irrespective of entrance qualifications (Sirhan & Reid, 2001). This perhaps highlights the importance of widening accessibility and promoting the utility and skills training of careers based on scientific foundations. In addition, Harden (1996) and Kilminster (1997) argue that vocational education should do more to enhance critical thinking, and general science education could have a role to play in this respect. Furthermore, it is often understood that the learning and application of scientific knowledge is a top priority for any nation to be economically viable (De Rennes, 1999); this perhaps supplementing the ethos of promoting continuing education in science.

In looking at learner perspectives, Connelly & Halliday (2001) have reported that the single most important reason for studying in FE, was to obtain a better job. The upgrading of skills, gaining formal qualifications, improving job prospects, social reasons, and improving personal confidence were other important motivational factors. Gallacher, Crossan, Leahy, Merrill, & Field (2000) noted that self-development, the overcoming of critical life incidents, and the opportunity "to get out of the house" were other reasons for studying in FE. Analysis of FE awards for the year 2011-12 shows that those 24 and under (at the start of the academic year, and including those under 16) make up 54% of enrolments (Scottish Funding Council, 2013b).

There are some reports of age mixing in college classrooms, (Howard, Short, & Clark, 1996; Merrill, 2001), however, McNair, Parry, Brooks, & Cole (2004) highlighted that this appeared to benefit younger students more than their older counterparts. They attributed this to life experiences that adults brought to the learning environment. IFF (2003) also noted that younger learners are less likely to have doubts about studying in the further education environment, however, given the expectation of the older workforce (Scottish Executive, 2003), this could provide the impetus for promoting the ethos of lifelong learning. Indeed, this is perhaps the power of FE colleges, in that they offer pathways into continuing education and lifelong learning.

## Nursing in Further Education

During the academic year 2011-12, there were over 14,000 enrolments in health, care and Access to Nursing courses (Scottish Funding Council, 2013b), and some of these courses will have guaranteed students entry to funded Bachelor of Nursing places at university. Thereafter, graduates will be able to register as qualified nurses with the Nursing and Midwifery Council (Nursing and Midwifery Council, 2013); the professional body that oversees and regulates nursing and midwifery professionals across the UK. Public bodies like NHS Education for Scotland (NHS Education for Scotland, 2013) and Health Education England (Health Education England, 2013) are responsible for ongoing training and continuing professional development – and such engagement will also support the professional body requirement for nurses to evidence their ongoing personal and professional development. Another institution of relative importance is the Royal College of Nursing (Royal College of Nursing, 2013) and they aim to offer a united voice to the nursing profession, and in influencing and supporting nurses and nursing.

## Is Science Difficult to Learn?

What is taught by teachers may not necessarily be learnt by pupils. Indeed, the literature has focussed heavily on the learning and teaching of science, and the evidence base is continuously growing. In addition, there is a growing perception for science education, like medicine, to become evidence-based and informed by research (e.g. Goldacre, 2013); although it is strongly argued that this notion is already present within the science teaching infrastructure across the UK (e.g. Taber, 2013).

Firstly, let's consider the nature of science, and in particular chemistry, which has been the subject of debate for some time. Chemistry has often been considered to be more difficult to learn (compared to other school subjects), but it will often feature and be a compulsory requirement amongst many courses (with examples including nursing, care, medicine and sports science). One way of looking at chemistry is viewing the subject through three lenses: (1) tangible and concrete experiences (2) representations of chemistry such as equations and graphs; and (3) mental, abstract visions of electrons and bonds (Tasker & Dalton, 2006). Such integrated exemplifications may be perfectly natural to experienced chemistry educators and experts, but they may be challenging to novice chemistry learners. This has been demonstrated to have implications in information processing capacity and working memory (Danili & Reid, 2004; Johnstone & El-Banna, 1986).

For adults, and particularly so for those whom have been out of learning science for a long time, it will be even more important to illustrate the relevance of learning science. This notion aligns itself with the principles advocated for adult learners (Knowles, 1990; Kolb, 1984). There is a relentless need for adults to know why they need to undertake learning, and particularly so if initial perceptions suggest that engaging in learning does not give rise to meaning or relate to their own life experiences. In addition, there is a requirement that the learning fruitfully links to potential utility and application. This perhaps may well be the key link in facilitating meaningful learning for non-major science students. Further, Ang & van Reyk (2013) reported that prior knowledge (of chemistry) and previously perceived success, supported a sense of current mastery; this perhaps highlighting the need to link previously gained knowledge, meaning and understanding, to support current learning and application. Indeed, the literature has suggested that studying chemistry as part of non-major science courses may be challenging for students (e.g., Ang & van Reyk, 2013; El-Farargy, 2009a; Mahaffy, 2004).

Lastly, the language of science needs some consideration. Amongst native speakers, the unfamiliar language of science may pose problems (Cassels & Johnstone, 1984), but the situation is more cognitively demanding for students learning science in a second language (Tao, 1994). Empirical research suggests that pupils understanding of common scientific words differ between native speakers and those learning science in a second language (Johnstone & Selepeng, 2001). In addition, their research demonstrated that pupils lose at least 20% of their apparent working memory space when working in a second language. Given that some college entrants may be unfamiliar with the language of science, there may well be some implications in the FE college classroom and laboratory.

Thus, it appears that whilst chemistry may appear to be more cognitively challenging than other subjects in the curriculum, educators can embed a number of strategies to support its learning, assessment and application.

## Attitudes to Science

Attitudes towards chemistry for example, are often highly complex and multi-faceted – they will comprise of a group of evaluations relating to, for example, the relevance of chemistry, its career utility, personal enjoyment, the textbooks, teaching, and laboratories. Together, they involve the cognitive, affective and behavioral domains (Fishbein & Ajzen, 1975), and this intertwined belief manifests itself as behavior. For attitude change to occur there has to be some sort of internal or external disruption to this intertwined matrix – which could also involve personal dissatisfaction with the *status quo*. There could of course be independent reasons why attitudes held may be stable and resistant to change (e.g. Eagly & Chaiken, 1993). In addition, the concept of simply enjoying and having passion for what one learns, works and does (Anderson, 2004) will be equally applicable to students of science.

Osborne, Simon, & Collins (2003) argue that attitudes play an important role in the learning of science, and some of the reasons for the low uptake of science and technological subjects could be related to the negative attitudes held by young people towards science. Some attitudinal studies have looked into gender related patterns, highlighting that boys exhibit more positive attitudes towards science than girls (e.g. Barmby, Kind, & Jones, 2008; Stead, 1985). Other work has concentrated on factors affecting the uptake of the sciences, namely issues like, attitudes to science in society, attitudes to school science, and attitudes to scientific careers (e.g. Bennett & Hogarth, 2009; Bennett, Lubben, & Hampden-Thompson, 2013; Osborne et al.,

2003). Further, some research has discussed attitudes in relation to the curriculum delivered (El-Farargy, 2009a), teaching factors (Woolnough, 1993) and parental contributions to science education (Ing, 2013). In addition, Osborne et al. (2003) highlighted a mismatch between pupils and teachers in the potential utility of science: pupils viewed science with a technological lens, whereas teachers perceived science in terms of milestones.

Despite the large body of attitudinal implications in science education research, it should be highlighted that each educational context – with its associated workforce, scientific developments and outputs – will vary. This is clearly evident across the US, Europe and the rest of the world. Even within the UK, the place of science education in Scotland varies from that in England, Wales and Northern Ireland. Additionally, power over education is devolved to Scotland, resulting in a unique education system.

## Some UK Trends

In the UK, most students have to engage with at least one science subject up until the age of 16. However, for many, this may the last stage in which they encounter science. In England, there is a declining number of students deciding to choose science (as post-compulsory education), with more students interested in studying the arts and humanities (Osborne et al., 2003). Within the same context, Pell & Jarvis (2001) reported that a decline in positive attitudes towards science was also evident at primary school, and similarly, Jenkins & Nelson (2005) highlighted that most students by the age of 14-15 (for both boys and girls), have little appeal for scientific careers. Further, Reiss (2004) reported that enthusiasm for science diminished even at the higher elective stages of sixth form college. This perhaps reflects a need to further explore attitudinal factors affecting science, even at the elective stages.

In Scotland, there is also some evidence regarding the falling enrolments in the sciences (BBC News, 2006) and the falling student numbers graduating from universities in the physical sciences (Young, 2006). Nevertheless, there is a slightly different pattern, in that enrolment numbers for Scottish Higher Grade biology, chemistry and physics, are the highest for the elective subjects (e.g. Scottish Qualifications Authority, 2006). In addition, for physics at least, there is some evidence that more girls tend to continue studying the subject (Reid & Skryabina, 2003); and this is not generally seen elsewhere in the UK. Following more recent initiatives, reports of higher levels of science uptake in Scotland were published (e.g. Denholm, 2011). Moreover, the Royal Society advocated a reform of the English system, with suggestions in emulating the Scottish science scene (The Royal Society, 2011).

In a recent review of the UK Science, Technology, Engineering and Mathematics (STEM) infrastructure, Tomei, Dawson, & Dillon (2013) noted that whilst the UK as a whole is ranked sixth best in the world, there is still a gap between the higher and lower achievers in science. Perhaps this presents an area that needs addressing.

## Strengthening Science Education

The literature is plentiful of suggestions and strategies in strengthening science education and the following points attempt to summarise these:

- **Non Formal Science Learning:** For instance, this may include museum visits (Gilbert & Priest, 1997) which could provide rich insights into how science operated in the past. To support the consolidation of learning, such visits would need to be followed up by classroom work. However, further research is perhaps required to fully understand the factors involved (Osborne, 2006).
- **Laboratory Practicals:** The presence of high quality laboratories and practicals to help inspire learners of science (Roberts, 2002).
- **Learning Materials:** A number of research studies have demonstrated that bespoke learning materials, based on the research evidence, support learning in chemistry. These have been largely based on contextual and applications based curricula (e.g. El-Farargy, 2009a; Hussein & Reid, 2009).
- **Assessment Methodologies:** Processes that support both formative and summative assessment could prove successful (e.g. Bell & Cowie, 2001; Black & Wiliam, 1998; Black & Wiliam, 2005), however, in the FE context, summative assessment requires more attention (Torrance & Coultas, 2004). In addition, it is suggested that a mix of assessment formats are utilised in order to support all learner cognitive styles.
- **Student Interaction with Peers:** The constructivist school of thought strongly advocates exploratory and guided learning, in which peers and interactive materials could have a considerable role to play(El-Farargy, 2009b; Taber, 2011; Vygotsky, 1978).
- **The Applications of Science:** Presenting students with the relevance and contextual applications of science has been shown to promote positive attitudes, and increase the future aspiration of studying science further (El-Farargy, 2009a). In addition, textbooks that aim to support these notions are available within the market (e.g. Atkins, 2003; El-Farargy, 2008; Lister, 2006); although these are more notably available for more advanced students.
- **Qualified Science Teachers and Educators:** Enthusiastic and confident science teachers who promote the utility of science have been shown to promote more positive attitudes to science (Osborne et al., 2003). Other aspects of

teaching could also involve the incorporation of the history of science (Niaz, 2005), which could provide contextual backgrounds in the exploration and discovery of science. There is some evidence in the UK, however, that some science teaching is delivered by non-specialists, which may affect various aspects of teaching and learning. This perhaps reflects a drive to support the recruitment, training and supply of specialist teachers.

## CASE DESCRIPTION

### The FE Science Scene

The FE science sector varies widely in terms of the demographics and courses studied. From a study of over 800 FE learners from across 10 colleges in Scotland, 42% indicated that they had at least one science qualification, (at levels Intermediate 1 to Advanced Higher or their equivalent); demonstrating the great variety in background science knowledge. Younger students viewed school science more positively than their older counterparts, with school biology being viewed more positively than chemistry.

These students also indicated their preferred activities in science lessons (denoted as either biology or chemistry). Results are grouped as science major courses and non-major science courses (Table 1).

*Table 1. Students' perceptions into the most enjoyable topics in science (n=843). Note: students could tick as many as applied*

| | **Preferable activities in science lessons (%)** | | | | | | |
|---|---|---|---|---|---|---|---|
| | **Studying the theory** | **Doing practical work** | **Explaining events of everyday life** | **Studying how science can make our lives healthier** | **Studying about the human body** | **Studying making equipment** | **Solving everyday problems** |
| Service science (n=477) | 36 | 43 | 43 | 40 | 77 | 6 | 23 |
| Science major (n=366) | 52 | 61 | 53 | 37 | 64 | 10 | 38 |
| $X^2$ *(df1)* | 21.8 $p<0.001$ | 25.3 $p<0.001$ | 9.1 $p<0.01$ | 0.7 ns | 17.0 $p<0.001$ | 6.5 $p<0.05$ | 23.5 $p<0.001$ |

*Table 2. Healthcare students' perceptions into the most enjoyable topics in science (n=127, from 4 Scottish colleges). Note: students could tick as many as applied*

| | Preferable activities in science lessons (%) | | | | | | |
|---|---|---|---|---|---|---|---|
| | **Studying the theory** | **Doing practical work** | **Explaining events of everyday life** | **Studying how science can make our lives healthier** | **Studying about the human body** | **Studying making equipment** | **Solving everyday problems** |
| Healthcare students – chemistry component n=127 | 41 | 56 | 51 | 35 | 73 | 10 | 29 |

*Table 3. Students' perceptions of their own learning (n~843, percentages illustrated)*

| Intellectual Development Questions (Perry, 1999) | | | | | | | |
|---|---|---|---|---|---|---|---|
| **Osgood-semantic differential questions** | | | | | | | |
| In order to pass my course, I need to study just what the lecturer covered. | 12 | 14 | 19 | 22 | 20 | 12 | I do not have to rely totally on the lecturer. Part of learning is to work things out myself. |
| I cannot be wrong if I accept what the lecturer says. If I question anything I might end up failing. | 2 | 5 | 13 | 29 | 31 | 20 | I do not believe in just accepting what the lecturer says without question. Success involves thinking for myself. |
| I prefer not to work with other students because I might up wrong ideas. | 1 | 3 | 10 | 27 | 33 | 26 | It is good to work with other students because listening to their points of view, I can correct my ideas. |
| **Likert questions** | | | | | | | |
| **Statement** | **SA** | **A** | **N** | **D** | **SD** | | |
| Sometimes I learn more about a subject by discussing it with others than I do by revising at home. | 3 | 14 | 13 | 55 | 15 | | |
| I feel uncomfortable when I am left to express an opinion, not knowing the view the lecturer feels. | 3 | 21 | 28 | 40 | 8 | | |

At this stage, it can be seen that there are significant differences between those studying the sciences by choice and those students undertaking science as part of a compulsory course remit. Practical work and studying theoretical foundations are viewed more positively than non-major science students. Studying the human body generated the highest interest for both groups; demonstrating inherent relevancy and meaning. In addition, when the results are grouped by gender, traditional gender related patterns emerge.

For healthcare students learning chemistry (i.e. a non-major science course), similar results are found (Table 2). Interests in learning about the human body and in undertaking practical work emerge positively, with third place being related to everyday life. Once again, these results perhaps reinforce the need to align the learning and teaching of chemistry with the overall course requirements and to everyday life.

In looking at attitudes relating to students' perceptions of their learning, it can be seen (Table 3) that students are highly social in their learning, and have somewhat developed views with respect to their own learning and development.

Very weak correlations exist according to age for both groups of female students and science service students.

Similarly, with respect to healthcare students studying chemistry (n~127), it can be seen that students hold relatively developed views in regard to their own role in learning (Figure 1).

*Figure 1. Healthcare chemistry students' perceptions of their own learning (n~127, percentages illustrated)*

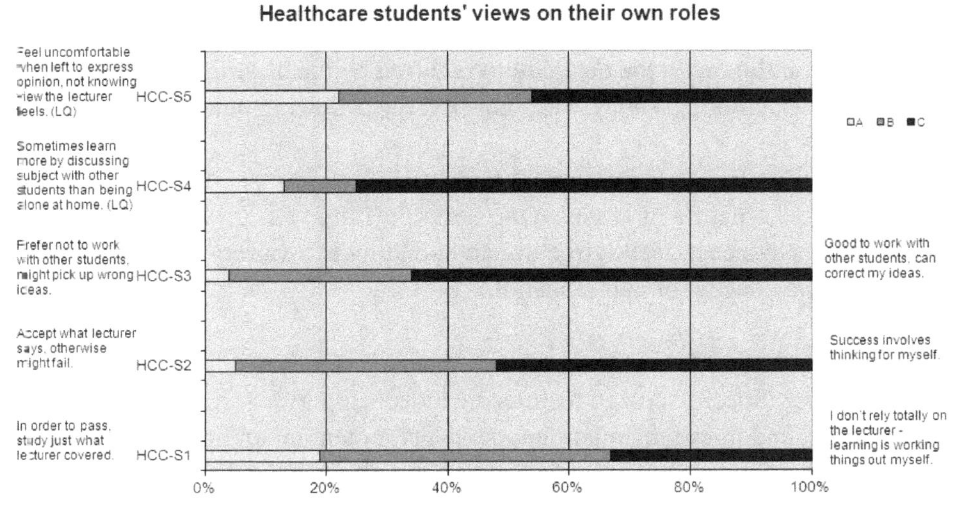

## APPLICATIONS-BASED LEARNING: A CURRICULAR CHANGE

18 students (2 males and 16 females), in one Scottish FE college completed a compulsory, newly revised, 18 week 'Introduction to Chemistry' module, as part of their health care studies (El-Farargy, 2009a). These students were taught by one lecturer, who was familiar with the course, over two concurrently running groups. Overall, the student cohort was of a mixed age (ranging from 17 to over 43), with varying abilities in chemistry. Five students indicated that they had never studied chemistry before, and only two students indicated that they were direct school leavers. Eight students were in employment prior to enrolling in college, in which they had enrolled due to wanting a better job.

The intervention was compared to a previous teaching group, who completed a traditional chemistry curriculum within the same college. This group consisted of 24 students (5 males and 19 females) of age range 17-42. Eight students indicated that they had never studied chemistry before, and four students indicated that they were direct school leavers. 12 students indicated that they enrolled in college because they wanted a better job.

The whole 'Introduction to Chemistry' course was redesigned such that students could be supported in appreciating the applications and context of chemistry in relation to their nursing studies (El-Farargy, 2009a). This removed the chemistry taught from the 'traditional' dimension to one that was relevant and meaningful to everyday life and to their studies. It was hoped that cognitive processes could be 'lightened' by presenting chemistry in such a way that linked to past learning and future applications. Load on the working memory was also reduced by 'chunking' related material together and making clear connections to the nursing situation.

Nine units, encompassing the full spectrum of basic and introductory chemistry, were devised, in line with the curriculum required by the awarding body (Scottish Qualifications Authority, 1993). They covered the following outcomes:

1. "Apply the concept of a chemical reaction in a variety of situations;
2. Relate the structure of atoms to the Periodic Table;
3. Apply the concept of the structure and bonding to a variety of substances;
4. Apply the concept of neutralisation."

In this way, the actual content was not changed; it was simply its presentation and sequencing. Specific design features included support for accessibility, a glossary of terms, and questions relating to students' potential applications. However, there was little room to change the laboratory practicals and the assessments could not be changed.

## What was Found?

In measuring the impact of the work, students expressed their views through questionnaires and focus groups; thus generating a mix of rich qualitative and quantitative data.

Across the range of data collection, it was found that there was overwhelming support for the new materials, as well as improved perceptions and attitudes:

- Improved perceptions relating to the organisation of the course.
- Assessment methods were perceived more positively, and students highlighted that they were growing intellectually and obtaining many new skills.
- All students indicated that they intended to study further at university, with more expressing more favorable attitudes to studying chemistry further.
- Notably there was an increase in positive attitudes in relation to relevance of the course in relation to health care.
- Students expressed increased personal enjoyment and interest in chemistry.

In addition, the participating lecturer felt that the new materials helped produce more positive attitudes to learning chemistry. However, given student perceptions, the need to perhaps revisit the practical laboratory component of the course (which was not changed during the intervention), to be aligned with the ethos of the newly revised curriculum was evident.

Students were also invited to reflect on what they enjoyed most in chemistry lessons (ranging over cognitive, affective and behavioural domains), (Table 4).

*Table 4. Students' perceptions into the most enjoyable topics in chemistry. Note: students could tick as many as applied. Frequencies are presented*

| | **What do you enjoy most in the chemistry lessons?** | | | | | | |
|---|---|---|---|---|---|---|---|
| | **Studying the theory** | **Doing practical work** | **Explaining events of everyday life** | **Studying how science can make our lives healthier** | **Studying about the human body** | **Studying making equipment** | **Solving everyday problems** |
| Traditional curriculum (n=24) | 7 | 9 | 10 | 5 | 19 | 1 | 5 |
| Revised curriculum (n= 18) | 3 | 11 | 7 | 8 | 13 | | 4 |
| Totals | 10 | 20 | 17 | 13 | 32 | 1 | 9 |

It can be seen that most students have a preference to learning about the human body, and in undertaking practical work; demonstrating needs in the behavioural realms of learning and teaching.

In addition, students' perceptions of their learning, assessment, teaching and nature of scientific knowledge yielded interesting results (El-Farargy, 2010). Students undertaking the new curriculum exhibited more developed views than those students undertaking the 'traditional curriculum' across these four domains. This perhaps demonstrates the impact of a curriculum change on learning chemistry for non-major chemistry students.

Students undertaking the curriculum intervention (n=18) were also invited to indicate topics that they felt were difficult. Results are presented in comparison with a cohort of other students who undertook the course using the traditional set of materials (n=11). However, given a difference in questionnaire distribution, this 'traditional' cohort is not the same as the one described above. It can be seen that many of the comments relate to the complex nature of chemistry and the need to understand its concepts and nature (Table 5). Nevertheless, those involved with the intervention supported the new developments and implementation of the curriculum.

Overall, in relating the intervention to the findings from the wider science FE scene (e.g. Tables 1-3 and Figure 1) the intervention appeared to meet the needs of students in relating chemistry to its applications and to everyday life.

## CURRENT CHALLENGES FACING THE ORGANIZATION

### Management and Organizational Concerns

The project ran its course smoothly; however, there are perhaps a number of key issues that need to be mentioned.

- The opportunity to distribute the 'difficulties' questionnaire to one cohort undertaking a traditional curriculum did not take place.
- There was no opportunity to change the assessment structure, and this is perhaps something that needs addressing for future work. It is understood that much assessment within FE is of short summative assessments, and there are questions whether students are gaining any 'real learning', i.e. there is a need to avoid the perceived culture of 'learn, pass exam, and forget'.
- There was no opportunity to revise the laboratory practicals to a great extent. This work has highlighted that students do indeed view practical laboratories as important.

*Table 5. Students' perceptions into the difficult topics within the 'Introduction to Chemistry' course: Note: students could tick as many as applied. Frequencies and comments are presented*

| Consider the following topics of Introduction to Chemistry. *Tick those subjects that you feel you had difficulty with: Please also indicate why you found them difficult (I still do not fully understand it)* | | | | |
|---|---|---|---|---|
| **Topics** | **'Traditional' Curriculum (n=11)** | | **Revised Curriculum (n=18)** | |
| | **Student Frequency** | **Comments** | **Student Frequency** | **Comments** |
| **Apply the concept of a chemical reaction in a variety of situations.** | | | | |
| (a) Physical/chemical changes | 2 | Initially hard to grasp<br>Found chemistry quite difficult to grasp but felt that I got there in the end and completed and passed all exams | | |
| (b) Rate of reaction: particle size, temperature, concentration | 1 | Found chemistry quite difficult to grasp but felt that I got there in the end and completed and passed all exams | | |
| **Relate the structure of atoms to the periodic table.** | 2 | Initially hard to grasp<br>Found chemistry quite difficult to grasp but felt that I got there in the end and completed and passed all exams | 1 | Just find it quite difficult/ the same as mathematics |
| (a) Mass, charge and position of electrons, protons and neutrons | 3 | It was hard to understand<br>Initially hard to grasp<br>Found chemistry quite difficult to grasp but felt that I got there in the end and completed and passed all exams | | |
| (b) Predicting position of element in periodic table, given electron arrangements | 2 | Initially hard to grasp<br>Found chemistry quite difficult to grasp but felt that I got there in the end and completed and passed all exams | 1 | Just find it quite difficult/ the same as mathematics |
| (c) Description of the properties of an element, given the electron arrangement | 1 | Found chemistry quite difficult to grasp but felt that I got there in the end and completed and passed all exams | 1 | Just find it quite difficult/ the same as mathematics |
| **Apply the concept of structure and bonding to a variety of substances.** | | | | |
| (a) Covalent bonding | 2 | Found them hard to understand<br>Found chemistry quite difficult to grasp but felt that I got there in the end and completed and passed all exams | | |
| (b) Ionic bonding | | Found them hard to understand<br>Found chemistry quite difficult to grasp but felt that I got there in the end and completed and passed all exams | | |

*continued on following page*

*Table 5. Continued*

| Consider the following topics of Introduction to Chemistry. *Tick those subjects that you feel you had difficulty with: Please also indicate why you found them difficult (I still do not fully understand it)* | | | | |
|---|---|---|---|---|
| Topics | 'Traditional' Curriculum (n=11) | | Revised Curriculum (n=18) | |
| | Student Frequency | Comments | Student Frequency | Comments |
| (c) Metallic bonding | | Found them hard to understand<br>Found chemistry quite difficult to grasp but felt that I got there in the end and completed and passed all exams | | |
| (d) Formulae | | Found them hard to understand<br>Initially hard to grasp<br>Found chemistry quite difficult to grasp but felt that I got there in the end and completed and passed all exams | | |
| **Apply the concept of neutrali-sation.** | | | | |
| (a) pH changes ($OH^-$ and $H^+$) | 3 | [No comment received]<br>Trying to remember about $OH^-$ and $H^+$ [!]<br>Found chemistry quite difficult to grasp but felt that I got there in the end and completed and passed all exams | | |
| (b) Acid and alkali word equations | 2 | Quite complex<br>Found chemistry quite difficult to grasp but felt that I got there in the end and completed and passed all exams | | |

Generally, the student FE experience in Scotland is very good; and supports peer interactivity, discussion and debate. In addition, students gave universal support to course organisational structures and the academic staff involved. Given that research has demonstrated the varying reasons for entering FE, these factors will no doubt have positive impacts on the students themselves – for example, increased confidence, a sense of achievement and intentions to study further.

On a more strategic scale, the FE sector in Scotland is currently undergoing major changes, with colleges being rationalized and integrated.

## SOLUTIONS AND RECOMMENDATIONS

This study perhaps demonstrates to educators, lecturers and teachers, some of the factors that could implicate meaningful and relevant learning. 'Getting to know your learners', e.g. through questionnaires and focus groups, could help teachers understand what the cohort is really interested in, gauge their prior knowledge, and to formulate and align strategies for meaningful and relevant learning, as set by awarding standards.

FE provision in Scotland is excellent and has a considerable role in supporting formal education, lifelong learning, and science education. Students consistently praise the work of the sector and how learning has opened fruitful doors to further development. Based on the research evidence, however, a review of laboratory practicals and assessment structures could be undertaken. This could help support an integrated and holistic infrastructure to courses and modules, which also align with the concepts of meaningful and relevant learning. In addition, with the requirement of many learning establishments to have a minimum e-Learning presence, a vast amount of online learner analytics will yield valuable information on learning, achievement, and success. Indeed, this is a growing area within the educational research arena. Other potential research strands could involve the role of work-placements and how they consolidate learning and application in the workplace. On a more ambitious scale, the whole of the FE nursing curriculum could be methodologically explored and adjusted as per the obtained research evidence. Indeed, these are exciting times for the FE sector in Scotland, with many opportunities to refresh and embed best practices that align with educational neuroscience and meaningful learning.

In summary, the following recommended principles are suggested for supporting non-major science students.

- A research informed, evidence based curriculum.
- A curriculum that supports relevancy and authentic learning.
- A mixture of assessment formats, including summative and formative processes.
- Constructivist approaches, through peer and material interaction.
- The laboratories provide opportunities to explore and 'confirm' knowledge.
- The opportunity to engage in exploratory, authentic or project based learning.

The incorporation of chemistry within nursing may have a considerable role to play, and given this importance, the presentation, sequencing and teaching of the subject can be delivered in such a way that it is perceived to be relevant and exciting. In turn, this delivery will also (implicitly) include universal science education and the opportunity to further develop students' scientific thinking skills, such as reviewing evidence, being objective, critiquing developments, and drawing conclusions. Taken together, these can all support the development of nurses for citizenship, employment and in the delivery of excellent and sustainable services.

## REFERENCES

Anderson, N. (2004). Work with passion: How to do what you love for a living (Revised and Expanded ed.). Novato, CA: New World Library.

Ang, K. C. S., & van Reyk, D. (2013). 'Teach Me Chemistry Like a Ladder and Make it Real' - Barriers and Motivations Students Face in Learning Chemistry for Bioscience. *International Journal of Innovation in Science and Mathematics Education*, *21*(2), 1–12.

Atkins, P. (2003). *Atkins' Molecules* (2nd ed.). Cambridge, UK: Cambridge University Press.

Barmby, P., Kind, P. M., & Jones, K. (2008). Examining Changing Attitudes in Secondary School Science. *International Journal of Science Education*, *30*(8), 1075–1093. doi:10.1080/09500690701344966

Bell, B., & Cowie, B. (2001). *Formative Assessment and Science Education*. Dordrecht, The Netherlands: Kluwer Academic Publishers.

Bennett, J., & Hogarth, S. (2009). Would You Want to Talk to a Scientist at a Party? High school students' attitudes to school science and to science. *International Journal of Science Education*, *31*(14), 1975–1998. doi:10.1080/09500690802425581

Bennett, J., Lubben, F., & Hampden-Thompson, G. (2013). Schools That Make a Difference to Post-Compulsory Uptake of Physical Science Subjects: Some comparative case studies in England. *International Journal of Science Education*, *35*(4), 663–689. doi:10.1080/09500693.2011.641131

Black, P., & Wiliam, D. (1998). Inside the Black Box: Raising Standards Through Classroom Assessment. *Phi Delta Kappan*, *80*(2), 139–148.

Black, P., & Wiliam, D. (2009). Developing the theory of formative assessment. *Educational Assessment, Evaluation and Accountability*, *21*(1), 5–31. doi:10.1007/s11092-008-9068-5

Canning, R. (1999). Post-16 Education in Scotland: Credentialism and inequality. *Journal of Vocational Education and Training*, *51*(2), 185–198. doi:10.1080/13636829900200083

Cassels, J. R. T., & Johnstone, A. H. (1984). The Effect of Language on Student Performance on Multiple Choice Tests in Chemistry. *Journal of Chemical Education*, *61*(7), 613–615. doi:10.1021/ed061p613

Childs, P. E., & Sheehan, M. (2009). What's difficult about chemistry? An Irish perspective. *Chemistry Education Research and Practice*, *10*(3), 204–218. doi:10.1039/b914499b

College Development Network. (2013). College Development Network. Retrieved 21 December 2013, from http://www.collegedevelopmentnetwork.ac.uk/cdn-homepage.html

Colleges Scotland. (2013). *Colleges Scotland.* Retrieved 21 December 2013, from http://www.collegesscotland.ac.uk/colleges-scotland-homepage.html

Connelly, G., & Halliday, J. (2001). Reasons for Choosing Further Education: The views of 700 new entrants. *Journal of Vocational Education and Training*, *53*(2), 181–192. doi:10.1080/13636820100200155

Crown Copyright. (1992). *Further and Higher Education (Scotland) Act 1992.* Retrieved 21 December 2013, from http://www.legislation.gov.uk/ukpga/1992/37/contents

Crown Copyright. (2005). *Further and Higher Education (Scotland) Act 2005.* Retrieved 21 December 2013, from http://www.legislation.gov.uk/asp/2005/6/contents

Danili, E., & Reid, N. (2004, November). Some strategies to improve performance in school chemistry, based on two cognitive factors. *Research in Science & Technological Education*, *22*(2), 203–226. doi:10.1080/0263514042000290903

De Rennes, J. (1999). *Maintaining Space for Adult Learners in Science.* Paper presented at SCUTREA 29th Annual Conference. Warwick, UK.

Denholm, A. (2011). Higher levels of uptake in science subjects. *The Herald.* Retrieved 21 December 2013, from http://www.heraldscotland.com/news/education/higher-levels-of-uptake-in-science-subjects.14629814

Donaldson, G. (2011). *Teaching Scotland's Future: Report of a review of teacher education in Scotland.* Edinburgh, UK: Scottish Government.

Eagly, A. H., & Chaiken, S. (1993). *The Psychology of Attitudes.* London: Harcourt, Brace Jovanovich College Publishers.

Education Scotland. (2013). *What is Curriculum for Excellence?* Retrieved 21 December 2013, from http://www.educationscotland.gov.uk/thecurriculum/whatiscurriculumforexcellence/

El-Farargy, N. (2008). Book review: Cutting Edge Chemistry. *Reviews: A guide to publications in the Physical Sciences, 9*(1), 11-12.

El-Farargy, N. (2009a). Chemistry for student nurses: Applications-based learning. *Chemistry Education Research and Practice*, *10*(3), 250–260. doi:10.1039/b914507a

El-Farargy, N. (2009b). Epistemological beliefs and intellectual development in the physical sciences. *New Directions in the Teaching of Physical Sciences,* (5), 1-6.

El-Farargy, N. (2010). Evaluation of a chemistry curriculum intervention using the Perry model of intellectual development. *Chemistry Education Research and Practice*, *11*(2), 98–106. doi:10.1039/c005353h

Fishbein, M., & Ajzen, I. (1975). *Belief, Attitude, Intention and Behavior: An Introduction to Theory and Research*. London: Addison-Wesley Publishing Company.

Gallacher, J., Crossan, B., Leahy, J., Merrill, B., & Field, J. (2000). Education for All? Further Education, Social Inclusion and Widening Access. Glasgow, UK: Centre for Research in Lifelong Learning (CRLL), Glasgow Caledonian University.

Gilbert, J., & Priest, M. (1997). Models and Discourse: A Primary School Science Class Visit to a Museum. *Science Education*, *81*(6), 749–762. doi:10.1002/(SICI)1098-237X(199711)81:6<749::AID-SCE10>3.0.CO;2-I

Goldacre, B. (2013). *Building Evidence into Education*. Retrieved 22 December 2013, from https://www.gov.uk/government/news/building-evidence-into-education

Griggs, R. (2012). *Report of the Review of Further Education Governance in Scotland*. Edinburgh, UK: Scottish Government.

Harden, J. (1996). Enlightenment, empowerment and emancipation: The case for critical pedagogy in nurse education. *Nurse Education Today*, *16*(1), 32–37. doi:10.1016/S0260-6917(96)80090-6 PMID:8700068

Health Education England. (2013). *Investing in people, for health and healthcare. Workforce Plan for England: Proposed Education and Training Commissions for 2014/15*. London: Health Education England.

Howard, J. R., Short, L. B., & Clark, S. M. (1996). Students' Participation in the Mixed-Age College Classroom. *Teaching Sociology*, *24*(1), 8–24. doi:10.2307/1318894

Humes, W., & Bryce, T. (2003). The Distinctiveness of Scottish Education. In T. G. K. Bryce, & W. M. Humes (Eds.), *Scottish Education post devolution* (2nd ed., pp. 108–118). Edinburgh, UK: Edinburgh University Press.

Hussein, F., & Reid, N. (2009). Working memory and difficulties in school chemistry. *Research in Science & Technological Education*, *27*(2), 161–185. doi:10.1080/02635140902853632

IFF. (2003). *Study of learners in Further Education (No. 1 84478 083 X)*. IFF Research Ltd.

Ing, M. (2013). Can Parents Influence Children's Mathematics Achievement and Persistence in STEM Careers? *Journal of Career Development*. doi: doi:10.1177/0894845313481672

Jenkins, E. W., & Nelson, N. W. (2005). Important but not for me: Students' Attitudes towards Secondary School Science in England. *Research in Science & Technological Education*, *23*(1), 41–57. doi:10.1080/02635140500068435

Johnstone, A. H., & El-Banna, H. (1986). Capacities, Demands and Processes: A Predictive Model for Science Education. *Education in Chemistry*, *23*(5), 80–84.

Johnstone, A. H., & Selepeng, D. (2001). A Language Problem Revisited. *Chemistry Education: Research and Practice in Europe*, *2*(1), 19–29.

Kilminster, S. (1997). *Vocational Education and Really Useful Knowledge*. Paper presented at the 27th Annual SCUTREA Conference. Crossing orders, Breaking Boundaries: Research in the Education of Adults Conference. London, UK.

Knowles, M. S. (1990). *The Adult Learner: A Neglected Species* (4th ed.). London: Gulf Publishing Company.

Kolb, D. A. (1984). *Experiential Learning: Experience as The Source of Learning and Development*. Upper Saddle River, NJ: Prentice Hall Inc.

Liddell, G., & Macpherson, S. (2013). *SPICe Briefing: Post-16 Education (Scotland) Bill*. Edinburgh, UK: Scottish Parliament.

Lister, T. (2006). *Cutting Edge Chemistry*. London: Royal Society of Chemistry.

Mahaffy, P. (2004). The Future Shape of Chemistry Education. *Chemistry Education: Research and Practice in Europe*, *5*(3), 229–245.

McNair, S., Parry, G., Brooks, R., & Cole, P. (2004). *Learning together: Age mixing in further education colleges*. Learning and Skills Research Centre.

Merrill, B. (2001). Learning and Teaching in Universities: Perspectives from adult learners and lecturers. *Teaching in Higher Education*, *6*(1), 5–17. doi:10.1080/13562510020029563

BBC News. (2006). *Decline in student science uptake*. Retrieved 20 December 2013, from http://news.bbc.co.uk/1/hi/scotland/5124108.stm

NHS Education for Scotland. (2013). *Nursing and Midwifery: Supporting Scotland's nurses and midwives*. Retrieved 21 December 2013, from http://www.nes.scot.nhs.uk/education-and-training/by-discipline/nursing-and-midwifery.aspx

Niaz, M. (2005). How to Facilitate Students' Conceptual Understanding of Chemistry? A History and Philosophy of Science Perspective. *Chemical Education International*, *6*(1), 1–5.

Nursing and Midwifery Council. (2013). *Our role*. Retrieved 21 December 2013, from http://www.nmc-uk.org/About-us/Our-role/

Osborne, J. (2006). *Communicating Science: A BAI Roundtable Summary*. Retrieved 22 December 2013, from http://cils.exploratorium.edu/pdfs/CILS%20BAI_Osborne_Communicating%20Science.pdf

Osborne, J., Simon, S., & Collins, S. (2003). Attitudes towards science: A review of the literature and its implications. *International Journal of Science Education*, *25*(9), 1049–1079. doi:10.1080/0950069032000032199

Paterson, L. (2003). *Scottish Education in the Twentieth Century*. Edinburgh, UK: Edinburgh University Press.

Pell, T., & Jarvis, C. (2001). Development of attitude to science scales for use with children of ages from five to eleven years. *International Journal of Science Education*, *23*(8), 847–862. doi:10.1080/09500690010016111

Perry, W. J. (1999). *Forms of Ethical and Intellectual Development in the College Years: A Scheme*. San Francisco: Jossey-Bass Inc.

Reid, N., & Skryabina, E. A. (2003). Gender and Physics. *International Journal of Science Education*, *25*(4), 509–536. doi:10.1080/0950069022000017270

Reiss, M. R. (2004). Students' attitudes towards science: A long-term perspective. *Mathematics and Technology Education*, *4*(1), 97–109. doi:10.1080/14926150409556599

Roberts, G. G. (2002). *SET for success. The supply of people with science, technology, engineering and mathematics skills. The report of Sir Gareth Roberts' Review*. London: HM Treasury.

Royal College of Nursing. (2013). *About us*. Retrieved 21 December 2013, from http://www.rcn.org.uk/aboutus

Scottish Executive. (2001). *A Science Strategy for Scotland*. Glasgow, UK: Scottish Executive.

Scottish Executive. (2003). *Life Through Learning, Learning Through Life: The Lifelong Learning Strategy for Scotland.* Glasgow, UK: Scottish Executive.

Scottish Funding Council. (2012). *Facts & Figures: The 2012 at a glance guide to the Scottish Funding Council.* Edinburgh, UK: Scottish Funding Council.

Scottish Funding Council. (2013a). *Colleges we fund.* Edinburgh, UK: Scottish Funding Council.

Scottish Funding Council. (2013b). *Infact Database.* Retrieved 21 December 2013, from https://stats.sfc.ac.uk/infact/index.htm

Scottish Funding Council. (2013c). *Learning for All.* Retrieved 21 December 2013, from http://www.sfc.ac.uk/funding/FundingOutcomes/Access/learningforall/LearningforAll.aspx

Scottish Funding Council. (2013d). *SFC Corporate Publication: Annual Report and Accounts 2012-13.* Edinburgh, UK: Scottish Funding Council.

Scottish Government. (2007). *Skills for Scotland: A Lifelong Skills Strategy.* Edinburgh, UK: Scottish Government.

Scottish Government. (2008). *Science for Scotland.* Edinburgh, UK: Scottish Government.

Scottish Government. (2010). *Science and Engineering 21 - Action Plan for Education for the 21st Century.* Edinburgh, UK: Scottish Government.

Scottish Government. (2011). *Putting Learners at the Centre: Delivering our Ambitions for Post-16 Education.* Edinburgh, UK: Scottish Government.

Scottish Government. (2012). *Supporting Scotland's STEM Education and Culture. Science and Engineering Education Advisory Group (SEEAG). Second Report: January 2012.* Edinburgh, UK: Scottish Government.

Scottish Government & Scottish Funding Council. (2011). *College Regionalisation: Proposals for implementing Putting Learners at the Centre.* Edinburgh, UK: Scottish Government & Scottish Funding Council.

Scottish Qualifications Authority. (1993). *National Certificate Module: Unit Specification General Information - Introduction to Chemistry.* Retrieved 22 December 2013, from http://www.sqa.org.uk/files/nu/3181213.pdf

Scottish Qualifications Authority. (2006). *Annual Statistical Report 2005.* Glasgow, UK: Scottish Qualifications Authority.

Scottish Qualifications Authority. (2013a). *About the Scottish Qualifications Authority (SQA)*. Retrieved 21 December 2013, from http://www.sqa.org.uk/sqa/5656.html

Scottish Qualifications Authority. (2013b). *Scottish Credit and Qualifications Framework*. Retrieved 21 December 2013, from http://www.sqa.org.uk/sqa/5659.html

Scottish Qualifications Authority. (2013c). *What We Do*. Retrieved 21 December 2013, from http://www.sqa.org.uk/sqa/5659.html

Sirhan, G., & Reid, N. (2001). Preparing the Mind of the Learner - Part 2. *University Chemistry Education*, *5*, 52–58.

Stead, K. (1985). An Exploration, Using Ajzen and Fishbein's Theory of Reasoned Action, of Students' Intentions to Study or not to Study Science. *Research in Science Education*, *15*(1), 76–85. doi:10.1007/BF02356528

Taber, K. S. (2011). Constructivism as Educational Theory: Contingency in learning, and optimally guided instruction. In J. Hassaskhah (Ed.), *Educational Theory* (pp. 39–61). New York: Nova.

Taber, K. S. (2013, May). The right medicine for educational research? *Education in Chemistry*.

Tao, P. K. (1994). Comprehension of Non-Technical Words in Science: The case of students using a 'foreign' language as the medium of instruction. *Research in Science Education*, *24*(1), 322–330. doi:10.1007/BF02356359

Tasker, R., & Dalton, R. (2006). Research into practice: Visualisation of the molecular world using animations. *Chemistry Education Research and Practice*, *7*(2), 141–159. doi:10.1039/b5rp90020d

The Royal Society. (2011). *Preparing for the transfer from school and college science and mathematics education to UK STEM higher education: A 'state of the nation' report*. London: The Royal Society.

Tomei, A., Dawson, E., & Dillon, J. (2013). *A study of science, technology, engineering and mathematics education in the United Kingdom, Consultant report: securing Australia's future: STEM: country comparisons*. Melbourne: Australian Council of Learned Academies.

Torrance, H., & Coultas, J. (2004). *Do summative assessment and testing have a positive or negative effect on post-16 learners' motivation in the learning and skills sector? A review of the research literature on assessment in post-compulsory education in the UK*. London: Learning and Skills Research Centre.

UK Commission for Employment and Skills. (2013). *How can education and training providers use NOS?* Retrieved 21 December 2013, from http://nos.ukces.org.uk/about-nos/Pages/Education-and-Training-Providers.aspx

Vygotsky, L. (1978). *Mind in Society*. Cambridge, MA: Harvard University Press.

Woolnough, B. E. (1993). Teachers' perception of reasons students choose for, or against, science and engineering. *The School Science Review*, *75*(270), 112–117.

Young, A. (2006). Why young people are turned off science. *The Herald*. Retrieved 20 December 2013, from http://www.heraldscotland.com/sport/spl/aberdeen/why-young-people-are-turned-off-by-science-1.21353

## ADDITIONAL READING

Adey, P. (1988). Cognitive Acceleration: Review and Prospects. *International Journal of Science Education*, *10*(2), 121–134. doi:10.1080/0950069880100201

Allport, G. (1935). Attitudes. In C. M. Murchison (Ed.), *Handbook of Social Psychology* (pp. 798–844). London: Open University Press.

Alsop, S., & Watts, M. (2003). Science Education and Affect. *International Journal of Science Education*, *25*(9), 1043–1047. doi:10.1080/0950069032000052180

Armitage, A., Bryant, R., Dunnill, R., Hayes, D., Hudson, A., & Kent, J. et al. (2003). *Teaching and Training in Post-Compulsory Education* (2nd ed.). Buckingham: Open University Press.

Ashcraft, M. H. (2002). *Cognition* (3rd ed.). Upper Saddle River, New Jersey: Prentice Hall Inc.

Bell, P., & Linn, M. C. (2004). Beliefs About Science: How Does Science Instruction Contribute? In B. K. Hofer, & P. R. Pintrich (Eds.), *Personal Epistemology: The Psychology of Beliefs About Knowledge and Knowing* (pp. 321–346). London: Lawrence Erlbaum Associates.

Caine, R. N., & Caine, G. (1990). Understanding a Brain-Based Approach to Learning and Teaching. *Educational Leadership*, *48*(2), 66–70.

Christoff, K. (2008). Applying Neuroscientific Findings to Education: The Good, the Tough, and the Hopeful. *Mind, Brain, and Education*, *2*(2), 55–58. doi:10.1111/j.1751-228X.2008.00031.x

El-Farargy, N. (2009). Book review: The concepts and practices of lifelong learning. *Reviews: A guide to publications in the Physical Sciences, 10*(1), 64-65.

Gilbert, J. K. (2006). On the Nature of "Context" in Chemical Education. *International Journal of Science Education*, *28*(9), 957–976. doi:10.1080/09500690600702470

Johnstone, A. H. (1991). Why is science difficult to learn? Things are seldom what they seem. *Journal of Computer Assisted Learning*, *7*(2), 75–83. doi:10.1111/j.1365-2729.1991.tb00230.x

Lising, L., & Elby, A. (2005). The Impact of Epistemology on Learning: A case study from introductory physics. *American Journal of Physics*, *73*(4), 372–382. doi:10.1119/1.1848115

Morgan-Klein, B., & Osborne, M. (2007). *The Concepts and Practices of Lifelong Learning*. Abingdon: Routledge.

Roberts, L. F., & Wassersug, R. J. (2009). Does Doing Scientific Research in High School Correlate with Students Staying in Science? A Half-Century Retrospective Study. *Research in Science Education*, *39*(2), 251–256. doi:10.1007/s11165-008-9083-z

Simpson, E., & Courtney, M. (2002). Critical thinking in nursing education: Literature review. *International Journal of Nursing Practice*, *8*(2), 89–98. doi:10.1046/j.1440-172x.2002.00340.x PMID:11993582

## KEY TERMS AND DEFINITIONS

**Applications-Based Learning:** A learning and teaching method in which the applications of a subject/course are firstly presented, after which the principles and underlying concepts are examined and understood. The applications can involve employment, workforce and everyday life examples, which outline the importance and relevance of the subject/course in question. In essence, it offers learners the context for the learning content.

**Attitudes:** A multi-faceted mixture of cognition and affect, manifesting as a favorable or dis-favorable evaluation of any object, event, environment, or person. Attitudes can be private or be manifested via behavior. They can be implicitly or explicitly formed, and may be entirely stable or unstable. Some may be subject to change through experience, knowledge accumulation, persuasive communication or affective appeal.

**Cognitive Characteristics:** A set of psychological (pertaining to knowing, learning, reasoning, perception, memory and judgment) or distinguishing qualities of the mind.

**Expertise:** A state of mind and behavior that characterizes mastery and sophistication of knowledge, skill, information processing, performance, judgment, ethics and decision making, in a particular area.

**Information Processing:** A way in which information is processed (e.g. in the mind). As a cognitive science, it researches and documents the gathering, storing, processing, classifying, linking, manipulating, retrieving and utilizing of information.

**Lifelong Learning:** A relentless pursuit of ongoing (mostly formal) voluntary learning throughout one's entire existence.

**Vocational Education:** An educational endeavor that is based upon, and prepares learners for, a specific trade or profession.

**Working Memory:** A suggested memory 'space' where e.g. the perception, retrieval and processing of immediate information takes place.

# Chapter 2
# Developing Scientific Literacy:
## Introducing Primary-Aged Children to Atomic-Molecular Theory

**Jennifer Donovan**
*University of Southern Queensland, Australia*

**Carole Haeusler**
*University of Southern Queensland, Australia*

## EXECUTIVE SUMMARY

*This chapter challenges existing school science curricula modes for teaching atomic-molecular structure and describes a current research project designed to provide supporting evidence for reviewing school science curricula. Using evidence from this project and other research studies, the chapter argues for the introduction of atomic-molecular structure in the curriculum at Year 3 or 4 and proposes that consideration be given to devising a spiral curriculum in which the macroscopic and microscopic properties of matter are taught concurrently rather than sequentially.*

DOI: 10.4018/978-1-4666-6375-6.ch002

## ORGANIZATION BACKGROUND

Three years ago, a former high school teacher responded to questions about matter and atoms from his young son. His son's interest and apparent capacity to grasp the concepts led to the teacher offering to teach the rest of his son's primary class. The apparent success of this early venture led to further development of the teaching and learning program and the backyard development of innovative hands-on models to better facilitate the learning. We are two science teachers, now University educators of preservice primary teachers, who became interested in this program. Our study seeks to verify whether the teacher's claims of success can be supported by research. Consequently, the research participants in this case are a diverse class of Year 4 children in a school new to the specialist science teacher. Our research examines the development in these children's understanding of atomic-molecular theory from their learning experiences with the specialist science teacher following 10 hours of instruction on atoms, molecules, and elements (1 hour per week over a 10-week period).

## SETTING THE STAGE

Commonly, the teaching of atomic-molecular structure begins in high school. For example, in the new Australian Curriculum: Science (Australian Curriculum, Assessment and Reporting Authority [ACARA], 2013) the first mention of 'atoms' is in Year 9, when most students are 14 years old. The new K-12 Next Generation Science Standards from USA (National Research Council [NRC], 2013) are based on disciplinary core ideas from their earlier framework (NRC, 2012). This K-12 Framework introduces particles at Grade 5, and then elaborates these as atoms at middle school level, Grade 6. By the end of Grade 8 students should know there are approximately 100 different types of atoms, but even in this bold new curriculum which aims to introduce core ideas in science, technology and engineering from students' earliest schooldays, the details of atomic-molecular structure and the Periodic Table are still not tackled until Grade 9. However, at least this progression attempts a spiral curriculum (pioneered by Bruner, 1960) by introducing the scientific language of atoms earlier and building upon this baseline. The new national science curriculum to be introduced in the United Kingdom from September 2014 appears at first glance to be conservative, but introduces the particle model and atoms from Key Stage 3, i.e. Year 7 and onwards (Department of Education, 2013). However, this is classed as high school and part of the secondary science curriculum; there is no mention of atoms in the primary science curriculum.

Yet an Australian researcher (Jakab, 2013) found that most of her participants aged 8 years or older could state some everyday knowledge of molecules when first asked, and some 11 year olds had sophisticated knowledge, one expressing the aspiration to become a particle physicist. This chapter will report on an independent innovative attempt to teach children of equivalent age about atoms, atomic-molecular theory and the Periodic Table.

This practice of leaving atomic-molecular structure to high school seems to be the consequence of the developmental stage theory of Piaget and others (Inhelder & Piaget, 1958). Interestingly, in the Australian context, this approach also seems to coincide with broad student resistance to, and lack of enthusiasm for, the learning of science. This is evidenced by measureable decline in the number of Australian secondary students who continue with the study of science, particularly the physical sciences, into the final years of high school and university (Goodrum, Druhan, & Abbs, 2011). Yet research reported by Tytler and Osborne (2012) has shown that students are highly interested in science at 10 years of age, and form their career aspirations by age 13 or 14. The importance of engaging students early in science education is supported by other studies: grade 8 students who expected to have a career in science are more likely to graduate with a science degree (Maltese & Tai, 2010; Tai, Lui, Maltese, & Fan, 2006) and 65% of a sample of scientists and graduate students had developed their interest in science before middle school (Maltese & Tai, 2010). Leaving the 'Big Ideas' of science until high school may be too late.

## The Problem with Piaget

The Piagetian model of developmental stages (Inhelder & Piaget, 1958) holds that children pass through four defined stages of cognitive development. Infants to age 2 years are in the sensorimotor stage, and from ages 2 to 7, children are in the pre-operational stage, during which they cannot conserve quantity nor think logically. Children aged 7 to 11 years are in the concrete operational stage in which they begin to think logically but only with practical aids, and from ages 11 to 16 years and onwards, children transition to the formal operational stage with the development of abstract thinking. It is on this basis that abstract concepts such as atoms are delayed in curricula until children are in the middle of the proposed transition to the formal operational stage.

Curiously, some curricula are inconsistent, in that some abstract concepts such as atoms and DNA are delayed, whereas other abstract concepts, such as energy, are not. For example, energy is introduced in Year 6 in the Australian curriculum (ACARA, 2013) and in Grade 4 in the new USA standards (NRC, 2013). However, the forthcoming UK curriculum is more consistent in that neither atoms nor energy concepts are mentioned in the primary curriculum (Department of Education, 2013).

Piaget's theory has been challenged by developmental psychology (Bidell & Fischer, 1992). Children's cognitive development is highly variable, and variability exists at all ages, in all areas of learning and at all points in learning (Siegler, 2007). Not only does variability exist between different people, it is also evident within an individual solving the same problem at two points close in time, or even within a performance on a single problem. Variability in thought and actions occurs in infants (Adolph, Bertenthal, Boker, Goldfield, & Gibson, 1997), toddlers (Chen & Siegler, 2000), pre-school children (Flynn, O'Malley, & Wood, 2004), older children, and adults. In a study of the development of scientific reasoning (Schauble, 1996), grade 5-6 children and non-science adults demonstrated significant variability in understanding of content and experimental strategies. The way people think is constructive, dynamic, and culturally embedded as are the organisation and pattern of their psychological structures (Fischer & Bidell, 2006). Rather than following distinct hierarchical stages, children's cognitive development shows variability in the age, synchronicity and sequence of acquisition of specific skills (Bidell & Fischer, 1992), and this variability is dependent upon factors such as the area of learning, cultural background, learning history and learning style.

Siegler's overlapping waves theory (Siegler, 1996,1998, 2006) also recognises the variability in cognitive development. For example, in solving problems, children choose adaptively among strategies, with some strategies becoming less frequent, others becoming more frequent; new strategies are discovered and others abandoned. A similar pattern of variability has been found in the age, synchronicity, and sequence of children's understanding of the concept of matter. Applying Fischers' dynamic skill theory (Bidell & Fischer, 1992) and Siegler's overlapping waves theory (Siegler, 1996, 1998) to the US sample from Third International Mathematics and Science Study (TIMSS) data set, Liu and Lesniak (2005) proposed a model of student matter concept development from elementary to high school which comprised a series of multiple successive and overlapping waves. A subsequent phenomenographic study by the same authors (Liu & Lesniak, 2006) of grade 1 to grade 10 students' conceptual progression patterns on matter confirmed that there was no clear conceptual leap between different grade levels.

## Children's Curiosity and Innate Capabilities

Piaget's theory underestimates children's capabilities. Many young children display uninhibited curiosity that has an affinity with the scientific method and philosophy itself. As the following examples will show, they are more than simplistic thinkers and are able to engage in quite sophisticated reasoning processes that are the foundations for scientific thinking (Fleer, 2009). Elementary aged children used the intuitive rule "everything comes to an end" when asked to consider the continual subdivision of both material and mathematical objects (Smith, Solomon, & Carey, 2005; Yair & Yair, 2004). In discussions about the process of evaporation (Tytler & Peterson, 2000), 5 year-old children used elementary conceptions of substance. Prior to instruction, children aged 7-10 were able to express naïve ideas of the particulate nature and behaviour of matter (Nakhleh & Samarapungavan, 1999). Similarly, Jakab (2013) describes how 6-11 year-old children were able to articulate ideas about the molecular nature of matter when offered the use of molecular artefacts such as symbols, diagrams, models and a website with interactive models.

## The Importance of Background Knowledge and the Quality of Instruction

The conclusions of cognitive psychology (Hirsch, 2006; Willingham, 2008) reveal that learning history and learning style are important factors in the conceptual development of children. Background knowledge is critical in providing contextual information enabling children to make sense of what they read, view, and absorb from the world around them. Therefore, both Willingham and Hirsch consider it integral to practice to expose children to background knowledge that may appear to be beyond their immediate full understanding but which helps to provide contextual information for future learning. In this, they follow in the footsteps of Bruner, who in 1960 suggested that no content should be off limits for school-age children. He said

*We begin with the hypothesis that any subject can be taught effectively in some intellectually honest form to any child at any stage of development. It is a bold hypothesis and an essential one in thinking about the nature of the curriculum. No evidence exists to contradict it; considerable evidence is being amassed that supports it. (Bruner, 1960, p. 33)*

Bruner went on to suggest that children are able to get an intuitive grasp of a complex concept before they have the background and maturity to deal with the same topic in a formal manner. More recently, Lehrer and Schauble's (2000) research showed that revisiting science ideas enables students to understand and apply concepts that they would not typically understand until several years later.

Murphy (2012) supports Vygotsky's contention that learning leads development, so teachers should always be challenging students rather than waiting for them to reach a predetermined developmental stage. Unfortunately, curricula do not always reflect these insights, and rarely give children the opportunity to engage with concepts beyond their current level of thinking or to revisit them periodically. Willingham (2008) points out:

*For children and adults, the understanding of any new concept is inevitably incomplete... If you wait until you are certain that the children will understand every nuance of a lesson, you will likely wait too long to present it. If they understand every nuance, you're probably presenting content that they've already learned elsewhere. (p. 39)*

It is the thinking of researchers such as Bruner, Willingham, and Vygotsky that encourages the earlier introduction of concepts, with concrete aids where possible. This aims to facilitate the transition of children through development in their cognition, whether or not such development occurs in set Piagetian stages or more gradually.

The conceptual understanding of children may be limited more by the quality of instruction than by any developmental process. In the 2007 National Academies report (Duschl, Schweingruber, & Shouse, 2007), *Taking Science to School: Learning and Teaching science in Grade K-8*, the authors reviewed the extant literature on cognitive and developmental psychology and science education. The conclusion from this review was that what young children are capable of is largely dependent on their prior opportunities to learn, and is not determined primarily by some fixed sequence of developmental stages. A student (or even a whole class) not understanding something does not mean that the task was developmentally inappropriate. Lack of understanding may indicate a lack of prerequisite knowledge or an ineffective way of presenting the material to make it easier to understand.

We note that the concept of teaching the 'Big Ideas' of science to younger children is not new. Other researchers have worked on ways of doing so; but thus far, curriculum policy has kept its distance from the outputs of such research. Effective teaching interventions can allow children to learn about atoms and molecules. Using role-play and building molecules with ball and stick molecular models can assist grade 5 students to learn about important molecules and their properties (Brown, Rushton, & Bencomo, 2008). Third grade students, exposed to a one-hour digital presentation of molecular models, were able to describe and draw accurate representations of molecules (Halpine, 2004). In 1993, Lee, Eichinger, Anderson, Berkheimer, and Blakeslee showed that addressing common misconceptions about matter and molecules improved Grade 6 students' understanding and application of the kinetic theory of matter to states of matter, changes of state, thermal expansion, and dissolving. The use of scientific modelling and argumentation in instruction is important in developing primary aged children's understanding of the atomic nature of matter (Schwarz et al., 2009). Acher, Arcà, and Sanmartí (2007) describe how 7-8 year old children used a "model of imaginary parts" (p. 401) built from their idea about the discrete materials to explain the behaviour of different materials. Extensive research by Nussbaum (1998) has demonstrated that in order to build students'understanding of atomic-molecular theory, they need to be engaged in cycles of model building and deep discussions about alternative theories and essential metaphysical and epistemological issues.

The recent development of learning progressions acknowledges that there are multiple pathways of conceptual change possible for student understanding of matter (Johnson & Tymms, 2011; Merritt, Krajcik, & Shwartz, 2008; Stevens, Delgado, & Krajcik, 2010; Wiser & Smith, 2008). However, most of these studies were based on existing curriculum models in which the macroscopic nature of matter is located in the primary curriculum and particulate models introduced in lower secondary years. Yet a longitudinal study of junior high school students in Grades 9 and 10 (Margel, Eylon, & Scherz, 2008) suggested that a long-term development of the particulate model requires building a strong foundation of knowledge about the microscopic structure of materials through a process of spiral instruction. In science, the judicious use of models, with clear explanations as to how they do and do not resemble the actual thing they are modeling, can be helpful in presenting abstract concepts to young children.

In earlier research (Donovan & Venville, 2004; Venville & Donovan, 2005), one of the authors and her colleague consulted expert geneticists for their opinions on essential genetics concepts students should acquire for everyday life, and on ways to teach these concepts. They recommended early introduction to vocabulary and use of pictorial and spatial models wherever possible. These findings led to the development of a simple wool model successfully used to introduce the essential vocabulary of DNA, gene, allele, and chromosome at a variety of age levels, the youngest being Year 2 students (aged 7 years). These students (Donovan & Venville 2005; Venville & Donovan, 2007, 2008) happened to be at an Islamic school and were all English-second-language students identified in Year 1 as requiring remedial assistance. At a subsequent post-test, these students demonstrated clear understanding that genes are made of DNA; that these molecules are responsible for our appearance being similar to that of our parents; and that identical twins would have the same DNA as each other. The model enabled them to learn some valuable genetics vocabulary and to link it with concepts of family identity. Consistent with Carey (2010), there is no claim that this fast mapping of the words 'genes' and 'DNA' enabled these students to develop full understanding of the words with all nuances of meaning. However, in current non-spiral curricula, which do not afford further exposure and opportunities for discussion and instruction, the extended mapping of these concepts, which Carey (2010) describes so clearly in the context of her research, will not occur. Opportunity has been lost. Thus we concur with Willingham's (2008, p. 39) notion that, "Without trivializing them, complex ideas can be introduced by making them concrete and through reference to children's experience."

Finally, support also comes from the field of neuroscience. It is now accepted that the brain is not fully developed early in life as was once thought. Instead, it has plasticity – structural and functional changes are possible throughout life. However, development is not linear. In very early life, the main plasticity involves the formation of new synapses, from 2,500 per cortical neuron at birth to 15,000 synapses per cortical neuron by age 3 (Gopnik, Meltzoff, & Kuhl, 1999). Adults have about half that number, so further development involves synaptic pruning. Neurons that are frequently used develop stronger connections; those rarely or never used eventually die. Learning may be defined as the ability to acquire new knowledge or skills through instruction or experience, memory as the process by which that knowledge is retained over time, and plasticity as the capacity of the brain to change with learning (Sousa, 2001). Information is initially placed into short-term memory, but over time is transferred into long-term memory, involving physical changes in the brain (Sousa, 2001). Drubach (2000) identified two types of such physical changes: a change in the internal structure of neurons, especially in the area of synapses; and an increase in the number of synapses between neurons. Further, recent neuroscience research suggests that ages 5-10 are years of heightened brain plasticity (Abdeldayem, 2012), during which the acquisition of science's 'Big Ideas' could be perfectly timed.

## Children's Prior Knowledge: The Influence of Media on Children's View of Science

The changing structure of the brain involved with learning results from the input of data. Children of today are surrounded by the mass media. A recent study conducted by one of the authors and her colleague (Donovan & Venville, 2012a, 2012b) of 141 children aged 10-12 years in four non-metropolitan areas in three Australian states reported an average level of exposure of 5 hours and 10 minutes per day. This averages 2 hours and 30 minutes per day less than for children in the USA (Rideout, Foehr, & Roberts, 2010). Television (TV) was the main contributor to this usage, averaging 800 hours per year. Children are thus exposed to considerable input of information.

Surprisingly, the study revealed very little research into the influence of this exposure to the mass media on children's academically relevant knowledge. Much is known of its influence on opinions, beliefs, attitudes and behaviours such as body image, risk-taking, and violence, but only a few studies had exposed children to specific TV programs or movies and probed how concepts presented were taken up by children. By contrast, the author's study considered the totality of children's voluntary exposure to entertainment mass media, and followed up these named examples for mentions of genes and DNA. These were cross-referenced to the understandings about genes and DNA expressed by 62 of the children in face-to-face interviews.

Although the study design cannot demonstrate causality, the evidence did indicate the likelihood that the participants' knowledge of genes and DNA (which, like atomic theory, is not taught in schools until children are aged 14 or 15) has been derived from their exposure to the mass media. The same genetics themes arose from the children, particularly DNA being used to solve crime and to resolve family relationships, as appeared prominently in the media examples they mentioned (Donovan & Venville, 2012a). Specifically, words used by children to describe how DNA is used to solve crime paralleled the way it is presented in crime shows that 79% of them reported viewing, despite these shows being rated for ages 15 years and older. For example, 12-year-old Annette (a pseudonym) said, "They use a special machine, and the machine will determine if it knows the DNA or if it's used that DNA before, and it will also show what the DNA looks like so you can compare it with other DNAs and find a culprit" (Donovan & Venville, 2012a, p. 25). Further evidence came from their relative lack of knowledge about the biological structure and function of DNA, which also paralleled the relative lack of this information in the mass media (Donovan & Venville, 2012a, 2012b). With 89% of the children knowing about DNA and 60% of them knowing about genes, this finding relates to the greater exposure of DNA compared with genes in the media. Collectively, this evidence indicates that, without formal teaching, primary children are capable of understanding more about genes and DNA than previously imagined and that the mass media are the most likely source of their information.

The children themselves (80% of them) acknowledged that TV was their major source of information, and were remarkably perceptive about which specific programs provided more information about DNA and genes (Donovan & Venville, 2012a). Furthermore, 27% of the participants had conducted their own research into genes and DNA and achieved sophisticated understandings. For example, 11-year-old Willis viewed few crime shows but had become interested in DNA from documentaries. He was able to describe in detail how DNA databases work, how to take a biopsy to test for cancerous cells, and knew that animals, humans and plants all have DNA. Thus, the participants in this study support Tytler & Osborne's (2012) findings that primary children are highly interested in science.

The favourite TV show nominated by participants in this study was *The Simpsons* (Donovan & Venville, 2012b). Searching *The Simpsons* wiki (http://simpsons.wikia.com/wiki/Simpsons_Wiki) indicates the show often mentions words related to atoms and atomic theory, with character Homer working in a nuclear power plant, outside of which is Nuclear Lake where waste is dumped. The local football team is The Springfield Atoms and the baseball team is The Springfield Isotopes. Many plotlines involve science and the show is far-reaching – even the eminent journal *Nature* was moved to select the staff's 10 favourite science moments in *The Simpsons* (Hopkin, 2007). However, it is not the only TV show to contain references to science. From classics such as *Dr Who* to the meteoric rise in popularity of *The Big Bang Theory*, today's children are bombarded with science as part of their daily entertainment.

Science is also found in other mass media. For example, 11-year-old John, one of Jakab's study children (Jakab, 2013) is very knowledgeable about molecules because he loves fantasy and science fiction books. He knew about methane from the plotline of a book that he has read. It would benefit primary school teachers to consider the sources of scientific vocabulary and concepts in the everyday worlds of the children they teach and 'add the science' to such encounters. At the very least, teachers should acknowledge that their children bring prior knowledge to the classroom, some of which may have been derived from their encounters with the mass media.

## Challenging the Paradigm

Science educators continue to express concern over the failure of traditional science curricula and traditional science pedagogy to engage students' interest in science (Tytler, Symington, & Smith, 2009). Wiser and Smith observe

*... science curricula treat knowledge as unproblematic facts; few students have any appreciation of the coherent nature of scientific theories or the role of ideas, models, and symbolisation, and cycles of hypothesis testing in their creation. (Wiser & Smith, 2008, p. 226)*

Margel, Eylon, and Scherz (2008) acknowledge that, despite the considerable time spent on instruction, existing traditional science curricula do not lead to robust particulate conceptions by the end of high school. Students' lack of understanding of matter and atomic-molecular theory continues to be reflected in many common misunderstandings (Özmen, 2004; Özmen & Ayas, 2003; Stein, Larrabee, & Barman, 2008; Vosniadou, 2012) even amongst senior high school students and college students of chemistry. Consequently, the argument that atomic-molecular theory should be introduced when students are 'developmentally' ready is flawed.

Johnson and Papageorgiou (2010) suggest that students' poor understanding of the particle theory of matter is a result of the 'solid, liquids, gases' context in which it is taught. Their work found that 9-10 year old children demonstrated greater understanding of the particle model when it was taught within the framework of a concept of substance. Wiser and Smith (2008) argue that atomic–molecular theory should be taught before students have a complete scientific theory of matter at the macroscopic level. How this is to be done has not been extensively explored. Our contention is that the elements of atomic-molecular theory should be introduced early in primary school, and continued within a spiral curriculum, revisited each year.

## CASE DESCRIPTION

### Research Participants

This pilot research project aims to verify claims made by a specialist high school science teacher that Year 3 and 4 children can learn atomic-molecular theory. Owing to mass media coverage of this teacher's innovative program, other schools have become interested in its uptake. Specifically, our research is being conducted in a metropolitan Catholic primary school in Queensland, Australia, because a parent of children at the school suggested to the Principal that their school could become involved. This made it an apparently ideal candidate to host the pilot research, as the specialist teacher has had no previous contact with the school that could confound the results. It is envisaged that future studies would expand the number of schools, contexts, and regions in which this program is offered to seek information about its generalisability to the Australian primary school population. Such broader research would also be more generalisable to the international scene. This chapter presents only preliminary results from the pilot study, in the hope of stimulating interest from potential collaborators to further this research.

Prior to commencing the research, ethics permission from both the Catholic education sector and our University was obtained, and the agreement of the Principal and classroom teacher. All participation in the research was with the written permission of parents and the continued willingness of the children to be involved, ascertained by asking them if they were happy to be interviewed each time. All names used in this chapter are pseudonyms from an appropriate cultural background.

The participants are thus a single class of 26 Year 4 children (average age 9 years 9 months) and one Year 1 child (Marcia, aged 6 years and present by the request of the parent). It is a diverse class. Three children (Kensei, Oliwia and Nadine) have English as their second language (ESL), with the latter two arriving late into the program from a holiday in their home country. Joel is another ESL student who also has Speech-Language Impairment (SLI). Edward has been designated as SLI and Intellectually Impaired (II), and requires an individualised learning program. Loughlin has Autism Spectrum Disorder (ASD) and Danisha is Hearing Impaired. The regular class teacher has welcomed the program as a professional development opportunity for herself as well as an extra learning opportunity for her students. At various times, a teacher aide and interested parents have joined the class to assist the children.

## Research Questions

Specifically, this pilot study sought to answer the following four research questions:

1. What do children aged 9 years believe science is, and what is their attitude towards science?
2. What prior knowledge about atoms, molecules, elements, and sub-atomic particles do children aged 9 years possess?
3. What knowledge about atoms, molecules, elements, and sub-atomic particles can children aged 9 years gain through an intervention designed by a specialist high school science teacher?
4. How can data obtained in this study inform the future development of the intervention?

## Research Methodology

The pilot project employs qualitative methodology involving the triangulation of three sources of data. The primary data set consists of the information gained from semi-structured interviews with individual children; this constitutes the main data presented here. Prior research experience with children of these ages yielded an expectation that repeating and paraphrasing questions in response to direct queries or body language would be necessary to achieve negotiated understanding of the questions. Consequently, a semi-structured interview protocol (Creswell, 2005) was the most appropriate method to yield rich qualitative data about children's conceptions about atoms. A secondary data set comprises the children's responses to classroom assessment tasks and the third data set is derived from teacher reflective journals.

The participant interviews are being conducted at three intervals – Stage 1: pre-instruction, Stage 2: post-instruction and Stage 3: approximately two months after post-instruction to assess children's retention of understanding. During the audio-recorded interviews, the children are able to draw or sketch how they visualise aspects of their thinking. In Stages 2 and 3, children have access to the models they have used in class in order to support their attempts to explain their understandings to the interviewer. At the time of writing, only Stages 1 and 2 interviews have been conducted. The authors (the researchers), without input from the specialist teacher, are conducting all interviews with the children to maintain appropriate distance and lack of bias.

The triangulation of data allows us to substantiate learning by matching the children's responses in interviews to responses on teacher devised assessment tasks (e.g., short response test items, investigation reports), and to teacher reflections on the learning processes. The children's responses to the interviews and assessment tasks are being analysed using a coding schema derived from Stevens, Delgado, and Krajcik (2010). The teacher reflective journals are analysed for teaching objectives, pedagogical strategies, and teachers' perceptions of student learning.

## What Was Taught during the 10 Hours of Instruction

In brief, the learning and teaching program covered the atomic nature of matter, properties of metals and non-metals, including conductivity, the structure of atoms, and the relationship of atomic-molecular structure to the properties of elements and their position on the Periodic Table. The children were taught how to read and interpret the Periodic Table in terms of the related properties of groups of elements such as the noble gases, the halogens, and the alkali metals. Valence electrons, covalent bonding, and the law of conservation of mass in simple chemical changes through conservation of atoms were also covered. The interview questions were drawn from the learning and teaching program but utilised different specific examples where possible. Consequently, a greater appreciation of what was covered in the teaching and learning program can be ascertained from the interview questions, and the marking scheme for scoring these questions, supplied in an appendix to this chapter.

In particular, the specialist teacher believes that the sequence of introduction of the concepts is critical, and this has been the subject of deep consideration in his development of the program. Also unique are samples and models that he has developed to support the learning. The samples include a set of 12 metals and 7 non-metals that the children can handle, including hydrogen and helium in balloons. The models include an atomic shell model to which children can add protons and electrons to build up the first 10 elements, and magnetic molecular models that accurately simulate the shapes of molecules, valency, and the sense of the involvement of energy in the making and breaking of bonds. Learning was also supported by worksheets and videos created by the specialist science teacher.

## Findings

In this chapter, we will present only the results of the comparison of the preliminary analysis of the repeated questions in the pre and post interviews. Analysis of the extra questions asked at the post-interview is still ongoing. The analysis is presented within two domains, the affective and cognitive domains.

## The Affective Domain

Asking the children if they liked science in the pre-interview showed this class was already very switched on to science, with 24 out of the 27 children reporting liking the subject. This was not necessarily an expectation of the researchers, as studies have shown that the teaching of primary science in Australia is patchy (Goodrum, Hackling, & Rennie, 2001). The remaining children were unsure. In the post-interviews, the number of unsure children dropped to one, Oliwia, who said she was "in the middle" and she "liked the activities, nothing much to dislike". English is a second language (ESL) for Oliwia and she arrived late into the program owing to a holiday in her home country. However, her sister Nadine, in the same circumstance, was more positive, enjoying science, and loving the experiments. One child, Merryn, said he did not like it now, as it was too hard. Interestingly his brother Tristan had a much more positive outlook, saying he liked it in both interviews, and looked ahead to the value of learning now about atoms for his future studies at high school. In the post-interview, Tristan expressed how much he enjoyed using chemicals and building molecules.

More differences were seen in the reasons why the children liked science. In the pre-interview, the main reason given was "fun", whereas in the post-interview the main reason given was the enjoyment of learning about atoms and molecules. Still a quarter of the class mentioned fun, and a quarter of the class now expressed a strong love for science, that it was their favourite subject, indicating their feelings had intensified since the pre-interview. Enjoying the experiments, activities, and models featured strongly in their responses, as did enjoying the challenge of learning about new things they did not know about before. Loughlin, the child with ASD, saw science as a means of making the world a better place, and Andrew, who also liked science, had been prompted to think deeply about the conflict between religion and the Big Bang because of the lessons.

Differences occurred in what the children thought science was, as seen in Table 1. Numbers refer to how many children mentioned each idea, but as children frequently mentioned several ideas, the numbers do not total to the number of children in the class.

Several trends are evident in Table 1. Before the pre-interview, the children had recently studied earth sciences, particularly volcanoes, accounting for the relative popularity of this answer, but it is apparent that not all children were constrained by this recent experience in their suggestions of what science is. Four children were already familiar with the Periodic Table, elements or atoms and atomic structure. Following the intervention, there was a large increase in the belief that science is about atoms and molecules, with more than half the class expressing this view, some

*Table 1. Ideas generated by children in response to the question: What do you think science is?*

| What do you think science is? | Number of children mentioning each idea | |
|---|---|---|
| | Pre | Post |
| experiments/data | 7 | 7 |
| chemical/mixing/explosions | 7 | 1 |
| the earth/volcanoes/rocks | 5 | 0 |
| discovering/finding or learning about new things or how things or the world works/inventing | 4 | 10 |
| space/sun/galaxies | 3 | 2 |
| gravity/push/pulls/friction | 3 | 0 |
| Periodic table/elements | 2 | 7 |
| engineering/technology | 2 | 1 |
| atoms/molecules/electrons/protons | 2 | 15 |
| cure diseases/cancer/germs/medicine | 2 | 3 |
| animals/nature/plants | 2 | 1 |
| dinosaurs/extinct animals | 2 | 1 |

of whom and others also mentioned the Periodic Table of elements. Again, children simply referring to what they had recently done could apparently explain this result. However, the numbers mentioning biological and space sciences changed only marginally, so recent experience does not entirely explain the new popularity of atoms and molecules. Experiments were still popular, but the 'flashy' idea that science is about chemicals and explosions gave way to more thoughtful interpretations of science, despite the program having included exploding a hydrogen-filled balloon. In particular, there was a large increase in the numbers of children believing that science is about discovering and learning about new things and how the world works.

At the end of the pre-interview we also asked the children where they had learned the science ideas they had spoken of during the interview. In descending order, their responses were school (11), parents (9), mass media – TV and movies (8), older sibling/cousin studying science (7), books (4), Periodic Table/element board (3), science show/museum (3), iPad game (1), and YouTube video (1). Again, many of the children cited more than one source of their information so the numbers do not match the number of participants.

## The Cognitive Domain

Notwithstanding the importance of the affective domain, our main interest nevertheless was in seeing what children had learned from participating in the program. Considering only the questions repeated in pre and post-interviews, referring to specific knowledge about atoms, molecules, elements, and sub-atomic particles, scores were assigned to the answers as indicated in the copy supplied in the appendix. Figure 1 shows the change in these scores (out of 50) between the pre and post-interviews.

Firstly, considering the pre-interview scores in Figure 1, it is clear that, while most children had minimal specific knowledge of atoms and molecules before the program, two children (numbers 9 and 27) had substantially more knowledge. These two children (Christian in Year 4 and Marcia in Year 1) are the children of the parent who had pressed for the program to be taken up by the school, which probably explains their pre-knowledge.

Secondly, considering the difference between the pre and post-interview scores, it is clear that every child in the class gained knowledge. In high school, a pass would be awarded to 14 year olds scoring 25 or better on a test incorporating such questions; on that criterion, 14 of these children passed. However, high school tests

*Figure 1. Individual participants' knowledge scores (out of 50) in pre and post-interviews*

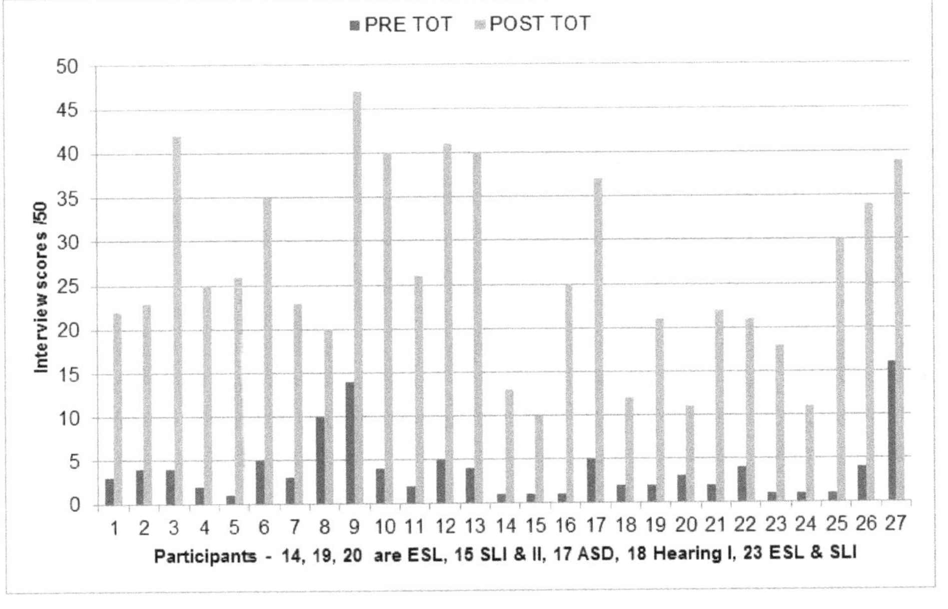

*Figure 2. The percentage increase in knowledge of individual participants as a result of the program*

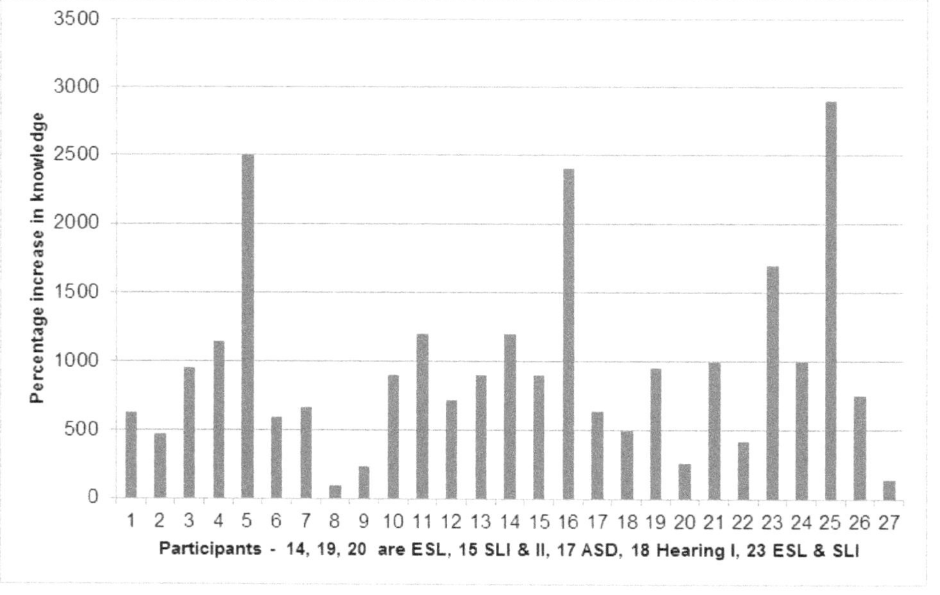


are often multiple-choice questions, an easier option than being asked face to face for an answer as these children were. Also, considering these children are only 9 years of age, an argument could be made that a score of 20/50 would indicate sound learning. On that criterion, 21 children passed. Given the diversity in this class, this is an outstanding result.

An alternative way of viewing their progress is shown in Figure 2, which maps the percentage increase in the knowledge of participants.

Figure 2 indicates that all participants experienced substantial increases in their knowledge. Children with the lowest percentage increases (children numbered 8, 9, and 27) were those with the highest starting knowledge. What is particularly telling is the gain made by children with special needs, as indicated in Figure 2. The intellectually impaired child, Edward (number 15), showed a 900% increase in knowledge. He was personally cognisant of this right from the start. One of the researchers, having completed the pre-interviews, sat in on the first lesson to observe. At the end of the lesson, Edward ran up to her and said, "I didn't know your questions the other day but now I know what an atom is!" His excitement was palpable. An ESL child, Kensei (number 14), showed a 1200% increase in knowledge as a result of the program, and Joel (ESL and SLI, number 23) showed a 1700% increase.

Further findings will be presented from analysis at a whole group level ($n$=27), for each subsection of knowledge examined: atoms, molecules, elements, and sub-atomic particles.

## Atoms

At the pre-interview, only three children had heard of atoms, but they knew very little else, other than two children knowing that atoms are very small. Not surprisingly, at the post-interview, all children had heard of atoms, all but one knew they are very small and most children offered several additional pieces of information about atoms. In all, 18 children knew that atoms make up everything, 11 could explain exactly how small atoms are, including five remembering a specific analogy used by the specialist teacher, five children launched into descriptions of the sub-atomic particles, and two thought to mention that elements have unique atoms. Only Matthew was unable to expand much on his claim of knowing the word.

When asked to draw an atom, 24 children made no attempt during the pre-interview, one drew a circle with flagella and dots in the middle, one drew a single circle, and one drew concentric circles. In the post-interview, only two children could not attempt a drawing, with one of these drawing the symbol for the element carbon. Ten children drew small dots, solid circles, single circles or circles side by side, two drew circles with connectors like the models, and one attempted to draw the fuzzy ball model of an atom, explaining that's what it was. The remaining 12 children came closer to drawing the internal structure of atoms, as one drew concentric circles with a nucleus, and six advanced on that by adding particles in the centre and on the rings. Five children drew atoms as concentric circles with positive protons in the nucleus and negative electrons on the rings, and could generally name the specific elements whose atoms they had drawn, as shown in Figure 3.

*Figure 3. Seb's drawing of an oxygen atom*

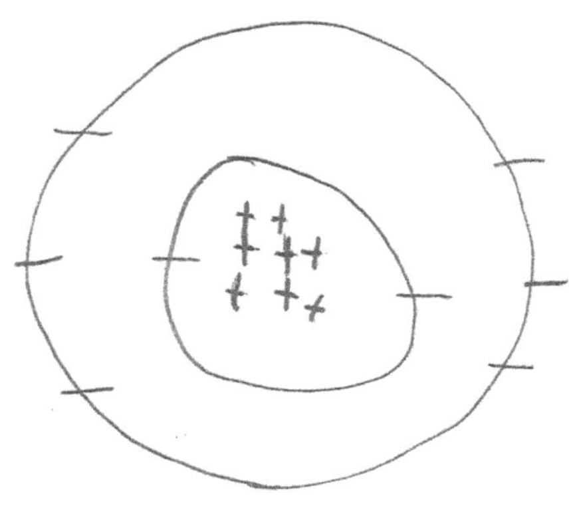

## Molecules

At the pre-interview, 14 children claimed to have heard of molecules, but that was all most had done, and they could not substantiate this with any other appropriate information. One child suggested it was something in a chemical, two thought it had something to do with liquids, one may have been thinking of models by suggesting it had to do with circles and toothpicks, one suggested germs, another suggested a machine and one said molecules help people survive. Only two children thought that molecules were bigger than atoms, none could name any molecules, and only two children attempted drawings. One drew linked circles and explained these were germs, whereas the other drew an oblong (the molecule) with smaller, filled in particles representing atoms inside.

At the post-interview, all except Edward, the intellectually impaired child, and Danisha, the hearing impaired child, now knew the word 'molecules'. The relatively larger size of molecules compared with atoms was known by 16 children, the others having forgotten or thought they were the same size. When asked for more information, seven children explained molecules were atoms joined together, while another eight children provided this information via a molecular or structural formula. Eleven children named appropriate molecules when asked, and these included water, carbon dioxide, hydrogen gas, methane, ethane, and acids. Seventeen children attempted a drawing, though one drew only a dot and one simply wrote $H_2O$. Loughlin (with ASD) drew the electron configurations of oxygen and hydrogen showing the sharing of electrons to make water, as seen in Figure 4.

The remaining molecules drawn ranged from simple ball-and-stick representations of water, $CO_2$, $H_2$ and $O_2$, to complex molecules such as $CH_3CH_2OH$, $CH_6ON_2$, $C_2H_3ON$, and $CH_3CH(OH)CH(OH)NH_2$ all drawn correctly as far as bonding and

*Figure 4. Loughlin's drawing of H and O forming water*

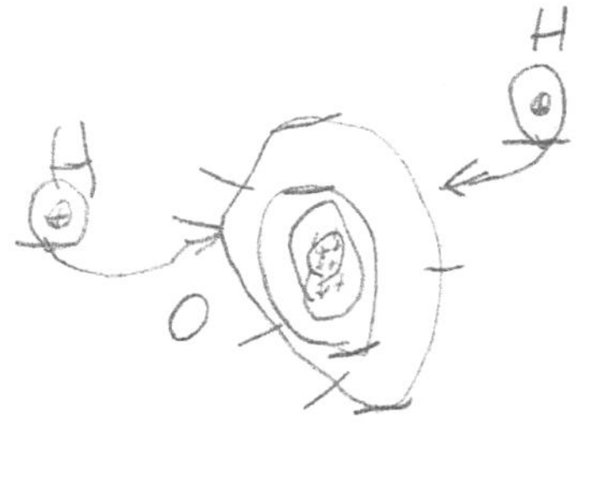

*Figure 5. Andrew's structural and molecular formula for ethanol*

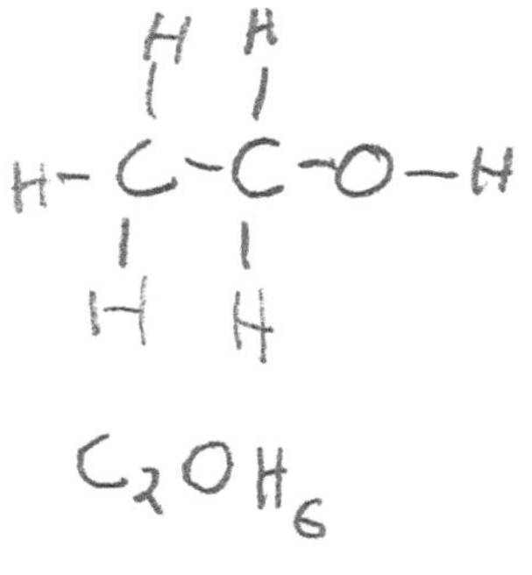

valency were concerned. The children who drew the last three molecules proudly declared they were "their own made-up molecules" and Victoria, who drew $C_2H_3ON$, needed to use the models to make it first before correctly drawing it. Of particular interest was Marcia who is in Year 1 (6 years old). She correctly drew $CO_2$ and explained the double bonds attaching each oxygen atom to the carbon. Some of the children's drawings were too pale to reproduce well, but Figures 5 and 6 show two of the more complex molecules drawn.

## Elements

In the pre-interviews, only four children said they had not heard the word 'element', but, when they were asked for more information, it became apparent that only three interpreted the word in its chemical sense. Marcia (the 6 year old) knew it was something with one type of atom, Olinda knew that two letters meant iron, and Christian mentioned that the element gold had gold atoms. Nine children spoke of earth, air, fire and water or variations on this, with two specifically mentioning they had seen this on TV. Tristan also referenced TV and referred to elemental powers, and Loughlin referenced the word 'element' as something you are good at, as in, 'You're in your element'. Others were unclear in their responses or said they had only heard the word and did not know more about it. Marcia was the only child who could name four elements, and she and her older sibling Christian were the only two who knew any letters representing elements (H, O, Fe, Ca, and Cu).

In post-interviews, only Oliwia claimed not to know the word 'element' because she was away at the time, although her sister Nadine had also been away on holiday but recognised the word. Oliwia and two others could offer no further information about elements, Edward and Kensei were unclear, and three children persisted with

*Figure 6. Oliwia (ESL) drew a complex made-up molecule correctly*

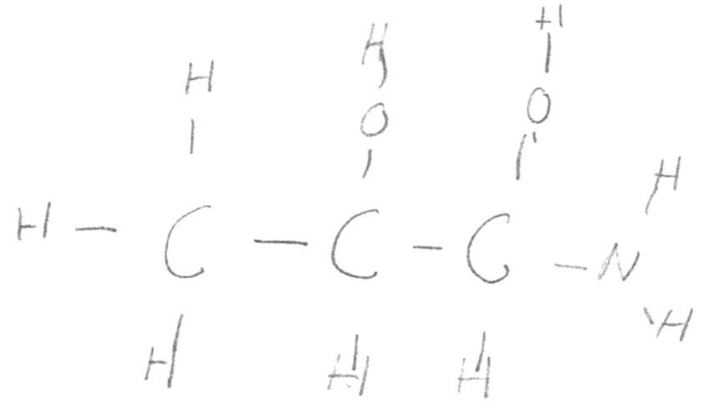

earth, air, fire, and water variations. Fifteen children specifically said that elements were made of one type of atom, eight mentioned the Periodic Table, examples were given, and three children knew there were 118 in all and that scientists had made some of these, whereas one child mentioned there were 92 natural ones. Other information supplied were that the elements were arranged from lightest to heaviest, and that the atomic number tells us what type it is, and Nathan volunteered that the left hand side of the Periodic Table were metals with loose electrons whereas the right hand side were non-metals with tight electrons. When asked to name elements, eight children could not do so correctly, whereas others began reciting the elements in order from hydrogen and helium, and others named anything from 3-15 different elements. Mark, who had answered the earlier question about what an element is with earth, air, fire and water, answered the question to name some elements with a long list, including titanium, vanadium, chromium, zinc, gold, silver, sulfur, silicon, iron, iridium, mercury, lawrencium, hafnium, samarium, and phosphorus. An equally long list of gold, argon, silver, tin, hydrogen, helium, beryllium, lithium, neon, carbon, oxygen, fluorine, sodium, plutonium, and silicon was given by 6-year-old Marcia, and Hanadi gave the second longest list: copper, iron, hydrogen, helium, lithium, beryllium, boron, carbon, fluorine, oxygen, neon, gold, silver and nitrogen. When asked to supply letter names for elements, only two children (Benedict and Evelyn) could not. Edward knew H is hydrogen, O is oxygen, and C is carbon despite his intellectual impairment and speech and language difficulties. Most children correctly gave the letters for several elements, often from the first 10 in the Periodic Table, with 13 children also knowing Au is gold, and Nathan even knew einsteinium is Es. There were very few errors.

## Sub-Atomic Particles

In the pre-interview, only four children had heard of protons, whereas eight had heard of electrons. However, their further answers indicated that they were conflating electrons with electricity and electronics, rather than referring directly to the sub-atomic particles themselves. The few suggestions regarding the size of protons and electrons were incorrect.

In the post-interview, all children had now heard of both protons and electrons, and all but three (Edward, Danisha, and Merryn, the child who said science was too hard) were now clearly referring to the sub-atomic particles. Six children clearly knew the correct charges and locations of both protons and electrons; and six had the right idea but confused the words 'protons' and 'electrons' either in location or in charge, indicating they had not consolidated the terminology. A further 10 children got either the location or the charge of protons and electrons correct but did not comment on the other criterion. Only two children (Oliwia and Benedict) made it clear that protons and electrons are parts of atoms but could not state the location in the atom of these particles or their charge. Seventeen children knew that both protons and electrons are smaller than atoms, and 17 children explicitly explained the octet rule (the first shell having two electrons and the second shell having eight).

The only 'extra' question asked in the post-interview commented upon here is the requirement to use the atom nucleus-shell model to make neon, as this informs our knowledge of their understandings of sub-atomic particles. The children were

*Figure 7. Atomic nucleus-shell model correctly depicting neon (designed and made by I. Stuart)*

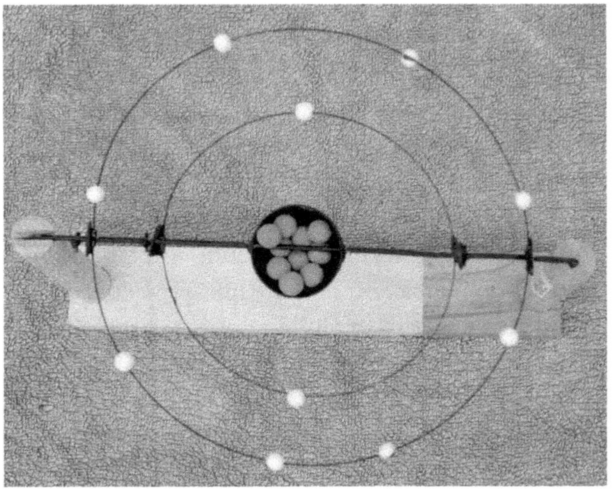

asked to find neon in the Periodic Table and then make it, so they had to work out that it was element 10 and what that meant in terms of the locations of protons and electrons. The model is shown correctly completed in Figure 3. Neutrons were not emphasised in the intervention and not included in the model.

The children put the correct heavier red balls (protons) in the central cup representing the nucleus, and located the lighter white balls (electrons) on the wire shells surrounding the nucleus, two on the inner shell, and eight on the outer shell. Every child except Edward was able to use the model to make neon correctly. This indicates that, although some children could not explicitly explain the octet rule, they knew the principle. This understanding was further demonstrated when responding to the question about neon's bonding capability. Most children knew it would not easily bond with other elements because the shells are full/there's no more room/its electrons are tight. Only five children thought it might be able to bond with other elements but could give no convincing reasons why. Interestingly, once they began using the model, only two children now confused the words 'protons' and 'electrons', so four children had self-corrected.

## DISCUSSION

We recognise this is a small-scale pilot study with just one class of children. Nonetheless, we find the results startling, especially when considering there were some factors operating against the successful implementation of the program in this context. Firstly, the specialist science teacher had no pre-existing collegial relationship with the classroom teacher, so he felt very much the visitor in her classroom. It also became apparent that she has a very different pedagogy, in that she rarely, if ever, addresses the whole class for instructional purposes. Instead, she moves and instructs each group in turn. As a former high school teacher, the specialist science teacher is used to being able to gain the attention of the whole class for instructional periods of at least ten minutes at a time, and it took him a while to realise this strategy was not successful in this group. He also felt constrained in terms of fully utilising the parents and aides and in using classroom tests to ascertain the individual unaided knowledge of each individual child. This was also a more diverse class in terms of children with special needs than would be typical of a high school science class; so again the specialist science teacher had to make some adaptations 'on the fly'. For every child to have gained as much knowledge as indicated in Figures 1 and 2 is truly remarkable in any circumstances, doubly so in this case.

The findings will be discussed in terms of the four research questions.

## 1. What Do Children Aged 9 Years Believe Science Is, and What Is Their Attitude Towards Science?

In the affective domain, children who already liked science generally liked it more, developed more sophisticated understandings of what science is, and appreciated the challenge of learning about atoms and molecules. Only one child thought it was too difficult. The positive response of children with special needs to the program is particularly gratifying.

## 2. What Prior Knowledge about Atoms, Molecules, Elements, and Sub-Atomic Particles Do Children Aged 9 Years Possess?

The results from the pre-interviews indicate that most of these 9-year-olds had relatively little prior knowledge of atoms and molecules, indicating this would be an opportune time to begin instruction, before misconceptions are acquired. Some had encountered the words 'atoms', 'molecules', and 'elements', showing these words are not beyond their sphere of reference, again indicative of this being an opportune age for exposure to this 'Big Idea' of science.

It is clear that children are exposed to some ideas about atoms and molecules from various sources, including the mass media. It is of concern that children referenced misconceptions about elements (earth, air, fire, and water) to television. This confirms the potential benefit of teachers deliberately drawing out the conceptions of children in their classrooms with consideration of knowledge they may have acquired from the mass media in order to expose children to the scientific use of these terms. The findings of this small-scale study also support the findings of Jakab (2013), in that more children claimed to have heard of molecules than had heard of atoms.

## 3. What Knowledge about Atoms, Molecules, Elements, and Sub-Atomic Particles Can Children Aged 9 Years Gain through an Intervention Designed by a Specialist High School Science Teacher?

Children were able to acquire a great deal of detailed and specific knowledge about all aspects of chemistry to which they were exposed. Children were now more aware that atoms are the building blocks of matter that make up everything, and had gained various degrees of understanding of atomic-molecular structure. Their understanding of molecules was wide-ranging, with fewer than expected being able to express confidently that molecules are atoms joined together, yet some were able to draw complex organic molecules. Confusions regarding the nature of elements were remedied in all but three children, with four other children lacking specific knowledge of what elements are. However, some children could recite long lists of elements, including some less common ones such as lawrencium, hafnium, and einsteinium. Most were accurate in their knowledge of the symbols used to represent elements, including some of those that are less obvious by not being the capital letter of the element's name, such as gold (Au), silver (Ag), and iron (Fe).

Children now had degrees of knowledge about sub-atomic particles, protons and electrons, although this terminology was not consolidated in all. Nonetheless, when children manipulated the model to make neon, only two children continued to confuse protons and electrons, indicating the importance of hands-on models to help children establish their understandings. That every child other than Edward (who is intellectually impaired) was able to manipulate the model to make neon with correct proton and electron arrangements is outstanding. As former high school teachers, we note that these concepts are often presented without hands-on models to high school students, in deference to their posited capacity to understand abstract concepts, and yet this approach is often unsuccessful in establishing sound understandings. We would suggest that such models would be beneficial whenever children first encounter these concepts, without regard to the Piagetian stage they are thought to be in. However, given the apparent capacity of 9 year-olds to comprehend these concepts with these models, we would suggest that starting at this age would be optimal, providing many opportunities to revisit these concepts over the following years.

The findings clearly indicate that, with appropriate instruction, children of this age are capable of dealing with the microscopic nature of atoms and sub-atomic particles. Such an understanding makes the macroscopic properties of matter, such as the shiny nature of metals, conductivity, and changes of state with temperature, eminently more explainable and comprehensible. We contend that teaching macroscopic and microscopic in tandem is likely to yield better results than the current approach of dealing only with macroscopic properties in primary school, delaying microscopic understandings to high school.

## 4. How Can Data Obtained in This Study Inform the Future Development of the Intervention?

Ten hours at one hour a week is not a lot of time to introduce such a wealth of information, nor does it provide ideal opportunities to consolidate this knowledge. The classroom teacher did do some consolidation activities, such as showing some of the specialist teacher's short explanatory videos, in between science classes. However, if tackled over a longer period of time, with more opportunity for diagnostic assessment of progress and consolidation of ideas, it would seem reasonable to suppose that even more dramatic gains in learning could be achieved. This study informs the future development of the intervention in that these data suggest:

- That Year 4, or possibly even Year 3 (before they become confused by what they see on TV or hear from other sources) are opportune times to introduce children to the concept of atoms,
- That taking the program more slowly, which probably means covering less information at this year level and leaving some to subsequent years, would be beneficial,
- The need to be more careful to consolidate the nature of molecules as compared with atoms, and
- The need to be more careful to consolidate the terminology of protons and electrons.

In addition, the specialist science teacher suggests that his introduction of magnets may have confused children's understanding of positive and negative charge and recommends omitting this in future.

A concern raised by the classroom teacher was whether the mathematics knowledge and capability of the children would hamper their understandings of how elements are constructed. However, that so many children grasped the octet rule indicates that at this level this is not an issue. With other classes, the specialist science teacher has introduced all the prefixes for smaller and smaller sizes, and has found children rather enjoy terms such as 'pico-', 'nano-', and 'yocto-', but the classroom teacher vetoed this approach with this class. In general, these concerns remind us that the mathematics capabilities of the children do need to be considered in consultation with the classroom teacher when implementing some aspects of this program.

## CURRENT CHALLENGES FACING THIS RESEARCH

The main challenge will be extending the research to a greater number of diverse schools in the first instance, consolidating our contention that children of this age can successfully learn atomic-molecular theory, and further refining the program. Following this, we would aim to develop a learning progression to introduce these concepts in a spiral curriculum over a number of years, and test the efficacy of this with a longitudinal study. The final research thrust would be to develop a professional development program that is effective in up-skilling existing primary teachers and a program for pre-service primary teachers so that they are confident in their ability to teach atomic-molecular theory.

## SOLUTIONS AND RECOMMENDATIONS

It is hoped that publication of even these preliminary findings will excite interest in this work. Further publications, particularly once the extra interview questions are fully analysed and the retention interviews have been conducted and analysed, will hopefully further engender interest that may translate to collaborations with schools nationally and perhaps internationally. Whilst being suitably cautious and cognisant of the small scale of this research, we conclude that the findings indicate that children have greater capability of understanding the microscopic aspects of atomic-molecular theory than was generally recognised previously. In this, these findings support those of other pioneering researchers mentioned in this chapter, such as Jakab, Liu and Lesniak, Nussbaum, Halpine, and Wiser and Smith.

We contend that appropriate instruction, including the thoughtful use of excellent hands-on models, is critical to children gaining understanding of this 'Big Idea' of science. It is clear that the models were particularly helpful to children in this study, and that they enjoyed using them. Furthermore, the children themselves judged the program appropriate for them; with only one believing it was too hard. They relished the opportunity to challenge their thinking and this furthered their interest in, and enjoyment of, science.

We argue that the Piagetian constructs for curriculum development should be discontinued. We suggest that primary curricula should include the 'Big Ideas' such as atomic-molecular theory at the time when children are encountering these concepts in the mass media, are cognitively ready and show interest in these ideas. Research indicates that if children were exposed to atomic-molecular theory in Years 3 and 4 they would be well primed to capitalise on their interest in genes and DNA in Years 5 and 6. Such would be the advantages of a spiral curriculum in which the macroscopic and microscopic properties of matter are taught concurrently rather than sequentially.

At the very least, science curricula should be sufficiently flexible for teachers to be able to take advantage of opportunities that present themselves. When children ask about atoms and elements, or genes and DNA, teachers should be able to take the time to capture and use this interest to establish science concepts, without stressing about how much set content there is to cover in the mandated curriculum.

## REFERENCES

Abdeldayem, H. (2012). *Brain plasticity: Slideshare presentation*. Retrieved on 14th June 2013 from http://www.slideshare.net/husseindayem/brain-plasticity-12531271

Acher, A., Arcà, M., & Sanmartí, N. (2007). Modeling as a teaching learning process for understanding materials: A case study in primary education. *Science Education, 91*(3), 398–418. doi:10.1002/sce.20196

Adolph, K. E., Bertenthal, B. I., Boker, S. M., Goldfield, E. C., & Gibson, E. J. (1997). Learning in the development of infant locomotion. *Monographs of the Society for Research in Child Development*, 162. PMID:9394468

Australian Curriculum, Assessment and Reporting Authority (ACARA). (2013). *Australian curriculum: Science*. Retrieved on 14th June 2013 from http://www.australiancurriculum.edu.au/Science/Curriculum/F-10

Bidell, T. R., & Fischer, K. W. (1992). Beyond the stage debate: Action, structure, and variability in Piagetian theory and research. In R. Sternberg, & C. Berg (Eds.), *Intellectual development* (pp. 100–140). New York: Cambridge University Press.

Brown, T., Rushton, G., & Bencomo, M. (2008). Mighty Molecule Models. *Science and Children, 45*(5), 33–37.

Bruner, J. (1960). *The process of education*. Cambridge, MA: Harvard University Press.

Carey, S. (2010). Beyond fast mapping. *Language Learning and Development, 6*(3), 184–205. doi:10.1080/15475441.2010.484379 PMID:21625404

Chen, Z., & Siegler, R. (2000). Across the great divide: Bridging the gap between understanding the thought of toddlers and older children. *Monographs of the Society for Research in Child Development, 65*(2).

Creswell, J. W. (2005). *Educational Research: Planning, conducting and evaluating quantitative and qualitative research* (2nd ed.). Upper Saddle River, NJ: Pearson Education.

Department of Education. (2013). *Science programmes of study: key stage 3*. National curriculum in England: Crown. Retrieved on 15th September 2013 from https://www.gov.uk/government/uploads/system/uploads/attachment_data/file/239134/SECONDARY_national_curriculum_-_Science.pdf

Donovan, J., & Venville, G. (2004). Genes and DNA: What kids and experts think. *SCIOS (Journal of the Science Teachers' Association of Western Australia), 40*(2), 26-32.

Donovan, J., & Venville, G. (2005). A concrete model for teaching about genes and DNA to young students. *Teaching Science, 51*(4), 29–31.

Donovan, J., & Venville, G. (2012a). Blood and bones: The influence of the mass media on Australian primary children's understandings of genes and DNA. *Science and Education.* DOI: 10.1007/s11191-012-9491-3

Donovan, J., & Venville, G. (2012b). Exploring the influence of the mass media on primary students' conceptual understanding of genetics. *Education, 40*(1), 75-95.

Drubach, D. (2000). *The Brain Explained.* Upper Saddle River, NJ: Prentice Hall.

Duschl, R. A., Schweingruber, H. A., & Shouse, A. W. (Eds.). (2007). *Taking science to school: Learning and teaching science in grades K-8.* Washington, DC: National Academies Press.

Fischer, K. W., & Bidell, T. R. (2006). Dynamic development of action and thought. In R. M. Lerner (Ed.), Handbook of child psychology: Vol. 1. *Theoretical models of human development* (6th ed., pp. 313–399). New York: Wiley.

Fleer, M. (2009). Understanding the Dialectical Relations Between Everyday Concepts and Scientific Concepts Within Play-Based Programs. *Research in Science Education, 39*(2), 281–306. doi:10.1007/s11165-008-9085-x

Flynn, E., O'Malley, C., & Wood, D. (2004). A longitudinal, microgenetic study of the emergence of false belief understanding and inhibition skills. *Developmental Science, 7*(1), 103–115. doi:10.1111/j.1467-7687.2004.00326.x PMID:15323122

Goodrum, D., Druhan, A., & Abbs, J. (2011). *The status and quality of Year 11 and 12 science in Australian schools.* Canberra, ACT: Australian Academy of Science.

Goodrum, D., Hackling, M., & Rennie, L. (2001). *The status and quality of teaching and learning of science in Australian schools.* Canberra, ACT: Department of Education, Training and Youth Affairs.

Gopnik, A., Meltzoff, A., & Kuhl, P. (1999). *The Scientist in the Crib: What Early Learning Tells Us About the Mind.* New York, NY: Harper Collins.

Halpine, S. M. (2004). Introducing molecular visualization to primary schools in California: The STArt! *teaching* science through art program. *Journal of Chemical Education, 81*(10), 1431–1436. doi:10.1021/ed081p1431

Hirsch, E. D., Jr. (2006). Building knowledge: The case for bringing content into the language arts block and for a knowledge-rich curriculum core for all children. *American Educator, Spring* [Electronic version]. Retrieved on October 16, 2010 from http://www.aft.org/newspubs/periodicals/ae/spring2006/editors.cfm

Inhelder, B., & Piaget, J. (1958). *The growth of logical thinking from childhood to adolescence*. New York: Basic Books. doi:10.1037/10034-000

Jakab, C. (2013). Small talk: Children's everyday 'molecule' ideas. *Research in Science Education*, *43*(4), 1307–1325. doi:10.1007/s11165-012-9305-2

Jean, A. (2007, July 26). Science in comedy: Mmm ... pi. *Nature*, *448*(7152), 404–405. doi:10.1038/448404a PMID:17653163

Johnson, P., & Papageorgiou, G. (2010). Rethinking the introduction of particle theory: A substance-based framework. *Journal of Research in Science Teaching*, *47*(2), 130–150.

Johnson, P., & Tymms, P. (2011). The emergence of a learning progression in middle school chemistry. *Journal of Research in Science Teaching*, *48*(8), 849–877. doi:10.1002/tea.20433

Lee, O., Eichinger, D. C., Anderson, C. W., Berkheimer, G. D., & Blakeslee, T. S. (1993). Changing middle school students' conceptions of matter and molecules. *Journal of Research in Science Teaching*, *30*(3), 249–270. doi:10.1002/tea.3660300304

Lehrer, R., & Schauble, L. (2000). Modeling in mathematics and science. In R. Glaser (Ed.), *Advances in instructional psychology* (Vol. 5, pp. 101–159). Mahwah, NJ: Lawrence Erlbaum Associates.

Liu, X., & Lesniak, K. (2006). Progression in children's understanding of the matter concept from elementary to high school. *Journal of Research in Science Teaching*, *43*(3), 320–347. doi:10.1002/tea.20114

Liu, X., & Lesniak, K. M. (2005). Students' progression of understanding the matter concept from elementary to high school. *Science Education*, *89*(3), 433–450. doi:10.1002/sce.20056

Maltese, A. V., & Tai, R. H. (2010). Eyeballs in the fridge: Sources of early interest in science. *International Journal of Science Education*, *32*(5), 669–685. doi:10.1080/09500690902792385

Margel, H., Eylon, B. S., & Scherz, Z. (2008). A longitudinal study of junior high school students' conceptions of the structure of materials. *Journal of Research in Science Teaching*, *45*(1), 132–152. doi:10.1002/tea.20214

Merritt, J. D., Krajcik, J., & Shwartz, Y. (2008, June). Development of a learning progression for the particle model of matter. In *Proceedings of the 8th International conference for the Learning Sciences* (Vol. 2, pp. 75-81). Utrecht, The Netherlands: International Society of the Learning Sciences.

Murphy, C. (2012). Vygotsky and primary science. In B. J. Fraser, K. G. Tobin, & C. J. McRobbie (Eds.), *Second international handbook of science education* (pp. 177–187). Dordrecht, The Netherlands: Springer. doi:10.1007/978-1-4020-9041-7_14

Nakhleh, M. B., & Samarapungavan, A. (1999). Elementary school children's beliefs about matter. *Journal of Research in Science Teaching*, *36*(7), 777–805. doi:10.1002/(SICI)1098-2736(199909)36:7<777::AID-TEA4>3.0.CO;2-Z

National Research Council (NRC). (2012). *A Framework for K-12 Science Education: Practices, Crosscutting Concepts, and Core Ideas*. Washington, DC: The National Academies Press. Retrieved on 15th September 2013 from http://www.nap.edu/openbook.php?record_id=13165&page=1

National Research Council (NRC). (2013). *Next Generation Science Standards: For States, By States*. Washington, DC: The National Academies Press. Retrieved on 15th September 2013 from http://www.nap.edu/NGSS/

Nussbaum, J. (1998). History and philosophy of science and the preparation for constructivist teaching: The case of particle theory. In J. J. Mintzes, J. H. Wandersee, & J. D. Novak (Eds.), *Teaching science for understanding–A human constructivist view* (pp. 165–194). San Diego, CA: Academic Press.

Özmen, H. (2004). Some student misconceptions in chemistry: A literature review of chemical bonding. *Journal of Science Education and Technology*, *13*(2), 147–159. doi:10.1023/B:JOST.0000031255.92943.6d

Özmen, H., & Ayas, A. (2003). Students' difficulties in understanding of the conservation of matter in open and closed-system chemical reactions. *Chemistry Education: Research and Practice*, *4*(3), 279–290.

Rideout, V. J., Foehr, U. G., & Roberts, D. F. (2010). *Generation M 2: Media in the lives of 8-to-18-year-olds*. Menlo Park, CA: The Henry J. Kaiser Family Foundation. Last retrieved July 12, 2011 from http://www.kff.org/entmedia/mh012010pkg.cfm

Schauble, L. (1996). The development of scientific reasoning in knowledge-rich contexts. *Developmental Psychology*, *32*(1), 102–119. doi:10.1037/0012-1649.32.1.102

Schwarz, C. V., Reiser, B. J., Davis, E. A., Kenyon, L., Achér, A., & Fortus, D. et al. (2009). Developing a learning progression for scientific modeling: Making scientific modeling accessible and meaningful for learners. *Journal of Research in Science Teaching*, *46*(6), 632–654. doi:10.1002/tea.20311

Siegler, R. S. (1996). *Emerging minds: The process of change in children's thinking*. New York: Oxford University Press.

Siegler, R. S. (1998). *Children's thinking* (3rd ed.). Upper Saddle River, NJ: Prentice Hall.

Siegler, R. S. (2006). Microgenetic analyses of learning. In D. Kuhn & R. S. Siegler (Eds.), Handbook of child psychology: Volume 2: Cognition, perception, and language (6th ed., pp. 464-510). Hoboken, NJ: Wiley.

Siegler, R. S. (2007). Cognitive variability. *Developmental Science*, *10*(1), 104–109. doi:10.1111/j.1467-7687.2007.00571.x PMID:17181707

Smith, C. L., Solomon, G. E., & Carey, S. (2005). Never getting to zero: Elementary school students' understanding of the infinite divisibility of number and matter. *Cognitive Psychology*, *51*(2), 101–140. doi:10.1016/j.cogpsych.2005.03.001 PMID:16081058

Sousa, D. A. (2001). *How the Brain Learns* (2nd ed.). Thousand Oaks, CA: Corwin Press.

Stein, M., Larrabee, T. G., & Barman, C. R. (2008). A study of common beliefs and misconceptions in physical science. *Journal of Elementary Science Education*, *20*(2), 1–11. doi:10.1007/BF03173666

Stevens, S. Y., Delgado, C., & Krajcik, J. S. (2010). Developing a hypothetical multidimensional learning progression for the nature of matter. *Journal of Research in Science Teaching*, *47*(6), 687–715. doi:10.1002/tea.20324

Tai, R. H., Liu, C. Q., Maltese, A. V., & Fan, X. T. (2006). Planning early for careers in science. *Science*, *312*(5777), 1143–1144. doi:10.1126/science.1128690 PMID:16728620

*The Simpsons* wiki (2013). Retrieved on 14th June 2013 from http://simpsons.wikia.com/wiki/index.php?search=atom&fulltext=Search

Tytler, R., & Osborne, J. (2012). Student attitudes and aspirations towards science. In B. J. Fraser, K. G. Tobin, & C. J. McRobbie (Eds.), *Second international handbook of science education* (pp. 597–625). Dordrecht, The Netherlands: Springer. doi:10.1007/978-1-4020-9041-7_41

Tytler, R., & Peterson, S. (2000). Deconstructing learning in science—Young children's responses to a classroom sequence on evaporation. *Research in Science Education*, *30*(4), 339–355. doi:10.1007/BF02461555

Tytler, R., Symington, D., & Smith, C. (2009). A curriculum innovation framework for science, technology and mathematics education. *Research in Science Education*, 1–20.

Venville, G., & Donovan, J. (2005, Spring). Searching for clarity to teach the complexity of the gene concept. *Teaching Science*, *51*(3), 20–24.

Venville, G., & Donovan, J. (2007). Developing Year 2 students' theory of biology with the concepts of gene and DNA. *International Journal of Science Education*, *29*(9), 1111–1131. doi:10.1080/09500690600931079

Venville, G., & Donovan, J. (2008). How pupils use a model for abstract concepts in genetics. *Journal of Biological Education*, *43*(1), 6–14. doi:10.1080/00219266.2008.9656143

Vosniadou, S. (2012). Reframing the classical approach to conceptual change: Preconceptions, misconceptions and synthetic models. Second International Handbook of Science Education, 1, 119-130.

Willingham, D. T. (2008). What Is Developmentally Appropriate Practice? *American Educator*, *32*(2), 34.

Wiser, M., & Smith, C. L. (2008). Learning and teaching about matter in grades K-8: When should the atomic-molecular theory be introduced. In International handbook of research on conceptual change, (pp. 205–239). Academic Press.

Yair, Y., & Yair, Y. (2004). "Everything comes to an end": An intuitive rule in physics and mathematics. *Science Education*, *88*(4), 594–609. doi:10.1002/sce.10142

## KEY TERMS AND DEFINITIONS

**Affective Domain:** The field of study concerning perceptions, beliefs, and attitudes about a topic.

**Atomic-Molecular Theory:** One of the 'Big Ideas' of science; all matter is made of atoms, many of which are joined to make molecules.

**Cognitive Domain:** The field of study concerning knowledge held about a topic.

**Learning Progressions:** Sequences of concepts increasing in sophistication designed to be taught each year so that learning progresses over time; such progressions are integral components of a spiral curriculum.

**National Science Curriculum:** Detailed plans for learning and teaching of science developed for implementation across a nation, including such curricula for Australia, the USA and the UK.

**Piagetian Model of Developmental Stages:** This idea, developed by Jean Piaget and other psychologists, contends that children experience distinct phases of development in terms of their cognitive capacities.

**Primary Children:** Children who attend primary school: in Australia, this includes children from ages 6 to 11 years.

**Scientific Literacy:** The capacity of people to understand science sufficiently to make informed decisions about scientific issues.

**Spiral Curriculum:** An idea developed by Bruner and others, that concepts are best presented early to create foundational knowledge, and then revisited often and built upon over successive years.

# Chapter 3
# Implementing the Understanding by Design Framework in Higher Education

**Judy Alhamisi**
*Marygrove College, USA*

**Blanche Jackson Glimps**
*Tennessee State University, USA*

**Chukwunyere E. Okezie**
*Marygrove College, USA*

## EXECUTIVE SUMMARY

*This chapter describes an organizational initiative to develop and implement the Understanding by Design (UbD) curriculum-planning framework to improve learning outcomes for teacher candidates and their students during clinical experiences and in their future classrooms. This case study explores a pedagogical approach that has met with success in working with teacher candidates. The focus is on a narrow range of knowledge, skills, and dispositions related to effective teaching in science education: the ability to design, plan, and implement curriculum. Curriculum design using the Understanding by Design (UbD) Framework is a high priority when moving from simply covering subject matter to ensuring deep understanding. Using "Backward Design" helped many teacher candidates develop skills to plan effective science curriculum, units, and lessons. The experiences of two teacher education programs in building teacher candidates' skills in planning and implementing science education curriculum using the UbD Framework are presented in this case study.*

DOI: 10.4018/978-1-4666-6375-6.ch003

## ORGANIZATION BACKGROUND

Colleges and universities, in responding to political, social, economic, and technological pressures, are becoming more responsive to teacher candidates' needs and are more concerned about how well they are being prepared to assume future roles as teachers. Faculty are feeling the pressure to change their teaching strategies by developing rigorous curriculum, lecturing less, making learning environments more interactive, integrating technology into the classroom, and using collaborative learning strategies. This chapter describes two institutions of higher education and their experiences in scaffolding teacher candidates regarding the use of UbD as a framework for unit and lesson planning, including assessment.

These urban institutions are identified as University A and University B. University A, a Historically Black College/University (HBCU), is located in the southern part of the United States and has a long tradition of educating students who have been historically underrepresented. The mission of University A is to build a cadre of graduates who are prepared to conduct scholarly inquiry and research, become life-long learners, and are committed to service. Currently, University A offers bachelor's degree programs, master's programs, and awards doctoral degrees in several disciplines. University A is comprised of eight colleges and schools. The program for preparing teachers, at University A, is located within the College of Education. The program prepares teachers for elementary and secondary classrooms in specific content areas including science. Education faculty teach courses in programs, such as: curriculum and instruction, special education, reading, science education, and math education. Content area specialization for majors in secondary area is provided by faculty in different colleges and schools, including: math, science, history, humanities, music, art, and physical education. Students enrolled in the teacher certification programs take traditional courses to prepare them to design curriculum, develop assessments, engage all learners, and become reflective practitioners. In terms of professional development, University A faculty in the education department do not have access to funds to support attendance at conferences or bring well-known educational researchers and experts to the campus.

University B is a four-year, co-educational private liberal arts college. The institution moved to its current location in 1927 to serve young women who would otherwise be unable to obtain a college degree. While University B has evolved to meet the changing needs of its students throughout its 86-year history in the city, the institution has remained committed to the city and to the education of those disadvantaged by gender, race, economic circumstances, or social limitations. The mission of University B is to educate each student to become intellectually and professionally competent; ensure career flexibility through grounding in the liberal arts; and develop active compassion and commitment. Through excellent teaching in its undergraduate, graduate, and continuing education programs, the university provides a personalized learning environment for its students. The Institution, in an open, caring, nurturing, and friendly environment, provides learning experiences and opportunities for students to demonstrate leadership and develop confidence and self-reliance. The individual and collective excellence for which University B stands will continue to be measured by the quality of its graduates and their successes in serving society.

The Education Department of University B offers undergraduate and graduate programs to prepare students for careers as certified teachers at both the elementary and secondary levels. Perhaps no program at University B has focused more directly on the commitment to enabling positive change than the teacher certification program. Like most small, private liberal arts colleges, teacher training has been a curricular centerpiece, graduating students who staffed and led schools in the city, throughout the metropolitan area, and in many other states. Teachers from University B continue the legacy of service and leadership. University B is committed to professional development, with faculty receiving financial support to attend conferences and workshops. In addition, the Education Department hosts an annual conference that brings nationally known researchers and educators to the campus.

This chapter presents the Understanding by Design (UbD) Framework and its use in preparing teacher candidates in science and other curricular areas. In this framework, learning is about the acquisition of factual knowledge and the fluent patterning of behavior, as well as about making meaning. While these are elements of learning, the "goal is to teach enduring understandings that should be the central focus of curriculum building" (Wang & Allen, 2003, p. 37).The chapter focuses on elaborating steps embedded in UbD that target deep understanding. The goal is to acquaint readers about the UbD framework as an effective tool for developing understanding, particularly in science education. The steps followed to introduce students and guide them through creating science planning using UbD are presented.

## SETTING THE STAGE

The National Science Teachers Association (NSTA, 2006) supports strong performance-based science teacher education programs and state licensure standards for teacher candidates for both elementary and secondary schools. The NSTA recommended that teacher education programs should use a curriculum framework based on National Science Education Standards that provide teacher candidates with deep science knowledge and skills at the grade levels for which they are teaching.

UbD challenges teacher candidates to design, develop, and implement units and lessons to assist their students in acquiring science literacy. To achieve this objective, faculty must engage their teacher candidates in the learning process. Craven and Penick (2001) asserted that "students not engaged in the learning process leave with little more than shallow understandings, weak connections between big ideas, trivial knowledge, unchallenged naive conceptions of how the natural world operates, and an inability to apply knowledge in new settings" (para. 2). Although improvements have been found in student performance in science, a need exists to improve science education (National Science Foundation, 2006). According to Craven and Penick (2001), teacher educators must assist their teacher candidates move from planning units and lessons based on traditional curriculum models used to teach and learn science. UbD provides a tool to help achieve this objective.

Grant Wiggins and Jay McTighe (1998, 2013) developed UbD as a curriculum framework for improving student achievement through specific elements, including standards-driven curriculum, instructional design, assessment, and professional development. In practice, UbD uses a three-stage backward planning curriculum design process, a set of design standards with rubrics, and a comprehensive training program to help teachers design, edit, critique, peer-review, share, and improve their lessons and assessments. Teachers, using backward planning, identify essential questions that students must be able to answer by the end of the unit. Jones, Vermette, and Jones (2009) asserted that with the end in mind, "teachers then design the assessment of those understandings, followed by carefully crafted lessons to achieve this set of objectives" (p. 357). Wiggins and McTighe (1998, 2013) identified six facets of understanding within the UbD framework that should be evident in teachers' work with students, including: (a) explaining, (b) interpreting, (c) applying, (d) shifting perspective, (e) empathizing, and (f) self-assessing. UbD details the three-stage backward planning curriculum design process.

*Stage One – Desired Results.* This stage considers which transfer and content goals need to be met and what big ideas students should understand. The essential questions that students will explore and address and what knowledge and skills with which students will leave are included in this stage. Table 1, an example of this stage, presents the title and brief lesson summary, produced by a teacher candidate at University B, including state content standards of a first grade integrated science unit on the Life Cycle of an Apple Tree.

The student teacher reviewed the science standards needed to address the science topic at the first grade level. One set of science standards had to do with the Life Cycle of Apple Trees. The student teacher selected the topic based upon the first grade curriculum, as well as her own interest and motivation to teach this particular topic. She also observed during her clinical experience that students naturally learned across subject areas. As a result, she researched the state and national standards pertaining to the related disciplines of English Language Arts, Literacy, Art, and Math.

*Table 1. Title and Brief Lesson Summary Including Established Goals (State Standards)*

| |
|---|
| Life Cycle – Apple Trees is a two-week unit that teaches first graders to explore the stages in the life cycle of an apple tree. Students theorize, investigate, and discover how apple seeds sprout to become mature apple trees that reproduce after their own kind. Students engage in various activities and assignments across subject areas within an overarching apple tree cycle/apple theme.<br>**English Language Arts** (ELA) Standards: Common Core State Standards (CCSS).<br>ELA – Literacy.W.1.2; CCSS.ELA – Literacy.W.1.3; CCSS.ELA-Literacy.SL.1.2; CCSS.ELA-Literacy.SL.1.5<br>Science Standards: S.IP.E1; S.IA.E.1; SRS.E.1; L.OL.E.2<br>Art Standards: ART.VA.II.1.; ART.VA.III.1.1; ART.VA.V.1.4<br>Math Standards: N.ME.01.08; N.MR.01.11; D.RE.01.03 |

*Table 2. Transfer Goals (Student Learning Outcomes)*

| |
|---|
| Independently, students will be able to use their learning to . . .<br>• Describe the life cycle of a plant – the apple tree. (L.OL.01.21)<br>• Plan and conduct simple investigations. (S.IP.01.13) |

Next, the student teacher translated the standards into transfer goals, shown in Table 2, indicating the student's ability to transfer learning from one context to others.

She determined the Desired Results including what the students should know, understand, and be able to do at the end of the unit. The students will explore "big ideas" and related thought-provoking essential questions throughout the unit. In Table 3, examples are as follow:

Specific knowledge and skill statements linked to the standards, understanding and essential questions also are identified and are shown in Table 4.

*Stage Two – Determine Assessment Evidence.* At this stage, the student teacher determined the extent to which students have achieved the desired results in Stage 1. The student teacher decided which performances and products could reveal evidence of understanding, as well as other evidence that could be collected to reflect

*Table 3.*

| Understandings | Essential Questions |
|---|---|
| Students will understand that…<br>• Apple seeds develop into trees.<br>• Investigations entail coming up with questions, making predictions on outcomes, and doing an activity to see if the predictions were true and if the original questions were answered. | What essential questions will be considered?<br>• Where do apple trees come from?<br>• What can happen if you plant an apple seed?<br>• Why are apple trees important? |

*Table 4.*

| Knowledge | Skill |
| --- | --- |
| Students will know..<br>• The parts of the life cycle of an apple tree include seed form, sprout, sapling, blossoming tree, and apple producing tree.<br>• Apple seeds grow into sprouts, which develop into fruit bearing trees. | Students will be able to…<br>• Construct a plant cycle diagram using the pars given (cut-outs used) or draw them.<br>• Label pictures of the stages of the apple seed (a. Seed, b. Planted in dirt, c. Breaking through the soil, d. Small sprout, e. Tall sprout with long roots). |

desired results. When using UbD, assessment evidence must be aligned with the desired results detailed in the first stage. To accomplish this alignment, teachers and curriculum planners need to think like an assessor when designing specific units and lessons. In the Life-Cycle of Apple Trees Unit, the student teacher identified two transfer tasks that students need to complete that show their learning by doing as shown in Table 5.

*Stage Three – Learning Plan.* During this stage, the student teacher considered which activities, experiences, and lessons could lead to achievement of desired results (Stage 1) and success with assessment (Stage 2; Wiggins & McTighe, 1998). Graff (2011) believed that "teachers can approach unit planning in any order as long as they aim for coherence of all three components" (p. 155). The acronym WHERE-TO summarizes the key elements the student teacher considered as she designed the learning plan (See Table 6).

*Table 5.*

| **Transfer Task(s):** |
| --- |
| • Draw a picture diagram of how an apple seed goes to a sprout, to a sapling, to a grown tree that produces apples, to apples falling off the tree, and back to a seed germinating in the ground after the apple has disintegrated, with each part labeled.<br>• Plant their own apple seed in small containers of dirt or Miracle Gro, water them daily, and chart their growth over the course of two weeks. They will also write "How To" books on growing an apple seed.<br>Evaluative Criteria:<br>Performance is judged in terms of…<br>• Accuracy of diagram according to the model.<br>• Accuracy of Growth Chart according to the model and rubric<br>Sample Performance Tasks and Rubric:<br><u>Watering the Seed Charting Growth</u><br>3 Meets the standard of using 1 or 2 dropper full or no more than 1 Dixie cup of water to avoid spillage Accurately shows progress of the seed development using pictures and written details of observation according to the model<br>2 Uses the standard, but does not implement effectively (only using half a Dixie cup amount or not using the dropper's full capacity) Uses some aspect of the model to chart progress, with some aspects omitted<br>1 Ignores the standard by using other forms of water measurement or overflowing the plant with excess water Does not accurately chart progress or shows unrelated information |

*Table 6.*

| | |
|---|---|
| W | **Where are we going? Why? What is expected?** We are going to learn about where apples come from and how to get more of them. This is important to know because it is part of life science and explains where a source of nutritious food comes from. Students are expected to meet the state standard of being able to understand and describe the life cycle of a plant [apple tree] and produce an apple sprout from scientific experiment of planting and nurturing apple seeds. |
| H | **How will we hook the students?** Students will be hooked with lively videos and hands-on experiments, such as seed planting. |
| E | **How will we equip students for expected performances?** Students will be provided with appropriate scaffolds including many examples, visual aids, step-by-step instruction, and rubrics to guide them. |
| R | **How will we rethink or revise?** If planned lessons are not as grade level appropriate or conducive to the culture of the classroom due to limited schema as expected, then lessons and activities will be revised to be accessible to all students and meet them where they are (Zone of Proximal Development). As long as students come away with knowing more than they did before, the unit is successful. |
| E | **How will students self-evaluate and reflect their learning?** Students will self-evaluate by indicating their feelings and reflections of how they performed or how well they understood via communicating in writing/written journal entries, showing a thumbs up or happy face for understanding/doing well or a thumbs down for confusion/not doing so well, etc. |
| T | **How will we tailor learning to varied needs, interests, and learning styles?** Some students are considered low students and need more scaffolding than others. Thus, their lessons can consist more of verbal and pictorial expression. They can also point to pictures when asked about sequence and for labeling assessment. Students who are more visual will benefit from the videos and actual plants being used instead of just earing about the apple tree cycle. The use of dirt and actively planting seeds may appeal to kinesthetic learners. |
| O | **How will we organize the sequence of learning?** Young students should always start with the simplest understanding (basic knowledge) and build from there. A read-aloud introduction and videos set the stage for the scientific experiment of planting and documenting growth. |

Note: Adapted from Wiggins & McTighe, 2013

During stage three, the student teacher designed lesson plans for each teaching event for the two-week unit. The individual lesson plans addressed the essential questions and performance tasks, and were connected to the goals and assessments for the unit.

The big picture of the UbD approach (Meier, n.d.) illustrated the key instructional design questions, design considerations, filters (design criteria), and what the final design accomplishes. At stage one, the key instructional design question is what is worthy and requiring of understanding. The answer to this question is determined by an examination of national and state standards, as well as student learning outcomes. At this stage, the design criteria are enduring authentic ideas and discipline-based work. Stage two focuses on the evidence of understanding. The design criteria are the six facets of understanding in combination with a continuum of assessment types. Stage three considers what learning experiences and teaching promote understanding, interest, and excellence.

UbD is responsive to the No Child Left Behind legislation that emphasizes research-based programs to influence student achievement positively. UbD is not a program of scope and sequence of skills, as no prescribed teaching activities are delineated. No direct, causal evidence exists on the influence of UbD on student achievement (McTighe & Seif, 2003); however, UbD principles and practices reflect views of learning in cognitive psychology. UbD shifts views on effective learning from drill and practice to understanding and application. Learning, using the UbD framework, is guided by generalized principles that are applicable for most students. Superficial coverage of topics may be ineffective in helping students develop competencies that prepare them for future learning and work (McTighe & Seif, 2003). Feedback is fundamental to learning, with such opportunities often scarce in classrooms. Most standardized assessments measure factual knowledge, and fail to ask if students know when, where, and why to use that knowledge. In designing units and lessons with UbD principles, students can develop deep knowledge of a topic that can be used as a scaffold for future learning.

Constructivism is an approach to learning where prior experiences and current knowledge are integrated to form new understanding. Constructivism has implications for teaching and learning, using discovery, hands-on, experiential, collaborative, project-based, and task-based learning as strategies to improve academic achievement. Backward design, when compared to the traditional way of teaching, is a more constructivist route to teaching. UbD is a form of guided discovery that focuses on authentic learning of content by starting with the finished product and working backward. Teachers' use of creative, innovative, and interesting teaching strategies that relate to real-life situations result in students developing deeper understanding of the subject matter being taught.

UbD is a tool that can be used to facilitate a quality education for all students and is a framework for improving student achievement through standards-driven curriculum development, instructional design, assessment, and professional development. McTighe and Thomas (2003) suggested that for "backward design to work, educators need to identify desired results, analyze multiple sources of data, and determine appropriate action plans" (p. 53). These components of the three stage backward planning process were introduced to students at both universities during the study of UbD.

## CASE DESCRIPTION

The current case involves two urban universities and their attempt to introduce teacher candidates to UbD with the goal of stimulating innovation in preparation for their future classrooms. The National Science Teachers Association (NSTA) considers strong, performance-based science teacher education program and science teacher licensure standards to be essential for all science teachers (National Science Teacher Association, 2004). As they prepare teacher candidates for future science classrooms, faculty at both universities must be aware of continuously changing NSTA and state licensure standards.

University A and University B are terms used to distinguish the two institutions of higher education that comprise this case study. The universities each selected different routes for equipping teacher candidates with skills embedded in the UbD Framework. University A followed the traditional model of preparing teacher candidates for science and general education classroom. Consequently, at University A, only graduate students enrolled in a curriculum design course were exposed to UbD. Members of the course included students whose majors represented science education, English education, school psychology, and special education. This course introduced students to several models of curriculum design, including UbD. University A had no institutional support to follow any particular curriculum design paradigm. The choice of the paradigm to which students were introduced depended on the instructor. The faculty member responsible for the course at University A conducted a review of the literature (e.g., research articles, webinars, and podcasts) on different curriculum frameworks before deciding to learn about UbD in preparation for sharing the curriculum framework with students. Course evaluations, using the traditional student course evaluation system, were minimal as the majority of students did not complete the assessment.

At University B, UbD was embedded throughout the courses that students were required to complete in the teacher education sequence. A number of strategies for curriculum, instruction, and assessment had been used at the university for several years before the introduction of UbD as a curriculum framework. As the Education Department faculty began preparing for state and national accreditation, they began developing goals that included implementation of learning outcomes driven curricula and use of different approaches for engaging faculty and teacher candidates to achieve these goals. During the 2007-08 academic year, a university-wide accreditation review was completed by the Higher Learning Commission of the North Central Association. The results identified areas of strengths as well as challenges that needed to be addressed to improve program quality and academic excellence for all students, including teacher candidates.

The initial challenge was assessment, which led the Education Department (ED) faculty to begin a review of the curriculum. In an attempt to identify ways to improve the ED programs, the faculty began reviewing researched-based instructional models, approaches, and strategies. After examining several models, the education faculty decided to adopt the "Understanding by Design (UbD) Framework" in undergraduate and graduate teacher education programs. This case study focuses on the process that led to the integration of "Understanding by Design" as a framework to contextualize the teacher education programs. To provide a chronological review used to introduce, develop, and embed the UbD framework into the Teacher Education curriculum, the following steps were taken:

- On March 14, 2011 – the institution invited Grant Wiggins, an education expert, to deliver the keynote address at its two-day education conference, "Authentic Performance Assessment in Urban Education." During his presentation, Wiggins discussed the importance of assessment in developing units and lessons using the UbD curriculum framework. Understanding by Design (UbD) is an important, practical framework used locally, nationally, and internationally to improve student-learning outcomes in college programs and in K – 12 schools. Wiggins and McTighe's framework emphasizes teachers' roles as designers and planners of student learning experiences. Course planning using UbD also is called *backwards design* or *teaching with the end in mind*. Using this curriculum framework provides a way for teacher candidates to understand the big ideas by investigating, exploring, testing, and verifying important concepts. They are able to transfer knowledge using the previously learned concepts.
- As a result of the 2011 Education Conference and the emphasis placed upon UbD, the ED faculty began integrating UbD as the curriculum framework for Teacher Education. The ED began a curriculum mapping process, incorporating principles of UbD in redesigning the programs and courses in the Teacher Certification Program. Faculty began to map curriculum with the Backwards Design process used in UbD by focusing on what teacher candidates need to know at the end of the program/course/ unit. In addition, a course, Curriculum Instruction and Assessment, was designed to prepare teacher candidates to use Backward Design to create units and lessons for their students in elementary and secondary schools.
- In July 2011, the elementary and secondary education coordinators participated in two national UbD training programs, conducted by Grant Wiggins and Associates. The programs were "UbD Train the Trainers" and "Designing from the Standards." Participants included elementary and secondary teachers, higher education faculty, and administrators from across the United States and other countries. The attendees were at various stages in learning about and implementing UbD in their educational agencies and institutions.
- During the 2011-2012 academic year, the ED submitted its Elementary Education Program application to the state department of education to assure the program complied with state standards. In this application, the elementary teacher preparation curriculum was aligned to the elementary education stan-

dards, with the UbD framework included across all phases of the program. The ED also participated in the Teacher Education Accreditation Council (TEAC; now known as Council on Accreditation of Educator Preparation [CAEP]) national accreditation process of its teacher education programs. The UbD Framework was incorporated into the Inquiry Brief Proposal submitted to TEAC. Both the state and national reviews were positive, solidifying the Education Department's commitment to UbD as a viable curriculum framework.

- University B hired two adjunct faculty members who were experienced in UbD to assist with instruction and implementation across the Teacher Certification Programs. In 2012, the education coordinators and one of the newly hired adjunct faculty designed, planned, and delivered a three-session workshop for college supervisors, cooperating teachers, as well as the education and liberal arts faculty. The evaluation feedback was positive and indicated that additional training and support were needed to infuse UbD into the education programs adequately.
- During the 2012-13 academic year, education faculty began to embed the UbD Framework into the curriculum throughout the four phases of the Teacher Certification Program. In phase one, students became aware of the UbD Framework in an introductory course to the teaching profession. In phase two, pre-candidates began to develop and practice principles of basic unit and lesson planning using the framework of backward design in a curriculum, instruction, and assessment course. Pre-candidates also participated in microteaching that provided opportunities to practice teaching with friends and peers. The presentation was a 5 to 10 minute segment of a UbD lesson.

The teacher candidates complete their content major courses (integrated science, English language arts, social studies, and mathematics) that are offered through the liberal arts department. The methods courses for these subject areas are offered by the education department. While the liberal arts faculty attended some training on UbD, they had not fully adopted this curriculum framework into their courses.

In phase three, teacher candidates continue to develop and practice UbD units and lesson plans through a structured program of coordinated UbD theory, observation, and participation in field-based experiences. Teacher candidates apply ideas or procedures presented in their methods courses to K – 12 students in school settings, focusing on teaching integrated science, and using skills associated with classroom management, planning, and daily routines. They observe, participate in classroom activities, teach a science lesson using the UbD framework, and reflect on these experiences. During phase three, teacher candidates begin to develop an understanding of the daily routines that are part of the classroom environment. They develop a greater understanding of the need to know their students and learn to use this knowledge and understanding to inform the teaching-learning dyad, especially in a science classroom.

- Student teaching is the college-supervised instructional experience that occurs during the fourth phase of the teacher education program. Student teaching is the culminating field experience in the teacher education program. Student teachers design a unit and develop lessons plans based on the UbD Framework. Each student teacher videotapes a continuous 30-minute segment of a planned teaching lesson that includes a self-evaluation and a reflective assessment of the teaching event. At the elementary level, integrated science is one of the content-area subjects that is included in the teaching event. The successful completion of requirements of this 15-week experience typically leads to a degree and teacher certification
- The Education Faculty at University B meets on a regular basis to assess curricula, teaching, and assessment to ensure that key design issues are being addressed in course curriculum. During the current academic year, faculty will continue to update and revise curriculum to ensure that UbD is being implemented across disciplines, including science, in the education programs. The professional development team, including the elementary and secondary coordinators and adjunct professors, will continue to design training and consultation, along with follow-up activities to assist faculty in integrating UbD principles into their teaching and learning.

## CURRENT CHALLENGES FACING THE ORGANIZATION

Understanding by Design helps teachers to "provide a broader focus on scientific concepts and processes in a 'big picture' sense and not overemphasize the parts of the scientific concepts and processes" (Designing Curriculum for In-depth Understanding in Science, n.d., Slide 4). The challenge is to equip education and liberal arts faculty with the skills and resources to incorporate UbD into their teacher preparation courses. Effective implementation of UbD requires institutional support because of the time needed to provide professional development to faculty who are involved with instruction in the teacher education programs (Newton, 2003). Faculty who coordinate the elementary and secondary teacher education programs at University B are also required to be involved in scholarship and service, as well as teaching full course loads. These coordinators have been responsible for the adoption of UbD by faculty in ED, but have had partial success with liberal arts faculty.

Teacher candidates, at University B, understand the need for planning, designing, and implementing UbD into their units and lessons during the third and fourth phase of their programs. Class size in the third and fourth phase of the teacher education programs are small, allowing instructors to provide frequent and prompt feedback as teacher candidates seek to implement UbD. Teacher candidates, working closely with faculty, incorporate UbD into their professional tools.

In this time of economic distress for higher education, faculty development monies are not available for University A. Some professional development activities are funded by University B, however, the monies are limited. To provide training for UbD, selected faculty attended training sessions and then returned to their institutions to become trainers for other faculty. Time and funds must be made available to provide professional development to faculty in education and liberal arts programs to ensure that the adoption and implementation of the UbD curriculum framework is provided for teacher candidates in the university.

## SOLUTIONS AND RECOMMENDATIONS

Key personnel and shared vision were the overwhelmingly predominant supporting factors in adoption of UbD in teacher education programs. In the case of University A, the lead change agent was an individual faculty. However, the lack of program support limited the extent to which such change appeared within the program. The lead change agents at University B were two faculty from the Education Department. One faculty member from the department had a background in professional development and understood the level of professional development activities and ongoing support needed to produce organizational change.

Strategic vision is a predominant factor in supporting initiation of paradigm change in curriculum framework used in coursework for teacher candidates. As the courses are taught across disciplines (teacher education and liberal arts), a common vision is needed to adopt and implement curriculum to improve learning outcomes for teacher candidates and their students.

Sometimes change is not easy. Nevertheless, at University B, positive support for such a change was indicated by the numerous steps instituted to embed UbD across the teacher education program. At University A, there continues to be a lack of programmatic discussion of a need to introduce a common curricular planning tool. The lack of a common goal or vision for curriculum planning can be a major challenge in implementation.

Organizational structure was the predominant supporting factor in continuing the UbD curricular framework. Once the model was instituted at University B, faculty began to build the structure to support implementation and continuance of

UbD. Faculty met as a team and discussed the strengths and challenges involved in implementing UbD as the curriculum framework. At University A, the faculty member reported that she felt isolated in terms of exploring the use of UbD for curriculum design and development. Collaborative relationships and policy also can be challenges to the continuation of UbD at both universities. Personnel changes and expectations for preparing teachers are important challenges, at University B, in the continuing collaboration among faculty members and administration. Providing on-going training and support can ensure effective educational experiences in curriculum design and planning for teacher candidates.

In looking back at the activities involving both universities, assessment is an issue that needs to be addressed. Decisions about which curriculum framework should be presented to teacher candidates must be rooted in assessment. Data from program graduates who are practitioners and administrators need to be collected to determine the viability of using UbD as a curriculum-planning framework to improve student achievement. Data on teacher candidates' use of UbD and their students' achievement also need to be collected throughout the teacher preparation programs to examine progress in unit and lesson designing, planning, and implementing curriculum, especially in integrated science education.

## REFERENCES

Craven, J., & Penck, J. (2001). Preparing new teachers to teach science: The role of the science teacher educator. *Electronic Journal of Science Education, 6*(1). Retrieved from http://ejse.southwestern.edu/article/view/7670

Designing Curriculum for In-depth Understanding in Science. (n.d.). *PowerPoint presentation.* Retrieved from www.fass.cecs.ucf.edu/.../Designing%20Curriculum%20-%20M.%20Hay

Hambrick, D., & Cannella, A. Jr. (1989). Strategy Implementation as substance and selling. *The Academy of Management Executive*, *3*(4), 278–285. doi:10.5465/AME.1989.4277401

McTighe, J., & Seif, E. (2003). *A summary of underlying theory and research base for Understanding by Design*. Retrieved from assets.pearsonschool.com/asset_mgr/ubd_myworld_research.pdf

McTighe, J., & Thomas, R. (2003). Backward design for forward action. *Educational Leadership*, *60*(5), 52–55.

National Science Foundation. (2006). *America's Pressing Challenge — Building a Stronger Foundation: Mathematics and Science Achievement is Critical*. Retrieved from http://www.nsf.gov/statistics/nsb0602/

National Science Teachers Association. (2004). *NSTA Position Statement: Science Teacher Preparation*. Retrieved from http://www.nsta.org/about/positions/preparation.aspx?print=true

Newton, J. (2003). Implementing an institution-wide learning and teaching strategy: Lessons in managing change. *Studies in Higher Education*, *28*(4), 427–441. doi:10.1080/0307507032000122279

Prensky, M. (2006). Listen to the natives. *Educational Leadership*, *63*(4), 8–13.

STEM Smart Brief. (n.d.). *Preparing and support STEM educators*. Retrieved from http://successfulstemeducation.org/sites/successfulstemeducation.org/files/Preparing%20Supporting%20STEM%20Educators_FINAL.pdf

Wang, D., & Allen, M. (2003). Understanding by Design Meets Integrated Science. *Science Teacher (Normal, Ill.)*, *70*(7), 37–41.

Wiggins, G., & McTighe, J. (1998). *Understanding by Design*. Alexandria, VA: ASCD.

Wiggins, G., & McTighe, J. (2013). *Understanding by Design* (2nd ed.). Alexandria, VA: ASCD.

## ADDITIONAL READING

Alhamisi, J., Glimps, B., & Okezie, C. (2012). Understanding by design and culturally responsive teaching: Effective tools for instruction. *Tennessee Educational Leadership*, *39*(19), 21–28.

*Authentic education's understanding by design – related products*. (n.d.). Retrieved from http://www.authenticeducation.org

Graff, N. (2011). An "effective and agonizing way to learn": Backwards design and new teachers' preparation for planning curriculum. *Teacher Education Quarterly*, *38*(3), 151–168. Retrieved from http://www.teqjournal.org

Graffam, B. (2003). Constructivism and understanding: Implementing the teaching for understanding framework. *Journal of Secondary Gifted Education*, *15*(1), 13–22. doi: doi:10.4219/jsge-2003-439

Jones, K., Vermette, P., & Jones, J. (2009). An integration of "Backwards Planning" unit design with the "Two-Step" lesson planning framework. *Education, 130*(2), 357–360.

McTighe, J., & O'Connor, K. (2005). Seven practices for effective learning. *Educational Leadership, 63*(3), 10–17.

Shumway, S., & Berrett, J. (2004). Standards-based curriculum development for pre- service and in-service: A "partnering" approach using modified backwards design. *Technology Teacher, 64*(3), 26–29.

Staines, G. (2013). *The benefits of cloud computing in education are huge!* Retrieved from http://blogs.aspect.com/2013/06/27/the-benefits-of-cloud-computing-in-education-are-huge/

*The flipped classroom.* (n.d.). Retrieved from http://successfulstemeducation.org/sites/successfulstemeducation.org/files/Preparing%20Supporting%20STEM%20Educators_FINAL.pdf

## KEY TERMS AND DEFINITIONS

**Authentic Performance Assessment:** These are curriculum based measures that require students to construct responses on real world tasks.

**Backward Design:** A form of guided discovery that focuses on truly understanding the content by starting with the finished product and working backward.

**Curriculum:** The educational term for what students experience in school.

**Instructional Design:** The systematic process of designing and delivering instructional materials.

**Standards:** Expectations for what students should know and be able to do.

**Understanding by Design:** Wiggins and McTighe indicate that UbD is a framework for improving student achievement through standards-driven curriculum development, instructional design, assessment, and professional development.

# Chapter 4
# Martial Arts and Physics: A Multidisciplinary Approach to Increase Student Engagement and Interest in the Sciences

**Eugenie de Silva**
*Harvard University, USA*

## EXECUTIVE SUMMARY

*Many students perceive physics to be a difficult subject without any practical applications to their daily lives. Without the appropriate guidance, students will continue to lose interest in the sciences and will be hesitant to explore possible careers in the science disciplines. Accordingly, this research project examined the use of an annual physics day to promote active engagement amongst high school and college students in the study of physics, in addition to the success of the novel teaching of physics 100 classes through the martial arts. Both activities yielded high success rates that also proved that multidisciplinary teaching techniques could aid in raising the interest of students in physics.*

## INTRODUCTION

When students enter a physics class, they commonly have preconceived ideas of what they will be taught. For a majority of the students, their ideas are based on misconceptions of what learning physics constitutes. The failure to recognize the extent to which physics can be applied in an individual's daily life is undoubtedly

DOI: 10.4018/978-1-4666-6375-6.ch004

a reason why many students refrain from entering the field. Accordingly, this work was established to improve students' interest in physics, whilst explicating the connections of the field to their daily life circumstances. The aim was to show students the reasons why they should develop a career in physics.

In 2012, the United States (U.S.) dropped from being the 21st to the 24th in a ranking of top countries in science with Shanghai-China remaining the 1st (Weisenthal, 2013). If the U.S. is to move up in these charts, students within the nation need to be motivated to learn science and establish a future career in the discipline. Accordingly, what should be sought is a rigorous educational framework that places a spotlight on the teaching of science in a manner that focuses on the applicability of the field to individuals' daily lives.

Take for instance, when toddlers are being taught the alphabet, they are commonly shown a letter with a picture that begins with the letter that is being shown. Accordingly, this ensures that the toddlers can associate the letter with a known item, which helps them to learn the alphabet and understand how to apply it in their lives. In the same manner, when a student is learning a new subject in high school or college, abstract knowledge can be difficult to comprehend and apply. This is why the technique of multidisciplinary research is imperative in learning subjects such as physics. In this work, multidisciplinary refers to the inclusion of two or more fields to learn, teach, or discuss one specific discipline. Learning physics for the first time is comparable to learning a new language. Without a basis upon which novel information can be developed and ideas can be formed, it can be challenging to thoroughly grasp a new concept. Many students entering physics classes are faced with difficulties in recognizing the feasibility of applying concepts from the class to their daily lives. This inability to recognize the direct links between physics and one's life could also be a reason why students refrain from entering the science field.

There is also common misperception that when one enters a field, such as physics, career choices become limited to working in labs or being professors. While these career choices are certainly abundant, many students do not recognize the wide array of jobs that become attainable upon entering the science domain. Unless more experts and teachers/professors in the field of physics express that learning physics does not limit career opportunities to professions that are not entirely deemed exciting, it would not be possible to improve the number of students entering the field. In any science course, students should be exposed to not only scientific concepts and theories, but also to the links of their personal lives and the subject. Multidisciplinary research is the key to moving forward in the world and showing students that there is much more to physics and science than what meets the eye.

This chapter elaborates on the cumulative results of three surveys that were conducted to display the usefulness of the application of martial arts as a means of teaching physics. The analyses here explicate the ways in which one can utilize multidisciplinary means of teaching subjects to raise interest among students whilst also fostering an in-depth understanding of the topic at-hand. The work here explores the use of an annual physics day as a catalyst to heighten high school students' interest in the field of physics, in addition to the ways in which this event fostered an interest among the volunteer college students during the day. Moving forward, the work also indicates the results of using a novel syllabus in a college physics class as it pertains to the improvement of students' understanding and interest in science.

## MARTIAL ARTS AND PHYSICS

When an individual begins to learn martial arts, he/she embarks on a life journey. Over the course of time, the individual will begin to view life in a new light; they will view the world through the application of martial arts concepts and learned information of the discipline. Accordingly, when a student begins to learn physics, the individual begins to walk a new path in life. Similarly to martial arts, by learning physics, a student will be exposed to a plethora of novel facts that can and will ultimately alter the way in which the student lives his/her life. In these aspects, both fields have major commonalities; thus, the teaching of physics through martial arts can be easily facilitated.

Through years of teaching physics in this novel manner, de Silva (2007) established the aforementioned "Physics Day" that immediately gained popularity. This unconventional routine of teaching high school students about the applicability of martial arts, physics, and daily life activities ignited a noticeable flame of passion for science for those who attended. For the students attending the event, the use of martial arts as a means of learning physics was their first introduction to multidisciplinary research. Of course, for a learned physicist, the links between martial arts and physics may seem easily explainable. Au contraire, for students without a prior background in the sciences, these explanations can act as driving forces to learn more about science and the role it plays in their daily lives.

For some martial artists, the key to the discipline revolves around the philosophy of achieving "harmonious values by individuals who live by peace, wisdom, morals, love and self-discipline through intellectual means", whereas for other practitioners the art of martial arts is about self-defense and/or strength building ("Martial Art," 2008). The training involved in martial arts requires self-discipline and pure dedication to the sport; however, it could also be argued that a knowledge of physics is necessary in order for one to master martial arts techniques in a systematic manner.

On the other hand, advanced physics requires that individuals also have a background in mathematics. Many students have categorized physics as a difficult field, yet it can be simplified when taught in an organized manner through a multidisciplinary approach. The subject essentially provides an underlying foundation upon which students in the field can improve their daily lives while also establishing and improving their novel analyses of the field.

## SETTING THE STAGE

For a time period of seven years, annual "Physics Days" were held for local high school students in the states of Virginia, Kentucky, and Tennessee in the U.S. The students were brought to the event by their science teachers and were provided with a survey to be filled out prior to leaving at the end of the day. The high school students were predominately juniors or seniors; hence, they had previously taken various science courses and several of the students indicated that they had already established tentative plans with regard to their future careers and aspirations.

Upon arrival at the event, the students were taken to a consortium where they were given a background description of the events that would be taking place. Physics day was initiated to impart physics knowledge, while also explicating to the high school students that the subject can be applicable to a wide array of daily circumstances. Through these annual events, high school students were provided with a direct opportunity to take part in physics activities while being taught through a multidisciplinary method of incorporating martial arts and the sciences. This technique sought to raise awareness and possibly influence younger students who have yet to enter college in the direction of earning a higher degree in physics or any related science field.

During the development stage of the physics day for the local high school students, de Silva (2007) structured the day so that the students would first learn theoretical information, and then be allowed to apply the knowledge from a practical perspective. The morning session included the teaching of basic physics principles that mainly pertained to mechanics' sections. This component of physics was also taught through martial arts demonstrations for around two to three hours. The afternoon session then placed a focus on the utilization of a hands-on approach to learn the discipline through demonstrations of physics instruments and experiments.

Undergraduate college students who had volunteered to take part in the event conducted the physics experiments. The college students were not required to have backgrounds in the sciences; thus, many were majoring in other fields, such as nursing, child education, engineering, etc. The undergraduates who volunteered for the demonstrations did not have more than an introductory knowledge of physics. Therefore, they had to attend at least six-weeks of two-hour preparatory sessions. While high school students in the morning session were mere observers, the afternoon session gave them an opportunity to actively participate in physics demonstrations.

The first-year undergraduates were given an opportunity to teach/demonstrate what they learnt both in preparatory sessions and independent/directed learning through experiments that were performed for the high school students. This enabled them to improve their knowledge of physics and helped generate greater interest among undergraduates to learn more about physics and change their negative attitudes toward physics. The high school students on the other hand were given the opportunity to apply the concepts that had been taught in the morning session, during the afternoon, which further reinforced the knowledge of physics. The high school and undergraduate students were given surveys at the end of the day to assess their attitudes, interests, and opinions on learning physics (de Silva, 2007).

The second project included an introductory physics course taught through martial arts. Unlike traditional introductory physics classes, the physics 100 courses discussed here allowed the students to learn the provided information by applying martial arts concepts. The students were not required to have backgrounds in martial arts, since all necessary information was taught within the classroom. The objective of the novel syllabus that was introduced was to promote analytical thinking amongst the students whilst introducing them to the multidisciplinary nature of physics. The students were given a survey at the end of the semester to understand their perception of physics and describe their feelings toward the new physics class format.

## CASE DESCRIPTION

### Physics Day

As aforementioned, for Physics Day, the undergraduate students were selected from various classes to perform the demonstration for high school students who attended the occasion. Seventy-five percent of these volunteer students had not studied physics formally. They attended direct learning sessions over the course of six weeks, prior to the physics day event. They learned the material and produced posters to make their demonstrations and explanations effective. All material was thoroughly checked to ensure that all explanations and discussions were correct, yet the college students were allowed to express their own ideas and analyses through the poster sessions as long as they were scientifically correct. The creation of the posters also allowed the students to conduct their own research on the science topics and expand their own knowledge in a manner that provided them with a sturdy basis to present the topics and be confident in their understanding of the subject content.

The high school students were passive learners in the morning session, yet in the afternoon session during the experiments conducted by the undergraduate students, they took part in an interactive learning process. The results of 2000 high school students and 60 college students are given in this research. The following components have been included to provide a basic understanding of the material that were utilized, in addition to the surveys and data that were presented to the high school and college students on the physics day.

## Physics Day Demonstrations

### *Morning Session*

- **Physics and Philosophy.**
- **Scientific Method and Hypothesis.**
- **Free-Fall:** Forward roll and other demonstrations.
- **Force:** Martial Arts Techniques.
- **Velocity and Acceleration:** Application of force, momentum, and velocity in breaking tiles using hands and head.
- **Circular Motions:** Application of Martial Arts Techniques.
- **Momentum:** Breaking Beams.
- **Rotational Force:** Two, three, and five -finger one-hand pushups at 45-degree angles and breaking of wooden stick balanced on two wine glasses with a sword.
- **Inertia and Pressure:** Breaking concrete on the stomach while sandwiched between two beds of nails, breaking of concrete on the neck and other parts of the body.

### *Afternoon College Students' Demonstrations*

- Tesla Coil.
- Van de Graff.
- Smoke Ring Cannon.
- Bottle Rocket Launcher.
- Inertia, Balancing, Feather and Coin Demo.
- Rotation and Angular Momentum.
- Projectile Motion.
- Prism, Laser Pointers and Periscope.
- Pendulum and Potential Energy.
- Electricity Demonstrator and Solar LED Flashlight.
- Electromagnetic Flashlight.
- Plasma Globe.
- Magnetic Field Model and Levitating Globe.
- Perpetuator, Levitron.

- Monkey and Hunter.
- Elastic Launcher.
- Telescope.
- Frozen Gummy Bears and various Cryogenic Demonstrations.
- Levitation – Superconductivity.

### Survey Given to High School Students after Physics Day

1. What was the most interesting section of today's event? (Morning/Afternoon)
2. Which demonstrations did you like the most?
3. Have you taken a physics course before?
4. Did you consider physics as a difficult subject before today?
5. Has your perception on physics changed after today? (Please describe)
6. Would you recommend Physics Day to others? (Yes/No)
7. Would you now recognize more of the importance of physics in daily activities? (Yes/No; if No, please specify)
8. Would you be able to promote physics to non-scientist better as a result of this day? (Yes/No; if No, please specify)
9. Would you major in a science field as a result of Physics Day?
10. Any other comments?

### Survey Given to College Students after Physics Day

1. Have you taken a college physics course prior to participating in Physics Day? (Yes/No; If Yes, please provide the details)
2. Are you majoring in any science subjects? (Yes/No; if Yes, please provide the details)
3. If you are not majoring in any science fields, in what field would you major?
4. Did you consider physics difficult prior to Physics Day involvement? (Yes/No)
5. Why did you consider volunteering for Physics Day? (Please specify the reason)
6. Did you find your activity difficult to follow and learn? (Yes/No; please specify)
7. Has your perception of physics changed as a result of Physics Day? (Yes/No; please specify)
8. Would you recommend Physics Day to others? (Yes/No)
9. Would you now recognize more of the importance of physics in daily activities? (Yes/No; if No, please specify)
10. Would you be able to promote physics to non-scientist better as a result of this day? (Yes/No; if No, please specify)

## Physics 100 Course

The second project that has been analyzed herein this work included the teaching of 30 students per semester who were taking physics for the first time. Their majors were non-scientific subjects. About twenty percent of the class students were involved in sports, but not martial arts. These students underwent a study of one semester learning physics through martial arts. They were not required to perform any of the exercises or martial arts movements and were free to select their mini project that could include a non-martial arts subject, but it had to be multidisciplinary in nature. The martial arts links to physics within the classroom settings were completed through actual demonstrations, videos, and books. The survey results of three semesters are presented in the results and analysis of this chapter. In its entirety, the total number of students was 90.

### The Physics 100 Syllabus Combined with Martial Arts (© Eugene de Silva)

1. **Physics, Martial Arts, and Philosophy.**
2. **Scientific Method and Hypothesis:** Assessing the environment, avoidance, dissuasion, discouragement, and deflection techniques.
3. **Velocity and Acceleration:** Forward roll and break-falls, basic hand and leg movements and stances.
4. **Movement in One Dimension:** Free-fall – break-falls.
5. **Movement in Two Dimensions:** Punches and kicks.
6. **Force:** Martial arts techniques including breaking and inertia.
7. **Momentum:** Breaking tiles and beams.
8. **Circular Motions:** Demonstration of weapons and *kata*.
9. **Rotational Force:** Throws, locks, and blocks.
10. **Mini-Project Presentation.**

#### *Survey Given to Physics 100 Class Students*

1. Have you taken a college physics course prior to this Physics 100 course? (Yes/No; if Yes, please provide the details)
2. Are you majoring in any science subjects? (Yes/No; if Yes, please provide the details)
3. If you are not majoring in any science fields, in what field would you major?
4. Did you consider physics as a subject difficult prior to beginning this course? (Yes/No)
5. Why did you consider taking this physics course? (Please specify the reason)

6. Did you find this course difficult to follow and learn? (Yes/No; if Yes, please specify)
7. Has your perception of physics changed as a result of this course? (Yes/No; please specify)
8. Would you recommend this course to others? (Yes/No)
9. Would you now recognize more of the importance of physics in daily activities? (Yes/No; if No, please specify)
10. Would you be able to promote physics to non-scientist better as a result of this course? (Yes/No; if No, please specify)

## RESULTS AND ANALYSIS

The surveys were gathered and analyzed to determine the actual effectiveness of the physics day and the physics 100 courses. The results clearly showed a vast improvement in students' interest in physics due to the novel teaching methods using martial arts. To provide an overall, unbiased discussion of the results, a comparative statistical analysis was also conducted. (See Tables 1, 2, & 3)

Each of the survey results showed that the students considered physics to be a difficult subject prior to entering the physics 100 classes and physics day. However, the students explained that they later had changed their perceptions of physics. Many of the students felt that they no longer believed physics was a difficult topic and that they would enjoy learning the subject if it was taught through the multi-disciplinary approach of using martial arts. Even the college students who had volunteered for physics day stated that they would recommend other students to be volunteers.

The pass rate for all physics 100 classes was 100% and the class assessments included daily quizzes, homework, chapter quizzes, midterm, final exam and a mini-project. About 10% enrolled for higher-level physics classes, while majoring in non-science subjects.

## DISCUSSION/CHALLENGES FACED

Based on the provided analyses, it was clear that the students involved in physics and martial arts activities enjoyed the event and had gained valuable knowledge. The students, as per the results of the surveys, acknowledged the usefulness of physics in their daily lives. Accordingly, the results, as was explicated in more detail above, proved that when students are provided with an opportunity to be interactive and substantially engage in the physics activities that are being taught they commonly enjoyed the process more than otherwise.

*Table 1. Survey I: Given to high school students after Physics Day (Sample size = 2000)*

| Question | Q | Yes | % (Yes) | No | N/A or Did Not Answer | % (No and N/A or Did Not Answer) | Specified Answer |
|---|---|---|---|---|---|---|---|
| What was the most interesting section of today's event? Morning/ Afternoon | 1 | | 0.0% | | | 0.0% | 413 (Both), 1587 (Morning Session) |
| Which demonstrations did you like the most? | 2 | | 0.0% | | | 0.0% | 1848 (Breakings), 62 (Light), 69 (Electric and Magnetic), 21 (Other) |
| Have you taken a physics course before? | 3 | 227 | 11.4% | 1773 | | 88.7% | |
| Did you consider physics as a difficult subject before today? | 4 | 1752 | 87.6% | 224 | 24 | 12.4% | |
| Has your perception on physics changed after today? (Please describe) | 5 | 1906 | 95.3% | 77 | 17 | 4.7% | |
| Would you recommend Physics Day to others? Yes/No | 6 | 1994 | 99.7% | | 6 | 0.3% | |
| Would you now recognize more of the importance of physics in daily activities? Yes/No (If no, please specify) | 7 | 1823 | 91.2% | 21 | 156 | 8.9% | |
| Would you be able to promote physics to non-scientist better as a result of this day? Yes/No (If no, please specify) | 8 | 1476 | 73.8% | 317 | 207 | 26.2% | |
| Would you major in a science field as a result of Physics Day? | 9 | 1238 | 61.9% | 728 | 34 | 38.1% | |
| Any other comments? | 10 | | | | | 0.0% | |
| | | | | | 1048 | 52.4% | Wish it was taught in school like this |
| | | | | | 1356 | 67.8% | Like to know more about the subject |
| | | | | | 128 | 6.4% | Want to come this college |
| | | | | | 1792 | 89.6% | There must be Physics Days regularly |
| No comments | No comments | No comments | No comments | No comments | No comments | No comments | No comments |

*Table 2. Survey II: Given to college students after Physics Day (Sample size = 60)*

| Question | Q | Yes | % (Yes) | No | N/A or Did Not Answer | % (No and N/A or Did Not Answer) | Specified Answer |
|---|---|---|---|---|---|---|---|
| Have you taken a college physics course prior to participating in Physics Day? Yes/No (If Yes, please provide the details) | **1** | 7 | 11.7% | 53 | | 88.3% | High School Physics |
| Are you majoring in any science subjects? Yes/No (If Yes, please provide the details) | **2** | 8 | 13.3% | 52 | | 86.7% | |
| If you are not majoring in any science fields, in what field would you major? | **3** | | 0.0% | | | 0.0% | Business, Sports, Marketing |
| Did you consider physics as a subject difficult prior to beginning this course? Yes/ No | **4** | 52 | 86.7% | 5 | 3 | 13.3% | |
| Why did you volunteer to do the demonstrations? (Please specify the reason) | **5** | | 0.0% | | 60 | 100.0% | Extra credits |
| Did you find your demonstration difficult to follow and learn? Yes/ No (If Yes, Please specify) | **6** | 5 | 8.3% | 55 | | 91.7% | Difficult to understand some concepts |
| Has your perception of physics changed as a result of this learning? Yes/No (Please specify) | **7** | 58 | 96.7% | | 2 | 3.3% | |
| Would you recommend others to volunteer? Yes/No | **8** | 57 | 95.0% | | 3 | 5.0% | |
| Would you now recognize more of the importance of physics in daily activities? Yes/No (If No, please specify) | **9** | 54 | 90.0% | 3 | 3 | 10.0% | Somewhat |
| Would you be able to promote physics to non-scientist better as a result of this day? Yes/No (If No, please specify) | **10** | 44 | 73.3% | 5 | 11 | 26.7% | Need to review the work again/ Need more help |

For instance, 96.7% of the undergraduates claimed that they would recommend others to volunteer for the program. Additionally, 73% stated that they would be able to better promote physics to non-scientists as a result of the program. Along these lines, 95% of the high school students stated that their perceptions of physics have changed due to physics day. Moreover, 63% of the students said that they would

*Table 3. Survey III: Given to Physics 100 class students (Sample size = 90)*

| Question | Q | Yes | % (Yes) | No | N/A or Did Not Answer | % (No and N/A or Did Not Answer) | Specified Answer |
|---|---|---|---|---|---|---|---|
| Have you taken a college physics course prior to this Physics 100 course? Yes/ No (If Yes, please provide the details | 1 | 9 | 10.0% | 81 | | 90.0% | High School physical science |
| Are you majoring in any science subjects? Yes/No (If Yes, please provide the details) | 2 | 13 | 14.4% | 77 | | 85.6% | Biology Major |
| If you are not majoring in any science fields, in what field would you major? | 3 | | 0.0% | | | 0.0% | Business, Sports, Marketing |
| Did you consider physics as a subject difficult prior to beginning this course? Yes/ No | 4 | 80 | 88.9% | 7 | 3 | 11.1% | |
| Why did you consider taking this physics course? (Please specify the reason) | 5 | | 0.0% | | | 0.0% | Needs a science course, physics day, referred by a friend |
| Did you find this course difficult to follow and learn? Yes/ No (If Yes, Please specify) | 6 | 8 | 8.9% | 82 | | 91.1% | Need more time to do math related questions |
| Has your perception of physics changed as a result of this course? Yes/No (Please specify) | 7 | 86 | 95.6% | 2 | 2 | 4.4% | Physics was fun, was not difficult |
| Would you recommend this course to others? Yes/ No | 8 | 81 | 90.0% | 3 | 6 | 10.0% | |
| Would you now recognize more of the importance of physics in daily activities? Yes/No (If No, please specify) | 9 | 82 | 91.1% | 2 | 6 | 8.9% | Somewhat |
| Would you be able to promote physics to non-scientist better as a result of this course? Yes/No (If No, please specify) | 10 | 67 | 74.4% | 11 | 12 | 25.6% | |

major in a science field as a result of the day. Finally, 89% of these students stated that there should be regular physics days, and 52% claimed they wished physics was taught in this manner in their schools.

According to the comments that were provided through the surveys, it was clear that the students had first viewed physics as a difficult subject that did not provoke their interests. For a majority of the students, physics lacked excitement and/or direct links to their lives. Whilst the high school and college students had been exposed to learning about physics in previous courses, they had not genuinely understood or recognized the necessity of the discipline and the usefulness of linking the field to their lives. It seemed that they had been taught physics as an abstract subject that was to be learned and understood separately from other subjects. There was an overall failure in the students' abilities to quickly apply taught physics concepts to their own lives, until they had been provided with examples.

The study investigated students from areas in Northern Virginia, Kentucky, and Eastern Tennessee. This also provided the opportunity to establish the ways cultural norms affected the students' learning and understandings. Whereas a majority of the Virginia high school students who visited physics day were from gifted and talented schools, the students in Tennessee and Kentucky were members of the public school systems. From the perspective of the college students during the physics days, it was significantly much easier to teach the information to the gifted and talented students. Whereas the traditional students were also able to understand the presented data, the college students stated that teaching students who were more advanced than their peers made their duties easier. Nevertheless, in each of the instances, the students' reviews and survey answers proved that they felt much more comfortable in discussing physics concepts and holding academic-level conversations about the discussed topics. The different ways in which students engage with taught material and comprehend data were recognized as one of the challenges that are currently posed.

Multidisciplinary teaching is a notion that is commonly utilized in the twenty-first century; it has recently gained much popularity and its focus on the bringing together of diverse fields has made it the focus of research and the underlying basis of teaching in the future (de Silva, de Silva, Horner & Campbell, 2012). The concept of multidisciplinary teaching is a seemingly simple concept, yet when viewed through the lens of an educator, researcher, and even a learner, it becomes progressively clear that there are basic contextual issues that must be addressed. In the research that was conducted for this work, the students learned physics through the utilization of martial arts. Whereas other subjects could have been substituted for martial arts, the aim was to engage students in an exciting day that would also allow them to realize the role that physics play in their lives. In the classroom settings of the Physics 100 class, the students were taught the subject-content by allowing them to delve deeper into the subject through the multidisciplinary approach, in addition to a research-based method. The physics classroom was more interactive than traditional physics classrooms. For instance, in traditional settings, students are posed questions and in some cases are allowed to work in groups to develop an academically suitable answer (de Silva, Long, & Mennen, 2008). However, these classrooms do not expand on the topics in a manner that establishes a fundamental basis for students to actually remember the concepts being taught for longer periods.

In the physics 100 class, physics facts were taught by first discussing the actual meanings and backgrounds of the physics concepts. Consequently, the concepts were elaborated in terms of how they pertain to the general world and their applicability to the individual lives of the students in the class. Finally, the students were provided with opportunities to prove their knowledge and show what they had learned by conducting their own mini-research projects. The research that was conducted by the students was multidisciplinary in nature in that they decided to investigate specific physics concepts that pertained to their own future career goals. For instance, some students who wanted to have careers as veterinarians studied balance and center of mass as it related to obesity in animals (de Silva & DeBusk, 2008). By teaching students about the applicability of physics to martial arts within the classroom settings, the students were able to extrapolate the information to link the subject to their own fields. For instance, some students focused on reviews of osteopathic manipulation in Western countries, understanding the physics behind roping horses, and even the application of physics for a homerun in baseball (de Silva, Teitelbaum, & Long, 2008; de Silva & Garret, 2008; de Silva & Christian, 2009). Accordingly, the students were able to improve three major skills; physics, research, and multidisciplinary linking. Since the students had to learn physics concepts in order to complete their research and take part in the classroom activities, they began to genuinely learn about the taught materials. Moving forward, the students were able to improve their research skills, since they had to conduct research to develop an academic-level research paper that would incorporate not only physics, but also their selected fields. Finally, it was clear that the research and teaching within the classroom allowed the students to improve their multidisciplinary approaches to analyzing situations. Prior to entering the class, the students were only able to identify very few, limited ways in which physics could be applied to their lives. Even in these cases, it was clear that the students did not thoroughly understand why physics applied in the ways they were explaining; rather, it seemed that they were repeating information that had somehow been filtered to them. Upon completion of the course, the students were able to quickly establish links between physics and their lives, which in turn allowed them to be considerably more confident in explaining physics and holding conversations about the topic.

Analogously, the survey results from the high school and college students who participated in the physics days also indicated that they had substantially improved their knowledge and understanding of physics as it pertains to the application thereof in daily life matters. The students' answers to the surveys also showed that whilst they began the physics day with interests in the fields outside of physics, they had left the day with a genuine interest in physics. The students, in their comments and responses to the surveys, acknowledged that they would most probably be able to now discuss and describe the covered physics concepts to individuals who are not in the field. The survey results provided above clearly highlights the vast improvements that were made during the physics days and the physics 100 classes.

As is the case with any novel implementations that would require extensive changes, there are always challenges that arise. The educational systems of twenty-first century schools and universities have been undoubtedly shaped and formed based on decades of analysis, revision, and improvements in the educational arena. For most educators, their responsibilities have been established by standards that have been initiated to complement and correspondingly improve the standards of education. In this manner, students have also seemingly become complacent with their positions in the educational realm. Thus, alterations in the sense of changing the norm of educational systems require all-inclusive efforts by educators and students to adapt to the changes. The above-mentioned novel implementations require changes in terms of the ways subjects are taught and the opportunities that are provided to students.

Accordingly, the first challenge that may arise is the lack of training of educators to utilize the multidisciplinary approach to highlight the ways their subjects can be linked to the lives of the students. Although the educators may be enthusiastic about expanding their teaching horizons to include this method, the actual application of the method may pose major problems. For instance, those who are specialized in the sciences may have knowledge of the dynamics of their field, yet they may find it difficult to teach the links between their field and other subjects or activities due to limits in their academic and professional training. The difficulties are understandable, since educators are not commonly trained to tackle multiple subjects, but rather solely focus on their selected discipline. Nonetheless, this challenge could be overcome through the implementation of specified training courses that would include 6-week to 8-week courses. The instructors would have the freedom to select the courses that they feel would make them more comfortable in teaching through the multidisciplinary approach. Moving forward, the initiation of a multidisciplinary reviewer specialist positions could be established in schools and universities. The specialists would conduct annual or monthly assessments of the educators to determine their strengths and weaknesses. They could then provide suggestions with regard to the most useful courses that the instructors could complete to improve their techniques. This would further lead to higher professional development in the field of science.

The second challenge may pertain to the reaction of students when placed in an environment conducive to multidisciplinary learning. According to the results of the research conducted in this work, a majority of the students actively engaged on the day and felt comfortable in learning through this novel method; thus, they stated that they would prefer to be taught in this manner. However, there remains the possibility that students will feel overwhelmed by the novel teaching and learning method, and hence will fail to immerse themselves in the class. Therefore, in order to prevent these difficulties, teachers must take time, prior to teaching, to explain the way the subject will be taught, why, and how it can be useful. The students should be aware of how it can help them in their own lives; they should recognize the actual necessity of devoting themselves to the class and having fun in the classroom and actively engaging. Students should feel comfortable to answer questions, pose questions, take part in academic conversations within the classroom with their fellow classmates, and even question course materials. In this aspect, students should feel that they have a thorough understanding of why they should be interested in the course subject and how it will help them in the long run.

## CONCLUSION

Over the course of nine years, physics day has gained public interest in the media and has consistently had more and more students attending the event. The work here has presented the results of surveys that had been provided to attendees and volunteers throughout the years. The results ultimately prove that the annual physics days have improved student interest in physics, in addition to student engagement in the sciences. On the other hand, the novel methods that were utilized in the Physics 100 class also were accepted by a majority of the involved students. The results of each of the analyses proved that the students who actively participated in the classroom and physics day event felt that they had genuinely learned novel information and were confident in their newfound abilities to discuss the materials at an academic level.

The statistical results further showed that 96.7% of the undergraduates claimed that they would recommend others to volunteer for the program, whilst 73% stated that they would be able to better promote physics to non-scientists as a result of the program. To further explicate the effectiveness of the program, 95% of the high school students believed that their perceptions of physics have changed due to the physics day. Finally, 63% of the students noted that they would major in a science field as a result of the day. Finally, 89% of these students stated that there should be regular physics days, and 52% claimed they wished physics classes in their schools were taught in this same manner. The overarching theme here is that each of the students, even those who volunteered, benefited from the events by expanding their knowledge of physics. Moreover, the students claimed to have ultimately recognized the actual ways to apply physics to their daily life circumstances.

The research has paved the way for other specialists in the fields of physics or multidisciplinary research to further investigate the applicability of these methods in other environments. By applying the method to classroom settings and rigorously testing the effectiveness of these practices, it may be possible to further identify strategies to improve this approach. Accordingly, further research may provide detailed discussions with regard to the effects of cultural norms on the usefulness of the multidisciplinary approach of using martial arts to teach physics. Finally, as research progresses in this topic, it may be useful to expand these notions to apply them to other subjects and disciplines. By taking these steps, it may be possible to create a learning environment that is conducive to utilizing the multidisciplinary approach in students' academic endeavors.

## REFERENCES

de Silva, E. (2007, March). *Teaching Physics through Research.* Paper presented at the Annual Physics Day at Commonwealth Governor's School. Fredericksburg, VA.

de Silva, E., & Christian, S. (2009, October). *Application of physics for a home-run.* Paper presented at the Tennessee Academy of Science Annual Conference. Knoxville, TN.

de Silva, E., de Silva, E., Horner, J., Campbell, L., & McCamey, W. (2012, June). Multidisciplinary Research and Science Teaching. In *Proceedings presented at the Annual Conference on Chemical Science*. Colombo, Sri Lanka: Academic Press.

de Silva, E., & DeBusk, H. (2008, November). *A Comparative Study of Obesity and Animal Movements.* Paper presented at the Tennessee Academy of Science Annual Conference. Nashville, TN.

de Silva, E., & Garrett, T. (2008, November). *Understanding the Physics Behind the Performance of Roping Horses. Tennessee.* Paper presented at the Academy of Science Annual Conference. Nashville, TN.

de Silva, E., Long, C., & Mennen, M. (2008, October). *Teaching Science Through Multidisciplinary Research.* Paper presented at the Appalachian College Association's Annual Conference, Summit XI. Abingdon, VA.

de Silva, E., Teitelbaum, H., & Long, C. (2008, November). *A Review of Osteopathic Manipulation among western Physicians and Eastern indigenous practitioners.* Paper presented at the Tennessee Academy of Science Annual Conference. Nashville, TN.

*Martial Art Philosophy*. (n.d.). Retrieved March 29, 2014, from http://www.alljujitsu.com/martialartphilosophy.html

Weisenthal, J. (2013). Here's the New Ranking of Top Countries in Reading, Science, and Math. *Business Insider*. Accessed February 16, 2014 at http://www.businessinsider.com/pisa-rankings-2013-12

## ADDITIONAL READING

de Silva, E., & de Silva, E. (2012). *Multidisciplinary Research for College Students*. Ronkonkoma, NY: Linus Publications.

*Journal of Multidisciplinary Scientific Research Content*. (2014). Retrieved March 19, 2014 from http://jmsr.rstpublishers.com/issue_no_7.jws

*Multidisciplinary and Interdisciplinary Research*. (2010). Retrieved March 20, 2014, from http://www.oecd.org/site/innovationstrategy/45185249.pdf

*Multidisciplinary Research in Applied Science*. (2014). Retrieved March 20, 2014 from http://ws.edu/academics/research/

*Multidisciplinary Research Program of the URI*. (2013). Retrieved March 20, 2014 from http://www.onr.navy.mil/Science-Technology/Directorates/office-research-discovery-invention/Sponsored-Research/University-Research-Initiatives/MURI.aspx

*National Accrediting Commission for Martial Arts*. (2014). Retrieved March 20, 2014 from www.nacma.us

*The Higher Education Academy*. (2009). Retrieved March 20, 2014 from http://www.heacademy.ac.uk/assets/documents/resources/publications/developingundergraduate_final.pdf

*Virginia Research Institute*. (2014). Retrieved March 20, 2014 from www.virginiaresearchinstitute.org

## KEY TERMS AND DEFINITIONS

**Inter-Disciplinary:** The clear combination of two or more subjects that are commonly considered quite different.

**Martial Arts:** Combat practices that are used for self-defense, strength building, competition, and/or the improvement of one's physical wellbeing.

**Multidisciplinary:** The inclusion of two or more subjects to teach one given subject, without any clear boundaries. (Compare with Inter-Disciplinary)

**Physics Day:** A day devoted to the teaching of physics through the use of martial arts; teaching guides students through theoretical notions, in addition to practical experiments.

**Research-Based Methods:** The study or analysis of a subject or topic through the utilization of research.

**Traditional Education:** Education primarily held in classroom settings by following a strict syllabus and giving predetermined assessments.

# Chapter 5
# The Inclusion of Multidisciplinary Research in Science Teaching:
## A Novel Teaching Method

**Eugene de Silva**
*Virginia Research Institute, USA & MRAS – Walters State Community College, USA*

**Eugenie de Silva**
*Harvard University, USA*

**Jeffrey Horner**
*Walters State Community College, USA*

**Pamela Knox**
*Tennessee Board of Regents, USA*

## EXECUTIVE SUMMARY

*The educational process is described as a method whereby knowledge, skills, beliefs, values, and methods are transferred from one person to another. This chapter describes a series of research projects carried out from 1998 to 2013 that attempted to establish an effective process conducive to the transfer of chemistry and physics knowledge. The powerful combination of research and online studies with the latest technological tools are also discussed in this chapter. The chapter also provides the*

DOI: 10.4018/978-1-4666-6375-6.ch005

*START model that signifies how different contexts may actually influence core learning. This further emphasizes the importance of the inclusion of research in teaching and how it provides a fourth dimension to teaching. This work also elaborates the importance of the multidisciplinary research-based teaching and how it promotes independent thinking and flexibility among learners.*

## BACKGROUND

From 1990 – 2004, there was a gradual, yet very significant reduction in the number of students studying chemistry at the undergraduate level in the U.K. ("10th Report," 2006). This reduction compelled some universities to merge chemistry departments with other related departments to improve the numbers of students and the general economic viability (e.g. Manchester Metropolitan University merged the department of chemistry with materials science and the University of Manchester merged with UMIST). The reduction of chemistry students seemed to be related to the difficulty of chemistry students securing jobs after graduation, in addition to the disconnection of teaching of theories and how they relate to real-life applications ("Stem Briefing," 2009). The studies carried out by the Association of Graduate Recruiters (1995) showed that the employers believed that graduates could be better prepared for an employment environment by the university education. However, the graduate market as of 2013 still seems to be struggling with the same issues ("The Graduate Market," 2013). Duckett, Garratt, and Lowe (1999) suggested that graduate training included the skills that were most needed in employment for chemistry graduates. Moving forward, Brattan, Mason, and Rest (1999), in their findings, reported that most of the practical chemistry work was not geared to improving the key skills of students. Since a similar situation was also recognized previously by the research and surveys undertaken by Johnstone (1997), three possible implications were recognized. Firstly, after qualifying as chemistry graduates, their skills other than the subject knowledge may have failed to impress prospective employers of chemical industries since they lacked generic skills such as written and verbal communication, business awareness, pragmatic approaches, etc. Secondly, if the graduates of chemistry cannot impress employers from related industries, then they would not only have failed to secure jobs in the fields related to chemistry but invariably in other unrelated fields, since their knowledge was only concentrated on chemistry related fields. Thirdly, when the predicaments of chemistry graduates became apparent to other prospective undergraduates, then their perception of this subject as a choice of a future career would have also diminished to very low levels. These reasons were assumed as the logical reasons as to what eventually led not only to a reduction in the number of chemistry undergraduates, but also in the number of students taking Advanced level or equivalent studies in chemistry or in selecting chemistry as their majors in universities. As per the Education in Chemistry issue on Graduate Employability, employers still consider chemistry graduates as lacking the generic skills expected for employment ("Graduate Employability," 2009). Also, according to the statistics of chemistry education published by the Royal Society of Chemistry (2010), there was a reduction in Advanced Level chemistry and physics students from 1989 to 2010. The percentage of students for chemistry in 1989 was 6.8 and in 2010 it was 5.2, whereas the percentage of students in physics in 1989 was 7.1 and in 2010 was 3.6. This clearly shows a decrease in the number of students taking these vital courses. Whilst later reports may show slight increases, there have not been any significant changes in these trends.

According to Dewey (1998), the educational process involves the transfer of knowledge, skills, beliefs, values, and methods from one person(s) to another person(s). The traditional process involves this transference in a formalized, fixed framework. Under these conditions, traditional education is focused to deliver pre-determined subject-based knowledge, and the extent of this knowledge is determined by the level of education. In Bondelli's work that was published in 2007, focus was placed on the failure of the traditional educational system to recognize the personal discovery of meanings in subject matter at all levels of education. He further elaborated that the reliance on standardized testing in traditional school systems has limited the extent to which true learning takes place in schools (Bondelli, 2007). The emphasis that is placed on standardized testing compels the teachers to concentrate more on the material relevant to these tests since the effectiveness and funding of school teaching are determined by the test results. This is a trend that is currently seen in colleges and universities for the same reasons.

According to the Association of Supervision and Curriculum Development (1978), the, "process of human learning always has two parts: (a) confrontation with new information and experience (b) the learner's personal discovery of the meaning of that experience.

On another note, the book, "How People Learn: Brain, Mind, Experience, and School Committee on Developments in the Science of Learning" (Bransford, Brown, & Cocking, 2000), highlighted three key findings:

1. The students begin their learning with preconceptions. As a result, if a student's initial introduction to the subject is not engaged, then the students may fail to truly understand the novel topics. (Bransford, et. al., 2000). Otherwise, students may only learn the information to pass tests, and then later revert to their previous ideas and conceptions of the discipline (Bransford et al, 2000).
2. In order to develop competency in any area of inquiry:

a. The student should be provided with a thorough foundation that consists of factual knowledge (Bransford et al, 2000).
b. The student should understand facts and ideas with reference to a conceptual framework (Bransford et al, 2000).
c. The student should have the ability to systematically organize gathered knowledge in a manner that is conducive to retrieval and application (Bransford et at., 2000).

3. A "Metacognitive" approach to instruction helps students learn the ways in which they can control their learning through the establishment and identification of learning goals and the monitoring of progress by achieving the goals (Bransford et al, 2000).

The above three points should act as a guide to teaching and teacher preparation. It is important for a teacher to develop students' learning by building upon the students' preexisting knowledge. The teachers need to foster a deep understanding of the subject in students by adding relevant activities that provide an in-depth basis of factual knowledge. The importance of integration of metacognitive skills in the curriculum has been elaborated by Bransford et al (2000) in their work. A learner-centered learning environment enables the educators to understand the perceptions of students, the level of learning, and the rate of progress.

## SETTING THE STAGE

In the 1960's, McMaster University in Canada, initiated a Problem Based Learning (PBL) which attempted to move away from the traditional teaching methods and attempted to address novel aspects in teaching in medical education (Barrows, 1996; Neville, 2009). The PBL can be described as a curriculum and a process where the learner is provided with an opportunity to acquire vital information, analytical skills, application of knowledge, problem solving, self-driven learning, and team collaboration (Barrows & Kelson, 1980). The PBL method has eventually expanded to other fields of education in the realization of its effectiveness when compared with traditional teaching methods.

The University of Delaware, in their first workshop on Teaching Effectiveness, realized that the students were dissatisfied with the lecture approach to teaching science in undergraduate courses (Groh, Williams, Allen, Duch, Mierson, & White, 1997). The students whose major was not science or those who followed lectures of different disciplines within the fields of science also found it difficult to be entirely engaged in the learning process when the teaching was purely based on lectures. The University of Delaware followed the Kaufman model on PBL, and then went on to introduce PBL/

research-based education for undergraduates (Kaufman, Mennin, Waterman, Duban, Hansbarger, Silverblatt, Obenshain, Kantrowitz, Becker, Samet, & Weise, 1989).

Nevertheless, instead of developing the teaching of science through these techniques, according to various reports, the proficiency standards have been set for students across the U.S. quite low, "perhaps because it causes less embarrassment when more students can make it across the proficiency bar" (Peterson & Kaplan, 2013). It could be construed that the lowering of the barriers is representative of lowering the extent to which students should be actively involved; accordingly, it may also signify a form of complacency within the educational arena.

In his work, de Silva (1999) focused his research to understand the perceptions of chemistry learners as the first project. Consequent to the results gained from this first project, de Silva took the initiative to include mini-research projects in his teaching in 2000. In 2004, de Silva included research-based teaching in physics and chemistry curricula. Whilst the Kaufman (1989) model set the foundation for de Silva's work both in the U.K. and the U.S., and added some aspects of research-based learning, this chapter elucidates a more rigorous multidisciplinary research-based teaching system developed by de Silva (1989 – 2013) as an alternative to traditional teaching.

## CASES DESCRIPTION

### Project 1: Investigating the Perception of Chemistry Learners 1998 – 1999 (de Silva, 1999)

This project was a triangulation work which incorporated investigative/action/grounded-theory research carried out among a group of adult learners within a Science Foundation course and an Access course in the U.K. An Access course prepares students who are over the age of 21 for higher education at the university level. Additionally, the Science Foundation was the prerequisite to the Access course.

A questionnaire was provided to the selected study groups in which their reasons for selecting chemistry as a subject were questioned. The results were analyzed to understand the learners' expectations within the courses.

After the results were identified, the groups of adult learners were invited to a lecture which highlighted the advantages of selecting chemistry as a main subject and the importance in the use of analytical chemistry in many other fields of scientific study. A survey immediately after the lecture showed an increase in those who selected chemistry. Regular reminders and examples in the uses of analytical chemistry were provided throughout subsequent lessons. A final survey that was carried out after the applications were made to Universities and Colleges Admission

Service (UCAS), showed an increase in interest among students within this class. Ultimately, this number was above the national average for that year.

At the beginning, two groups of students in the Science Foundation and Access courses were selected for the cohort study. The age range of the group studying in the foundation science course was 16-25 years. There were 40 students in this group. The second group studying the Access Science consisted of 19 – 50 year olds. There were a total of 16 students in this access course.

The two lectures after the first survey included the following points:

1. Chemistry is widely known as the central science.
2. The analytical techniques used by other sciences such as biology, physics, microbiology, materials, etc. are all combined in analytical chemistry.
3. The key skills like application of number, IT, and communication can be improved immensely through the study of chemistry.
4. In general, chemists are widely employable in other professions such as banking, business, marketing, teaching, politics, etc.
5. The stereotype thinking where a chemist is considered as someone wearing glasses in a white coat and is mixing chemicals in a lab is a wrong impression. In fact, contrary to this type of image, they are actually well respected leaders, managers, and business people with a sense of responsibility and a good analytical approach.
6. Continuous practice as a chemist would lead to a professionally recognized prestigious status such as the chartered chemist for the Royal Society of Chemistry, U.K.
7. The analytical techniques and approaches to problem solving can be applied to daily life situations which enable one to achieve goals easily.
8. The training acquired through studying analytical chemistry may help one to secure quick promotions in the management sector and become great leaders (Belt, Clarke, and Phipps, 1999).
9. The study of chemistry should be embraced as a move beyond the laboratory settings that assists one to broaden one's horizons in different fields. This also helps one to change one's field and to move up in life, because chemistry as a central science provides holistic training.

Most of the initial replies did not show any academic interest in the subject of chemistry for their selection. They selected chemistry because it was a compulsory subject to enter the university system to embark on a science degree. Initially only about 2% in the group had selected chemistry as an undergraduate subject. After the initial lecture the total number increased to 16%. In the final survey, the students' number had increased to 50% for the access course. The students obviously had selected chemistry since it was taught as a central science. It appeared that students had realized that chemistry was a subject with many career opportunities.

The selection of cohort groups was not random because all the participants had to be enrolled in the Access program. The lecturers of chemistry and physics were also the ones who conducted the surveys and this may have influenced the answers by the students. Nevertheless, the selection of chemistry as their major at the end was verified by the number of students who declared the same in their application forms at the beginning of both years. The cohort group included mostly foreign students and the major biasness caused for the study was unavoidable. This was because foreign students showed the tendency to listen and follow the lecturers more closely than the local U.K. students, due to cultural differences under which they were raised. Nevertheless, this approach to teaching chemistry was never tested under these settings previously. The purpose of the research was to see a trend under certain conditions. The overall aim was to initiate an underlying basis upon which further analyses or models could be developed.

The learning process is an on-going process that begins at birth and continues throughout one's lifetime. In general, adult learners initiate the learning process for major reasons whereas youth traditionally follow the path that has been set for them. In this manner, adult learners take the initiative, while youth follow the common trends. Personal characteristics or personal variables may be considered as being continuous, representing a gradual growth towards maturity along three dimensions; the physical, the psychological and the socio-cultural (Cross, 1981). Adult educators have taken the view that adult education is markedly different from school. One school of thought is that youth education mainly revolves around the subject and is aimed at traditional future expectations, whereas adult education, on the other hand, is problem-centered (Verner, 1962). Adults in general attempt to mainly improve their skills relevant to their selected career or any common skill useful for them. It is also important that tutors be more person-centered. The "teach the person, not just the subject" approach from lecturers is beneficial to both adult and youth learners. It is important that tutors have an understanding of different types of learning theories in order to effectively cater the learning process for different levels. The learning theories can be mainly divided into cognitive, affective, and psychomotor categories ("Learning Domains," n.d.). Based on this information, it may be prudent to recognize that adult learning is mainly a combination of each of these domains.

There are several different theories related to cognitive styles of individual learning which are considered as the ways in which individuals organize and process information, or as the ways in which individuals conceptually organize the environment (Goldstein, 1978). Another consideration is that they represent individual's typical modes of information processing as the individual engages in perceiving, remembering, thinking, and problem solving (Knox, 1977).

Irrespective of an individual student's tendency to a learning style, the nature of the subject matter to be learned becomes primarily important (Knox, 1977). In all situations, the material that is to be learned should be presented in a manner that emphasizes the characteristics to be learned, and do so in a way that is as meaningful as possible to the learner. Yet, some cognitive styles of learning theories do not seem to acknowledge that there may be inherent differences in the demands made upon students in different disciplines.

This first project was directed at highlighting the advantage in studying chemistry where ample opportunities were available to improve key skills and also to have a core connection to other science subjects. Once this awareness was raised among the learners a noticeable change in their perception and attitude toward chemistry was noticed.

Consequent to the findings of the above research, the following projects were introduced successfully.

## Project 2: Goal-Oriented Approach to Teaching Chemistry and Physics 2000 – 2004 (de Silva, 2006)

The above approach was introduced at a college conducting both Advanced Level (A/L) and undergraduate studies from 2000 – 2004. The first cohort was selected from a group of A/L students. Each class consisted of 20 students. The approach was through a series of lectures and assignments where students were constantly reminded of their goals and the connection of these goals to their chemistry/physics classes. Most of the students in this cohort were taking these classes to get into the field of medicine. They had clear goals, but similarly to the cohort groups in previous studies, these students were not motivated to study physics or chemistry. They merely wanted a reasonable grade that would enable them to enter the medical school. As a result, they were searching for easy methods to do well in examinations. One method used was to continuously do past papers and memorize the types of questions and answers. Some students sought to study only sub-sections within sections in different areas of the subject to get high scores. Interestingly, most of the students were repeating the course for the third time. They were oblivious to the fact that their methods of study were not working well. In order to change their mindset, modus operandi, and motivation, this project was begun by showing them a broader aspect of physics and chemistry. Daily lesson plans always included 10 minutes of review that discussed previous lessons from the perspective of each student's goal. This was broadly done by asking them to write one or two sentences that related the previous lesson to their own ambitions. This generated the students' interests in these subjects. During the lesson, the students always related their work to their own ambitions. They were seeking connections and relevance. Accordingly, the students were more enthusiastic throughout the whole course and they were clearly making great improvements in their studies. They were also given essay types of questions within the lessons, and the students displayed better connections to their goals in their submissions. A vast improvement in the grades of these students was noticed. The average percentages of the chemistry A/L grades within the three years prior to 2000 were 50% C grades, 5% was B grades, 5% A grades, and the rest of the 40% was a combination of D and F grades. Since the new approach was implemented in the year 2000, the average percentage grades for the subsequent three years improved to 10% A grades, 10% B grades, 70% C grades and the rest 10% between D and F grades. Out of the majority of students who intended to enter the field of medicine, most of the students ultimately went on to study chemistry/physics or combined undergraduate programs. The same approach was initiated with undergraduate classes in 2003 and a small, but significant trend in the number of graduates with higher grades was noticed. The project ended in 2003, since de Silva left the U.K. and moved to the U.S. where further research was initiated.

## Project 3: Mini-Research Project Oriented Approach to Teaching Chemistry and Physics 2004 – 2007 (de Silva, 2006; de Silva & de Silva, 2009a)

In 2004, the above project was initiated in the U.S.A. for a group of gifted and talented students who were taking early undergraduate chemistry and physics classes during their school years. The cohort consisted of 200 students. They were introduced to mini-research projects during chemistry and physics lessons. Each lesson started with a question that led to a mini-project for the selected sections. Each lesson had a specific project that related to the section. During the teaching session, the fundamental principles taught were discussed through projects.

The students were then given the opportunity to select a project from a given a list of topics or create a project of their own choice. In addition, the students were given extra lessons about the methods to conduct independent research. An introduction of a research module as a four-year project also began during this time period to support the students to develop their research skills. There was an average of 99% pass rate of students with 30% A grades, 40% B grades, and the rest with C grades. This showed a clear success in this approach. The only deviation of this selected cohort from an average class was that this group of students was a group of gifted and talented students.

## Project 4: Teaching Chemistry and Physics through Multidisciplinary Research 2007 – 2010 (de Silva, 2006; de Silva & de Silva, 2009b)

Building upon the success of previous work, the teaching was extended to include multidisciplinary research in undergraduate chemistry and physics classes in 2006 and was implemented rigorously in 2007. The students were given several choices of research projects. The students in groups selected some of the projects relevant to their career choices. The students read the chapters ahead of the class, and highlighted the sections that were relevant to their projects. During the teaching, the students brought examples, discussion points, and questions from their chosen research projects. Teaching science through research turned out to be a very challenging task. The students were also advised to take classes in research methods that supported them to produce successful projects. This work was presented at several conferences including AP conferences and as workshops at the Appalachian Colleges Annual Conferences (de Silva, Long, and Mennen, 2008). As a result of this approach, the students also went on to win the first, second, and third places at the Tennessee Academy of Sciences Annual Conference from 2007- 2010. The undergraduate students carried out research projects in their first/second year of the university program because of this approach in teaching. This approach generated a greater interest among students to study chemistry/physics. The multidisciplinary research approach to teaching enabled the students to directly link the subjects to their chosen careers, and this generated a higher enrollment in chemistry and physics courses. The fact that the subject lessons always discussed different projects and how the students could use their subject knowledge to conduct a successful research project also established confidence among students and led to higher pass rates. The class discussions involved the students coming prepared to display the way they link knowledge to their projects, at least once every two weeks. Most of the projects followed the multidisciplinary approach to give students the opportunity to master more than one subject. This also made them realize the importance of studying different subjects and the usefulness of various aspects within those subjects when conducting successful research. This approach increased the pass rate to 99% with more students taking chemistry and physics classes. This trend continued steadily from 2007 until 2010. The project concluded in 2010.

## Project 5: START Model and Multidisciplinary Research Module Online 2008 – 2013 (de Silva, 2006; de Silva, Long, & Mennen, 2008)

Expanding on the results gained during the above mentioned projects, de Silva (2006) developed a universal model; START(©Eugene de Silva 2006), to be used in the integration of research with science teaching. In this model, the inclusion of multidisciplinary research in the teaching of science provided the opportunity to direct the students to recognize research as an active, diligent, and systematic process. This approach to "teaching science through research" (de Silva, 2006) as an effective method recognized the following:

- Research should focus on gathering, interpreting, improving/updating, or identifying data. Therefore, by conducting research, students developed their abilities in all of the above aspects.
- In the class, research was not separated to basic or applied/practical research. All research topics clearly had the practical aspects highlighted, and the selections were limited to readily available resources and limited timelines.
- At this level, pure basic research which mainly concentrates on the advancement of knowledge was least explored. The students' knowledge level made it difficult to embark on such projects, although there were some instances where students managed to complete exceptional basic research projects.
- The research projects made students realize that research can be applied to many situations and people. It made them see the relevance of the study of different subjects in various fields.
- The results generally covered a broad field or theory of knowledge. The work highlighted the vast knowledge gained by students within a very short time period.
- The applied research work conducted by students always attempted to improve a system or an application or to solve a specific question. This pragmatic approach in solving problems, paved the way for students to see how a combination of different study fields could provide a great deal of solutions to many problems.
- Research recognized as an active learning process enabled the students to be actively involved in the learning process.
- Research enabled the teachers to play both an active and a passive role giving the opportunity for students to improve their own abilities.
- Research projects delivered a forum for various learning processes and freedom of access to students.

- The students were given the opportunity to use visual, audio, and kinesthetic learning approaches.
- By teaching subjects through research, teachers were allowed to bring in a variety of resources, experts, and connections to classrooms.
- The time constraints and homework were not an issue. The in-class work and out-of-class work were part of the students' work schedule which was covered by their projects.

The multidisciplinary research approach to teaching has proved to be a very effective and novel approach in the realm of science teaching.

## START MODEL (©EUGENE DE SILVA 2006)

The START model identifies five key areas under which the teaching of science through multidisciplinary research was developed.

The above acronym was derived according to the steps an educator must follow, prior to introducing the process of teaching through research:

- **S:** Syllabus Analysis.
- **T:** Tracking System Covering Objectives.
- **A:** Assessment Methods.
- **R:** Research Component Inclusion.
- **T:** Testing Through Research.

Therefore, the first step is to analyze the syllabus of the subject to be taught. Two examples of analysis of syllabi are given below.

An analysis of a U.S. first year college chemistry syllabus highlighted the following:

- **Structure of Matter:** 20%
- **States of Matter:** 20%
- **Reactions:** 35 – 40%
- **Descriptive Chemistry:** 10 – 15%
- **Laboratory:** 5 – 10%
- **Chemical Calculations:** Included in all of the above sections.

An analysis of a U.S. first year college physics syllabus highlighted the following:

- **Sound:** 12%
- **Mechanics:** 52%

- **Light:** 7%
- **Electricity:** 11%
- **Magnetics:** 4%
- **Heat:** 6%
- **Lab:** 8%

Based on an analysis of the syllabus, a tracking system is generated with letters and numbers within the syllabus to cover the objectives. For example, the common objective, such as understanding force is covered in the sections that include mechanics, electricity, and magnetism. Therefore, these common objectives can be taught as one objective by combining all three sections when teaching force and related calculations. Therefore, at this step, common objectives are identified within different sections of the syllabus. The next step is to determine and finalize a suitable assessment method. These assessments become part of the research assessment process. Moving forward, the final steps would be to produce one or more multidisciplinary research projects to cover the objectives and the testing of knowledge required within the course.

An application of START program in the analysis of Physics 211 is given in Table 1.

*Table 1. START model exemplified (@Eugene de Silva 2006)*

| Physics 211 | S | T | A | R | T |
|---|---|---|---|---|---|
| **Objective 1** | Application of Force | 2a. Mechanics<br>3b. Electric<br>4a. Mechanics | Several calculations of force values through equations and lab work to calculate unknown forces. | Project<br>4. Breaking Boards<br>5. Balancing electric and magnetic forces. Novel transport systems. | 5-10 Calculations where final force is determined for both cases. |
| **Objective 2** | Sound Waves | 2a. Mechanics<br>5a. Sound | Lab work and wave mechanics calculations. | Project<br>6. Development of different tunes using various materials. | 5-10 Calculations to find wave lengths and frequencies to match tunes. |

## Examples of Projects

A common example of a project used for chemistry was the "Determination of the Effectiveness of Water Purifiers." Another common project used for physics was the "Determination of Safety Speed Limits on Roads and Safety Precautions."

Each student who undertook a research project presented the work at conferences and focused on putting forth each step of the research process, including the data collection, analysis, discussion, and conclusion.

Some other examples of actual projects conducted by students included:

- The effect of the angle theta while measuring blood flow velocities using a duplex Doppler.
- A comparative study of the impact of four different arrows and four different arrow tips.
- A study of consistent precise electrocardiograph lead placement for 12-lead ECG for more accurate diagnosis.
- The effectiveness of booster seats for children aged four years to nine.
- Physics and educational research.
- The effects of alcohol on the perception of pain.

*Table 2. Research rubric*

| Points (Variable according to grading rubric) | Zero points | Minimum Points | Medium Points | Highest Points |
|---|---|---|---|---|
| **Organization** | Lacks organization | Minimum organized approach | Work demonstrates more than minimum organization | Work is clearly organized |
| **Description (including relevant formulae)** | Lacks description | Minimum description | Work demonstrates more than minimum description | Excellent description with relevant formulae |
| **Calculations and application of knowledge** | Lacks calculations and application of knowledge | Covers the minimum requirements in calculations and application of knowledge | Covers more than minimum requirements in calculations and application of knowledge | Exceeds the criteria for calculations and application of knowledge |
| **Multidisciplinary Approach** | Lacks multidisciplinary approach | Minimum multidisciplinary approach | Covers more than the minimum multidisciplinary approach | Significantly proven multidisciplinary approach |

The testing through research that is linked to assessments used the common rubric given in Table 2.

The presented START model complemented and completed the multidisciplinary research module that had been developed in 2006 and was consequently modified as an online module in 2012.

The multidisciplinary research module was developed with the following expectations:

- Research should be introduced as a separate subject.
- All research topics should be linked to all subjects.
- Standard research projects should be introduced to include interdisciplinary studies.
- Achievements should be measured through research.
- Researching and finding data and evaluating and exchanging information, in addition to effectively presenting the findings should be linked to key skills.
- Appointment of a research coordinator at school, college, university, and state levels.

The developed research module included the following topics:

- Nature of Research.
- Research Process.
- Selecting the Research Project.
- Ethical Issues.
- Literature Review.
- Research Design and Planning.
- Research Method.
- Data Collection.
- Data Analysis.
- Writing-up the Research.

## Online Teaching

In 2009, the development of the first online physics programs for the State of Tennessee provided an opportunity to develop a course combined with mini-research projects and questions that could be conducted independently. The investigative work was provided through discussion boards, simulated labs, and projects. Online learning had thus promoted self-learning, self-regulation, self-assessment, self-assurance, and self-achievement. The technological advances in the Information Technology have also paved the way for online learners to connect with other learners beyond the traditional classroom setting to create international connections to broaden not only the subject knowledge, but also the key skills discussed earlier.

A few examples of discussion questions of an online college physics course are given below to highlight the promotion of key skills, research, scholarship, and independent learning.

### Section Taught: Free Fall

#### *Discussion Question*

Explain the scientific theory behind the free fall of an acorn and a pumpkin to an atheist who is also a non-scientist. You may visit the website: http://digital.library.upenn.edu/women//finch/1713/mp-atheist.html to read more on the poem, "The Atheist and the Acorn."

### Section Taught: Vectors

#### *Discussion Question*

"All roads lead to Rome" - Analyze this saying in terms of vectors, scalars, displacement, and distance, and provide examples.

### Section Taught: Energy

#### *Discussion Question*

The great Roman Poet, Lucretius wrote, "Things cannot be born from nothing, cannot when begottend brought back to nothing." This statement became an accepted theory as a result of the work of the chemist, Lavoisier in the 18th century. This is the law of conservation of matter. Similarly, the law of conservation of energy is an important law in physics. Can you discuss the validity of the conservation of energy in terms of potential (gravitational or elastic) and kinetic energy by providing real-life applications?

## CURRENT CHALLENGES

There are several challenges in the introduction of multidisciplinary research in schools and colleges. The lack of time, personnel and resources, initial misinterpretation of the suggested model, absence of incentives to educators, and the educators' resistance to change due to their many years of training in the traditional teaching methods, seem to act as obstacles in the initial progress of improving multidisciplinary-based research models in schools and colleges. Nevertheless, once the training is completed for responsible educators, it is hoped that most of these challenges would be eliminated. The schools that are being selected for the pilot project should begin to show signs of improvements within the next three years. The extent to which the schools would continue with this novel practice once the initial introductory, observational period is over is currently uncertain. It is hoped that a substantial policy change within the educational system would secure the practice of multidisciplinary research at schools.

## CONCLUSION

The changes over the past three decades have established the future trends in teaching. The START model has paved the way to integrate multidisciplinary research in any subject at a college level to prepare students for broader opportunities in their chosen career. Employers and universities have shown concern over the past decade at the level of key skills of students. The six key skills recognized by the Qualifications and Curriculum Authority (2004) are Communications, Application of Number, Information Technology (IT), Improvement of Own Learning and Performance, Working with Others, and Problem Solving. According to Delker (1974) "all adult education involves adult learning, but all adult learning is not adult education." This means that adult education refers to systematic and step-wise experiences that are established to fulfill the goals of adults. In addition, adult education should also satisfy the requirements set up by the educational department. Therefore, adults who engage in education while attempting to satisfy the educational requirements should also seek to improve their skills in order to benefit their chosen career pathways. According to the results published by The Independent (2000) in association with UCAS, of those who studied chemistry, only 37% entered careers related to chemistry or other sciences; nearly 20% entered management, marketing and administrative work; over 4% went into computing; around 3% became accountants; more than 35% went on to do further study, including teacher training; the rest became bank managers, army officers, and sales professionals. Therefore, it could be assumed that the transferable skills involved in chemistry enabled them to change their careers and secure employment in other fields. Similarly, the inclusion of multidisciplinary research in any subject would provide a sturdy foundation to be competent in allied fields, while also enhancing the opportunities for students to develop transferable skills. Accordingly, students would also receive the training to seek pragmatic solutions to common problems through interaction, interconnection, and integration with other fields/subjects. One of the main objectives of including multidisciplinary research in science subjects should be to provide an understanding on how scientific ideas are presented, evaluated and disseminated. In addition, it should also introduce students to the ways to interpret empirical evidence and make them understand the power and thresholds of science in addressing industrial, social, and environmental questions. The students at this stage would be taught how to plan experiments, where to obtain information, key factors involved in collecting evidence, and techniques involved in sampling and in the use of equipment, etc. They are further taught how to present evidence and appropriately conduct evaluations of the collected data. Hence, multidisciplinary research establishes links among subjects and prepares the students for practical work at a specialist level. Online teaching would continue to grow exponentially with the inclusion of advanced technology, such as holograms. These trends would steadily grow as more colleges favorably respond to online science teaching combined with multidisciplinary research. The promotion of independent learning would ultimately generate fully-fledged graduates who are competent and knowledgeable in their subjects and have a deeper understanding of true applications in real life. A concerted effort by the secondary school systems and the colleges/universities within each state would further pave the way to build a generation of students with sound, solid, and superior knowledge in science and its applications. If the suggested novel methods are introduced and closely monitored, followed, and supported by a group of academics selected both from schools and colleges/ universities, then an improvement in our science standards should be evident within the next decade.

## REFERENCES

Anonymous. (2004). *The Key Skills Qualifications Standards and Guidance*. Qualifications and Curriculum Authority. Retrieved from http://www.ocr.org.uk/Images/71651-key-skills-qualifications-standards-and-guidance-2004.pdf

Anonymous. (2009). *Graduate Employability*. The Royal Society of Chemistry. Retrieved from http://www.rsc.org/Education/EiC/issues/2009May/Graduateemployability.asp

Anonymous. (2009). *Stem Briefing*. The Russell Group of Universities. Retrieved from http://www.russellgroup.ac.uk/uploads/STEM-briefing.pdf

Anonymous. (2010). *Statistics of Chemistry Education*. The Royal Society of Chemistry. Retrieved from http://www.rsc.org/images/Statistics_of_Chemistry_Education_2010_tcm18-192310.pdf

Anonymous. (2013). *The Graduate Market in 2013*. High Fliers Research Limited. Retrieved from http://www.highfliers.co.uk/download/GMReport13.pdf

Anonymous. (n.d.). *Learning Domains or Bloom's Taxonomy*. The University of Dayton School of Law. Retrieved from http://academic.udayton.edu/health/syllabi/health/unit01/lesson01b.htm

Association of Graduate Recruiters Report. (1995). *Skills for Graduates in the 21st Century*. Cambridge, UK: Association of Graduate Recruiters.

Barrows, H., & Kelson, A. (1980). *Problem-based learning: A total approach to education*. New York: Springer.

Barrows, H. S. (1996). Problem-based learning in medicine and beyond: A brief overview. *New Directions for Teaching and Learning*, (68), 3-12.

Belt, S. T., Clarke, M. J., & Phipps, L. E. (1999). Exercises for chemists involving time management, judgment and initiative. *University Chemistry Education*, *3*(2), 52–58.

Bondelli, K. (2007). *An Evaluation of the Traditional Education System*. Scribd. Retrieved from http://www.scribd.com/doc/38418/An-Evaluation-of-the-Traditional-Education-System-by-Kevin-Bondelli

Bransford, J. D., Brown, A. L., & Cocking, R. R. (2000). *How People Learn: Brain, Mind, Experience, and School Committee on Developments in the Science of Learning*. Washington, DC: National Academy Press.

Brattan, D., Mason, D., & Rest, A. J. (1999). Changing the nature of physical chemistry practical work. *University Chemistry Education*, *3*(2), 59–63.

Cross, K. P. (1981). *Adults as Learners: Increasing Participation and Facilitating Learning*. London: Jossey-Bass.

de Silva, E. (1999). *Understanding the perception of chemistry undergraduates to increase the number of learners*. (Post-graduate Certificate in Education Thesis). University of Manchester, Manchester, UK.

de Silva, E. (2006). *Proceedings from Annual AP Conference: Applied Research in Science and Future Teaching*. College Board.

de Silva, E., & de Silva, E. (2009a). *Proceedings from Tennessee Association of Science's Annual Conference: COMSOL as a tool for teaching and research in physics and chemistry*. Knoxville, TN: Tennessee Academy of Sciences.

de Silva, E., & de Silva, E. (2009b). *Proceeding from Tennessee Academy of Sciences East Tennessee Collegiate Division Conference: Teaching Chemistry and Physics Through Software Programs*. Tennessee Academy of Sciences.

de Silva, E., Long, C., & Mennen, M. (2008). *Proceedings from Appalachian College Association's Annual Conference Summit XI: Teaching Science Through Multidisciplinary Research*. Appalachian College Association.

Delker, P. V. (1974). Governmental roles in lifelong learning. *Journal of Research and Development in Education*, *7*(4), 24–33.

Dewey, J. (1998). *Experience and Education Kappa Delta Pi*. Kappa Delta Pi.

Duckett, S., Garratt, B., Lowe, J., & Nigel, D. (1999). Key Skills: What do Chemistry Graduates think? *University Chemistry Education*, *3*(1), 1–7.

Goldstein, K. M., & Blackman, J. (1978). *Cognitive Style: Five approaches and Relevant Research*. New York: Wiley.

Groh, S. E., Williams, B. A., Allen, D. E., Duch, B. J., Mierson, S., & White, H. B. (1997). Institutional change in science education: A case study. In A. P. McNeal & C. D'Avanzo (Eds.), Student-active science: Models of innovation in college science teaching (pp. 83-94). Philadelphia: Saunders College Publishing.

Johnstone, A. H. (1997). …And Some Fell on Good Ground. *University Chemistry Education*, *1*, 8–13.

Kaufman, A., Mennin, R. S., Waterman, S., Duban, C., Hansbarger, H., & Silverblatt, S. et al. (1989). The New Mexico experiment. *Academic Medicine*, *64*(6), 285–294. doi:10.1097/00001888-198906000-00001 PMID:2719785

Knox, A. (1977). *Adult Development and Learning*. San Francisco, CA: Jossey Bass.

Neville, A. J. (2009). Problem-Based Learning and Medical Education Forty Years on. *Medical Principles and Practice*, *18*(1), 1–9. doi:10.1159/000163038 PMID:19060483

Peterson, P. E., & Kaplan, P. (2013). Despite Common Core, States Still Lack Common Standards. *Education Next*, *13*(4), 44–49.

Report of the ASCD Working Group on Humanistic Education. (1978). *Humanistic Education: Objectives and Assessment*. Washington, DC: Association of Supervision and Curriculum Development.

Report of the project on liberal education and the sciences. (1990). *The liberal art of science: Agenda for action*. Washington, DC: American Association for the Advancement of Science.

Science and Technology Committee. (2006). *10th Report of Session 2005/2006*. London: Authority of the House of Lords.

The Independent. (2000). "Report." Report Produced in Association with UCAS. *University Chemistry Education.*, *3*(1), 1–7.

Verner, C. (1962). *Conceptual scheme for the identification and classification of processes of adult education association*. Washington, DC: Adult Education Association.

## ADDITIONAL READING

*Council on Undergraduate Research (CUR)*. (n.d.). Retrieved from http://www.cur.org/

*Council on Undergraduate Research (CUR) Publication List*. (n.d.). Retrieved from http://www.cur.org/publications/publication_listings/

de Silva, E., & de Silva, E. (2012). *Multidisciplinary Research for College Students*. Ronkonkoma, NY: Linus Publications.

*Higher Education Academy*. (n.d.). Retrieved from http://www.heacademy.ac.uk/assets/documents/resources/publications/developingundergraduate_final.pdf

*Multidisciplinary Research in Applied Science (MRSA)*. (n.d.). Retrieved from http://ws.edu/academics/research/

*National Accrediting Commission for Martial Arts (NACMA)*. (n.d.). Retrieved from www.nacma.us

*Virginia Research Institute*. (n.d.). Retrieved from www.virginiaresearchinstitute.org

## KEY TERMS AND DEFINITIONS

**Goal-Oriented Approach:** An approach to learning in which the students' goals are the focus.

**Higher Education:** Any education after the completion of any certificate equivalent to a high school diploma. In the United Kingdom, this would be any education consequent to the G.C.E. (O/L).

**Holistic:** The process whereby all subjects are incorporated.

**Inter-Disciplinary:** The clear combination of two or more subjects that are commonly considered quite different.

**Multidisciplinary:** The inclusion of two or more subjects to teach one given subject, without any clear boundaries. (Compare with Inter-Disciplinary)

**Problem-Based Learning:** A process in which a subject is taught through the lens of a given problem(s).

**Traditional Education:** Education primarily held in classroom settings by following a strict syllabus and giving predetermined assessments.

# Section 2
# Content Presentation

# Chapter 6
# Developing a Research-Informed Teaching Module for Learning about Electrical Circuits at Lower Secondary School Level:
## Supporting Personal Learning about Science and the Nature of Science

**Keith S Taber**
*University of Cambridge, UK*

**Neil Mercer**
*University of Cambridge, UK*

**Kenneth Ruthven**
*University of Cambridge, UK*

**Fran Riga**
*University of Cambridge, UK*

**Christine Howe**
*University of Cambridge, UK*

**Riikka Hofmann**
*University of Cambridge, UK*

**Stefanie Luthman**
*University of Cambridge, UK*

## EXECUTIVE SUMMARY

*This chapter discusses the design and development of a teaching module on electrical circuits for lower secondary students (11-14 year olds) studying in the context of*

DOI: 10.4018/978-1-4666-6375-6.ch006

*the English National Curriculum. The module was developed as part of a project: "Effecting Principled Improvement in STEM Education" (epiSTEMe). The electricity module was designed according to general principles adopted across epiSTEMe, drawing upon research and recommendations of good practice offered in curriculum guidance and the advice offered by classroom practitioners who tested out activities in their own classrooms. The module design was informed by the constructivist perspective that each individual has to construct their own personal knowledge and so rejects notions that teaching can be understood as transfer of knowledge from a teacher or text to learners. However, the version of constructivism adopted acknowledged the central importance of social mediation of learning, both in terms of the role of a more experienced other (such as a teacher) in channeling and scaffolding the learning of students and the potential for peer mediation of learning through dialogue that requires learners to engage with enquiry processes and interrogate and critique their own understanding.*

## BACKGROUND

### Introduction

This chapter describes the development of a research-informed teaching module on electrical circuits for early secondary level (in particular aimed at 11-12 year olds) developed as part of the project 'Effecting Principled Improvement in STEM Education' (epiSTEMe). The principles informing the design of the module will be discussed, and the way those principles were applied in module development will be explored. Three levels of context for appreciating module development will be provided relating to issues of (i) research into student thinking and learning in the topic, (ii) the context of the epiSTEMe project more generally, and (iii) the wider curriculum context in which the work took place.

### Student Thinking and Learning about Electrical Circuits

There is an extensive body of research exploring student learning and thinking in various science topics (Duit, 2009; Taber, 2009), including electricity and electric circuits (Driver, Squires, Rushworth, & Wood-Robinson, 1994; Shipstone et al., 1988). Learning difficulties relating to the topic of electrical circuits are well established, and these are found across the secondary age range. A common problem concerns students not appreciating how current will be constant around a series circuit. A naive view would be that this could be countered by demonstration: simply showing learners a series circuit and measuring the current at various points.

A somewhat more informed view – informed by research into science learning (considered below) – might suggest that something more than this is needed: to first help learners make explicit their intuitive ideas about what would happen in the circuit, and then counter these by providing the evidence that their intuitions do not match what actually happens. This might be expected to lead to cognitive dissonance, and so motivate learning to make sense of the discrepant observations (Driver & Oldham, 1986).

This approach is commonly recommended because human beings generally manage to perceive the world as fitting expectations (finding matches between what is sensed and existing implicit knowledge elements such that perception is biased to fit existing cognitive structures) - what is sometimes known as confirmation bias (Nickerson, 1998). Driver noted how students put in open-ended discovery learning situations with minimal 'scaffolding' from teaching tended to fail to spot the patterns that it was hoped they would find salient and seldom 'discover' the scientific principles hoped for (Driver, 1983). Much more recent work has reinforced how rarely students take away from school science practical work the ideas such activities are intended to motivate or illustrate (Abrahams, 2011).

One pedagogic approach intended to address this issue is known as P-O-E, which stands for Predict-Observe-Explain (White & Gunstone, 1992). The principal assumption drawn upon here is that by first having students make predictions they would then be primed to extract the desired 'figure' from the 'ground' of sensory data - to borrow terms from the Gestalt psychologists (Koffka, 1967) - and also have some investment in observing a particular pattern or outcome. Where expectations are confounded, the potential cognitive dissonance (Cooper, 2007) is harnessed by asking students to explain what they have observed – thus reinforcing the outcome and requiring the learner to actively seek to make sense of the unexpected observations. This is considered important because research in science education suggests that learners commonly revert to alternative conceptions supported by their intuitions despite teaching events, once those events cease to be recent (Taber, 2003).

Interestingly, some research suggests that even employing the P-O-E strategy may be insufficient to overcome students' expectations about what goes on in electric circuits. A study showed that when a class of 14-year olds was asked to predict how current would vary round a simple series circuit most of the students predicted current would diminish around the circuit as previous studies had suggested (Gauld, 1986, 1989). This prediction can be tested by ammeters, or by using lamp brightness as an indicator (as long as similar lamps are used at different points in a circuit). After seeing the demonstration students accepted current did not change around the circuit, seeming to have changed their ideas about current in circuits. However when the same students were interviewed three months later many had reverted to their initial thinking – that current diminishes around a series circuit.

These students often remembered the demonstration, but now thought what they had seen fitted their initial predictions. So even when learners' prior thinking is made explicit, AND they are shown their predictions are wrong, AND they accept they were wrong and seem to change their minds, this may not be sufficient to bring about long-term conceptual change.

Human cognition has inherent drives for coherence and consistency (Jolliffe & Baron-Cohen, 1999; Parkin, 1993). It would seem that in Gauld's study, the observation of a confounding outcome was sufficient to lead students to accept a new way of thinking that matches the unexpected outcome, but without sufficient reinforcement of the new learning (Vertes, 2004) many students worked towards coherence by modifying their memory of the observations, rather than their preferred mental models of electrical current flow around circuits.

## Electricity as a Challenging Topic

It is perhaps not surprising that electrical circuits is a topic which many students find difficult (Shipstone et al., 1988). Whilst students can observe and manipulate simple circuits, and indeed many students seem to enjoy this type of practical work, the ideas involved are challenging. Electrical circuits are explained in terms of abstract ideas, in particular current and potential difference (p.d. or 'voltage') that link to the concepts of charge and energy respectively. Energy is acknowledged as a highly abstract topic – for example by Nobel laureate physicist Richard Feynman (1965) - which students commonly struggle with (Brook & Driver, 1984; Solomon, 1992; Watts, 1983). Current as a flow of charge can potentially be visualised, but to apply this idea to circuits students have to shift from considering the observable phenomena at the macroscopic 'bench' scale, to think about a process occurring at a submicroscopic scale. At this scale the apparently solid metal wires students observe are understood as a fixed lattice arrangement of atomic cores bound by electrical forces to a fluid-like (Buddle, Niedderer, Scott, & Leach, 2002) ensemble of delocalised electrons (see Figure 1). However, this model is not usually explicitly taught until much later in secondary education.

It has long been recognised that part of the challenge of school science learning relates to how learners are asked to cope with presentations at several 'levels' at once (Johnstone, 1982, 1991). In particular, students are often presented with two distinct re-descriptions or re-conceptualisations of phenomena they can observe, framed in terms of the technical symbolic language, theoretical concepts, and explanatory models used in science. In many topics students not only have to learn how an observable phenomena is categorised and conceptualised in formal terms (say a candle flame in terms of categories of chemical reaction and combustion)

*Figure 1. Conceptualising electrical circuits at two levels (Adapted from Taber, 2013)*

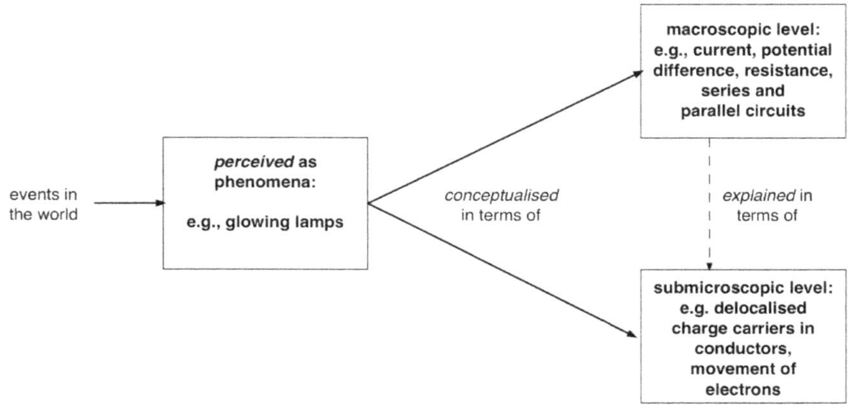


*Figure 2. Science teaching as moving between different levels or domains (Adapted from Taber, 2013)*

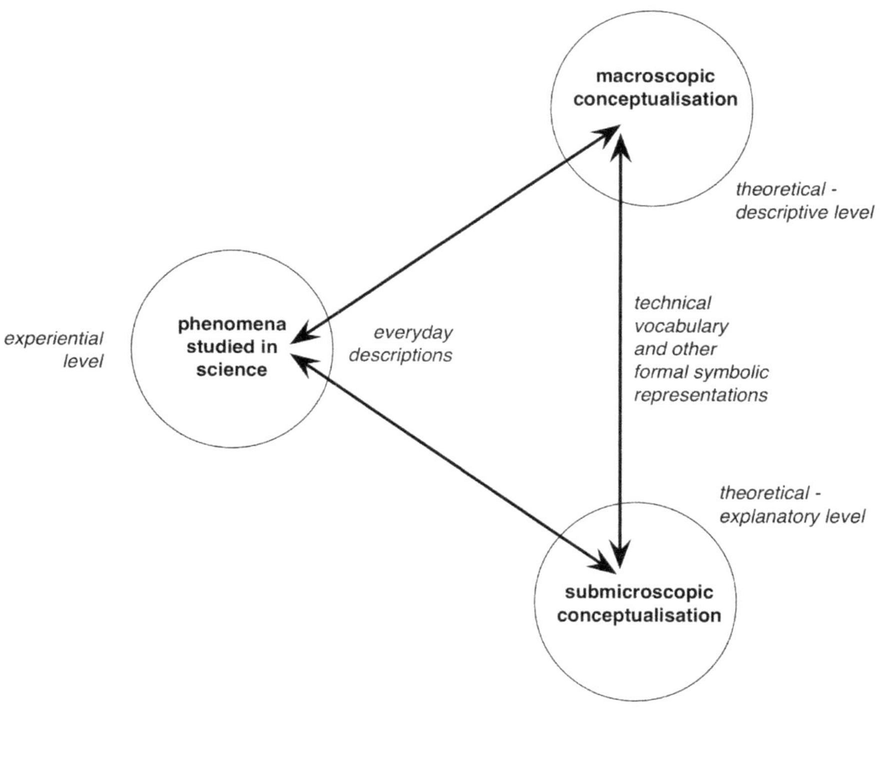

but also how scientists explain the phenomena in terms of models of the structure of matter at submicroscopic scales (see Figure 2).

## Circuit Diagrams

A key part of the challenge of learning about electrical circuits is the use of a scientific formalism to represent circuits - circuit diagrams. To the experienced physicist or science teacher a circuit diagram has probably come to be perceived as quite like a circuit: it has the important affordances of both offering 1:1 correspondence between components and their symbols, and also clearly reflecting the topology of the circuit, and so how different components are connected.

However, learners faced with circuit diagrams may find they less obviously reflect actual circuits, and can find it very difficult to build circuits from circuit diagrams. Students have to both identify specific components from different symbols, and appreciate how the formalism of straight lines and sharp corners can represent the key aspects of the arrangement of various leads - inevitably taking up myriad configurations on the bench, but seldom appearing linear.

## Common Alternative Conceptions

Although current passing through wires is often made more accessible to students through the use of teaching models making analogies (discussed further below) with, for example, fluid flow through pipes, this offers limited explanatory power by itself. Secondary students commonly commence formal study of the topic of electricity with a vague notion of 'electricity' from everyday discourse, which is not differentiated between current, potential difference, energy and power (Arnold & Millar, 1987). Students commonly initially make sense of circuits in terms of intuitive ideas that lead to mental models that have been labelled as 'unipolar' (something comes from one side of the cell or battery to the component) or 'clashing current' (something different comes from each side of the cell or battery and meets at the component).

Generally then a major shift is needed to persuade learners to consider that something, charge in the form of a current, is flowing all around the circuit. However, this shift does not lead to a mental model of circuits that can support desired learning unless current as a flow of charge is clearly distinguished from current as a means of transferring energy. For one thing, current is conserved around circuits, despite work being done in lamps and other components. Moreover, the charge that is flowing does not really go anywhere (although in a direct current circuit individual electrons do slowly drift around the circuit) – in the sense that the electrons in the wires are simply replaced by other, entirely equivalent, electrons. The only structural difference between the current carrying wire, and the same wire when it is not carrying current, is that in the former case there is a very slight drift superimposed on the otherwise random patterns of electron movements in the metal. (Indeed in a.c. circuits, such as that used in house lighting, there is not even a net drift – and typically electrons may only shift a fraction of a millimetre before the direction of flow switches). The speed of electron drift in a simple circuit is extremely slow from a macroscopic perspective. Even in a physically small circuit it might take quite a few minutes for a particular electron to move through a distance equivalent to that from the cell terminals to a lamp: yet the circuit seems to work instantaneously. Students do not have to wait several minutes for lamps to glow or ammeters to register current.

## Circuits as Systems

The effect of current flow therefore can only be understood by coordinating ideas about current with something else – energy or electrical potential. The circuit can be understood as a device for transferring energy and the mobile charges that make up the current are in effect energy carriers. A circuit needs to be understood in terms of this process (i.e. systematically), but as Chi has reported, learners tend to conceptualise scientific processes in terms of substances, and it is then difficult to reassign the concept to a very different 'ontological tree' (Chi, 2008; Chi, Slotta, & de Leeuw, 1994). Of course this is not a tendency of school students in particular – the history of science offers many examples adopted by respected scientists – not just electrical fluid as used by Benjamin Franklin among others, but caloric, phlogiston, the ether, vital forces, etc. - of substances or pseudo-substances once mooted as elements of scientific explanations but now discredited.

Measurements of potential difference (i.e. 'voltages' in common parlance) are indicators of the amount of energy being transferred in sections of the circuit. Where students commonly expect current to diminish around the circuit, the scientific account suggests they should instead be paying attention to the voltmeter readings between different points around the circuit as this relates to where work is being done in different components. In a series circuit with two dissimilar lamps, apparent phenomenologically as glowing with different degrees of brightness, the same current flow will pass through both (being determined by the total p.d. across the circuit and its total resistance) but the p.d. across the two lamps will differ.

An electric circuit is therefore a system, and in a sense it is an emergent system, as the different specific components, and their configuration, need to be specified to understand what is going on at any point. Studies on student understanding of systems suggest school age learners often have limited basis for understanding systems and emergent phenomena (Wilensky & Resnick, 1999).

Taking these considerations together, it is perhaps less surprising that secondary age students tend to (i) focus their thinking about circuits on current (something they can visualise); thus (ii) think primarily in terms of a substance-like entity (rather than thinking in terms of process); (iii) consider the parts of a circuit sequentially attempting to understand each point locally (rather than as one part of an interacting system); and (iv) conceptualise the circuit in terms of current moving from a source (the battery) and being 'used up' around the circuit. This alternative conceptual framework is more accessible than the scientific alternative.

Given the very real barriers to effective learning about circuit concepts, and in particular the abstract understanding needed to make good sense of circuits, it might be questioned whether this is a suitable topic for teaching students at the start of secondary education – perhaps instead electronic circuits should only be taught at this age as part of technology classes so that students become familiar with components and their affordances, to provide a context for theoretical learning later in the school. However, in the English curriculum context discussed below (as many others) teaching and learning about circuits is prescribed for lower secondary level science.

## Challenges of Teaching Electrical Concepts through Simple Circuit Work

Research suggests that although many students enjoy practical work in school science, and some certainly develop competence in manipulative work (meeting educational objectives in the sensori-motor domain), such activities are often less successful in engaging students in using their observations to support conceptual learning. Many school practicals are meant to illustrate scientific principles (Millar, 2004): but in science the link from observation to theory is often not straight-forward (Kuhn, 1996; Lakatos, 1999), and expecting students to draw the 'right' conclusions without careful scaffolding is often unrealistic (Abrahams & Millar, 2008). Moreover, whereas research scientists practise and refine techniques they use on a regular basis, school students are generally operating with relatively unfamiliar apparatus and techniques. This adds to the excitement of lessons, but undermines learning in two ways.

Firstly scientific apparatus often needs nursing to 'work' as intended. Polanyi (1962) stressed how scientific work depends upon 'tacit' knowledge that scientists develop over time: implicit knowledge of how to get particular set-ups to work that relies on close familiarity with that kit and laboratory environment. In principle, scientific papers provide all the details for others to undertake the replications that are part of science - but in practice new experimental set-ups can sometimes only be transferred between research groups when scientists visit other labs so that the scientist with the specialist experience can model the processes to others (Collins, 2010).

Whilst the type of apparatus involved in school practical work in electricity is routine and far from the forefront of research, it is notorious for being problematic. Practical work can be spoiled for example by corroded switches and contacts; intermittent faults due to unseen breaks in insulated leads; old cells with high internal resistance (so the measured terminal p.d. drops significantly as soon as an external load is applied); lamps with partly evaporated filaments having very different power ratings to other nominally identical lamps; and poorly calibrated meters. These difficulties can only be avoided when technical support is available to carefully check all kit before each lesson - a time-intensive process - and the teacher or support staff are able to fault-find during student work.

A second problem concerns the limits of human working memory (Baddeley, 2003). People can only mentipulate a limited about of material at any one time. School practical work generally involves relatively novel aspects for learners (reducing the potential for 'chunking' to use working memory more effectively). Following instructions, collecting and manipulating apparatus, and recording observations may 'load' students' working memories in full. This will leave limited, if any, capacity for the kinds of reflection on what is being experienced in relation to (often recently introduced) concepts that is needed for what Abrahams (2011) refers to as 'minds-on', rather than just 'hands-on', practical work.

## SETTING THE STAGE

### Electrical Circuits in the English Lower Secondary Curriculum

The present chapter describes the process of developing a research-informed teaching module to support learners in developing a scientifically appropriate understanding of simple electrical circuits. The work reported here derives from the English context, where electricity is a major topic in the lower secondary school. At the time of developing the module as part of the epiSTEMe project the English National Curriculum for Science for 11-14 year old students had recently been revised (QCA, 2007). The new curriculum document might be considered 'content-lite' compared to the previous version of the curriculum (DfEE/QCA, 1999), in part as a deliberate attempt (a) to counter concerns about how in the previous curriculum excessive prescription of content was limiting depth of treatment and restricting the teacher's flexibility and creativity in meeting needs of particular students (Hacker & Rowe, 1997; Jenkins, 2000; Kind & Taber, 2005); and (b) to balance prescription of subject content with wider objectives relating to skill development and understanding the nature of science (QCA, 2005).

The physics content of the revised science curriculum for teaching across three years of study (for 11-14 year olds) was reduced to "The study of science should include energy, electricity and forces: (a) energy can be transferred usefully, stored, or dissipated, but cannot be created or destroyed; (b) forces are interactions between objects and can affect their shape; and motion; (c) electric current in circuits can produce a variety of effects" (QCA, 2007. p.210). The notes provided in the curriculum to explain the scope of the material to be taught about electricity (point c above) was limited to "Circuits: This includes current and voltage in series and parallel circuits" (QCA, 2007. p.210).

This provided limited guidance for teachers and a sharp change of approach. The previous much denser curriculum document had been supplemented by non-statutory schemes of work (QCA, 2000), and an extensive 'national strategy' for teaching science built around a comprehensive framework document suggesting how progression in understanding key concepts should be supported across the three years of the lower secondary phase (Key Stage 3 National Strategy, 2002).

## Teaching about 'How Science Works' as Part of the Science Curriculum

Part of the rationale for the new curriculum was to increase the emphasis within the secondary curriculum on the nature of science, or 'how science works' in the terminology used in the curriculum documents. The importance of teaching about the outcomes of science within a wider context has been recognised in many national contexts for some decades. In the 1980s there was an 'STS' movement that sought to prioritise teaching about 'science and technology in society' (McConnell, 1982). There has also been a widespread movement to teach more about the nature of science itself - in particular through informing school curricula with scholarship in the history and philosophy of science (Duschl, 2000; Hodson, 2009; Matthews, 1994).

When the UK government decided to introduce a national curriculum into English schools (Statutory Instrument, 1989), the original proposals for the science curriculum included an attainment target focused on the nature of science. However, later simplification of the proposals led to this aspect becoming largely implicit - leading to it having limited effect on practice (Donnelly, 2001). Several attempts were later made to address this concern through tweaks to the curriculum, guidance documentation, and the assessment regime (Taber, 2008).

The 2007 revision of the curriculum was more substantial, reducing specification of science content to be taught to brief topic descriptions such as those above, and setting this content as just one of several aspects of the curriculum: so 'range and content' (as it was headed) followed what was referred to as 'key concepts' and 'key processes' (QCA, 2007). "Key concepts that underpin the study of science and how science works" (p.208) included scientific thinking, applications and implications of science, cultural understanding and collaboration. 'Key processes' related to practical and enquiry skills, critical understanding of evidence, and communication (p.209). Under the heading of 'scientific thinking' students were expected to use "us[e] scientific ideas and models to explain phenomena and develop... them creatively to generate and test theories" and "critically analys[e] and evaluat[e] evidence from observations and experiments" (p.208).

## The TISME Initiative and the epiSTEMe Project

The epiSTEMe project was part of an overarching *Targeted Initiative on Science and Mathematics Education* (TISME). TISME is a programme of research funded by the UK's Economic and Social Research Council in partnership with the Gatsby Charitable Foundation, The Institute of Physics and the Association of Science Education. The aim of the initiative was to find new ways to encourage children and young people to greater participation, engagement, achievement and understanding of Science and Mathematics. The initiative funded a number of projects including one based at the University of Cambridge: *Effecting Principled Improvement in STEM Education* (epiSTEMe). EpiSTEMe was concerned with student engagement and learning in early secondary school physical science and mathematics.

The epiSTEMe project set out to develop classroom activities and supporting materials that drew upon research-based approaches in four lower school science and mathematics topics: probability and proportionality in mathematics and forces and electric circuits in science. Our aspiration however, was not simply to support the teaching of four topics, but to demonstrate how research-based pedagogy could be built into school teaching schemes. It was hoped that if schools used and saw the value of our modules these would provide experience in a particular teaching approach and offer models of effective classroom activities. We looked to frame classroom tasks that could help build students' abilities to think as mathematicians and scientists and support key conceptual advances in a topic. In particular, tasks were designed to trigger critical examination of common alternative conceptions. Lessons were planned around carefully crafted problem situations intended to appeal to shared student experiences and interests. As the intention of epiSTEMe was to adopt research-informed pedagogy, suitable existing classroom-tested activities were incorporated into the modules alongside newly designed activities.

A distinctive feature of the *epiSTEMe* approach is its use of dialogue – in small student groups and the whole class – to elicit and examine differing points of view on problem situations (Howe et al., 2007; Kleine-Staarman & Mercer, 2010). As it was recognised that students (and teachers) need to develop skills in working through such approaches, an introductory module was developed to build teacher and student understanding of the value of talk and dialogue in supporting subject thinking and learning, and to help teachers develops rules and processes to underpin effective small-group and whole-class discussion. As a result, two topic modules in each of science and mathematics were designed to stimulate and capitalise on talk and dialogue, based on the assumption that ground-rules and good working habits had been established through the introductory module.

The epiSTEMe team worked closely with teachers from several schools over an 18-month period to develop, trial and refine the intervention. The development process drew on the expertise of teachers and researchers, as well as on a synthesis of relevant research literature (Ruthven et al., 2010) and analysis of evidence from classroom trialling. Its aim was to generate resources for developing teachers and teaching students, as well as to improve understanding of teaching and learning processes in school science and mathematics.

Teachers from partner schools enrolled in the project attended project days with the university team to discuss the aims of the project, to explore the pedagogic approach, to critique (and sometimes try out) draft activities and to make suggestions for modifications or additional activities drawing on their own teaching repertoires. In particular the classroom practitioners were able to offer advice on how the constraints of their real teaching contexts should be considered in planning teaching and learning activities. Sometimes teachers were video-recorded trying out activities with their own classes to allow later review at a project day. Through this process, module materials were refined sufficiently to be suitable for testing in schools that had not been part of the development process.

## CASE DESCRIPTION

### Principles Adopted in Developing the Electricity Module

The module on electric circuits was informed by the general principles adopted in epiSTEMe, combined with specific considerations particular to the topic. A feature shared with the other topic modules was orchestration of lessons to permit shifts between student group work and teacher-led full classroom discussion; and which moved between eliciting and examining students' own thinking, and considering the canonical curriculum accounts reflecting scientific concepts and models. This is the type of approach discussed by Mortimer and Scott (2003) in their exploration of classroom science teaching. This is considered further below.

The perspective informing the development of the module was personal constructivism, in the sense of psychological or pedagogic constructivism (Glasersfeld, 1989; Sjøberg, 2010; Taber, 2009), which suggests that each person has to interpret their experiences to construct their own understanding of the world. The corollary of this principle is that all learning is contingent upon the interpretative frameworks available to a learner and so the teacher cannot assume that teaching will be understood as intended. Sometimes personal constructivism is presented as being in opposition to social constructivism or constructionism, but the version of constructivism adopted here fully acknowledged that human learning normally takes place in a social context, and that culture provides affordances and constraints on learning (Kleine-Staarman & Mercer, 2010; Scott, 1998). School learning is often highly contingent not only upon the student's prior learning, but also on features of the classroom context (Finkelstein, 2005): such as curriculum, teaching approach, teacher language, teaching models, and in particular learning activities and the opportunities for engagement with ideas these provide.

The module included an extended series of group practical activities of building and examining simple circuits. In selecting electricity as a project topic it would have been possible to have focused on building circuits with different transducers (lamps, buzzers, light dependent resistors, light emitting diodes etc) in response to problems that could have been contextualised in everyday situations. So, for example, students could have been asked to build a circuit that turned on a light if it was dark when someone (who could not see the light switch) whistled. This would have motivated problem-solving through everyday relevance (and would have matched the kind of approach used extensively in the other epiSTEMe topic modules). Such an approach could have treated circuit components as 'black boxes' and been based on how technological solutions are met by using logic gates and various transducers in different combinations.

However, as suggested above, the key problem for science educators in a curriculum context that expects learners to understand basic circuit principles is how to help learners to acquire a scientific model of current in circuits that distinguishes the flow of charge itself from the energy being transferred through the circuit. It was decided therefore to focus on these more fundamental abstract aspects of circuits rather than their technological applications. Given the problems, described above, that lower secondary students often experience in making sense of scientific models of circuits, there might be a case for arguing that theoretical understanding could be deferred to upper secondary level, and that it is more appropriate to provide experience of practical uses in the lower secondary school: but since the prescribed curriculum was set out in terms of the physical principles, these were addressed.

## Minds-On Practical Work

As suggested above, there are significant challenges in expecting students, especially those in the lower secondary school relatively unfamiliar with circuit work, to

*Figure 3. Predict-observe-explain was used to motivate dialogue within groups*

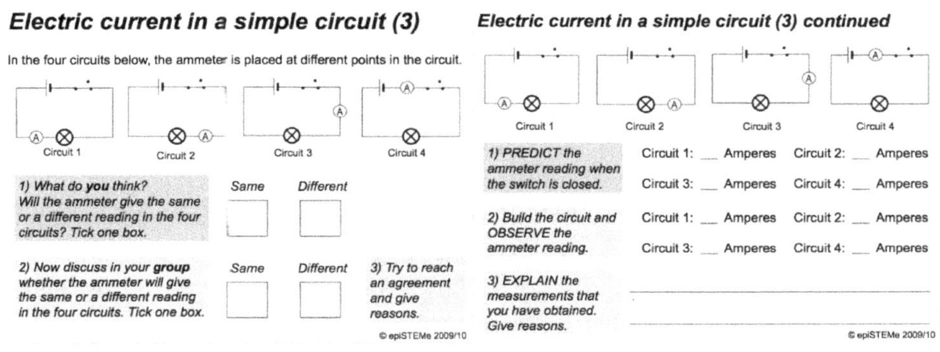


make the desired links between observations made when constructing circuits and the concepts they are expected to learn. In particular, these concepts will often be contrary to the mental models students develop from their intuitive ways of making sense of electrical circuits.

The core of the module was a sequence of practical activities organised around building simple series and parallel circuits. Despite the potential difficulties associated with practical work it was considered important to include 'hands-on' work to motivate students to see a need for modelling what was going on in circuits by presenting actual phenomena to be explained.

In order to ensure the work was also minds-on this was undertaken within a dialogic frame at two levels. Firstly, the circuit investigations were to be undertaken within groups where (a) students had been taught about effective group work in the epiSTEMe introductory module, and (b) the P-O-E technique was adopted so that circuit building would be undertaken with a view to testing particular ideas about what was going on in circuits (e.g. see Figure 3).

Secondly, the teaching and learning activities were designed to shift between group work and classroom-led discussion where the teacher was asked to work with students' ideas and explore their adequacy in relation to the empirical observations. It was also recommended that (given the potential for equipment failures to lead to anomalous results) the teacher should reinforce student findings by using either a large demonstration version of the circuits students were building, or projected computer simulations of the circuits, to ensure that the scientifically 'correct' observations were being discussed and recorded by students.

This central core to the module involved learners in a succession of similar activities as they built a sequence of circuits allowing comparisons to be made between different arrangements of circuit components. We were aware in designing the module that different classes would progress through the material at different rates, and that teachers saw the limited number of lessons they could commit to any particular topic as a major constraint. We therefore included some optional material, and wrote modules that allowed differentiation by giving teachers flexibility to choose to omit some activities for some groups of students.

The epiSTEMe electricity module allows learners to work their way through a series of closely related practical exercises to help them build up a conceptual understanding of phenomena - that is it offers an opportunity to experience a much more authentic form of scientific enquiry than a series of discrete stand alone practicals each related to a distinct scientific idea. This more authentic approach also helps counter the problems referred to above of working with unfamiliar kit which tends to lead to a major part of both time on task and working memory capacity being given over to manipulation, leaving less resource for mentipulation of the ideas the practical work is meant to link to.

## Building upon Existing Good Practice

The epiSTEMe project, then, sought to build upon, and develop design principles around, existing research and demonstrated good practice. Within the electricity module this was enacted in two ways. The common use of models and analogies in teaching this topic was developed and made a key focus of the module (see below). In addition it was decided by the research team that rather than just writing new activities, it was important to include existing research-informed teaching resources developed by other researchers. In particular we draw upon two existing sources. One of these is the UK's Institute of Physics' 'Supporting Physics Teaching 11-14' materials (Whitehouse, 2002). The other is guidance materials published as part of a government funded 'National Strategy' (The National Strategies Secondary, 2008). These in turn drew upon activities designed as part of a teaching scheme (Hind, Leach, Lewis, & Scott, not dated) developed at the University of Leeds during a funded project (the *Teaching and Learning Research Project* funded by the UK Economic and Social Research Council).

So for example, one of the activities included in the epiSTEMe module was 'the big circuit' - a teacher-led activity asking students about what would happen to a lamp when a switch is closed in a circuit that is set up around the full perimeter of the room for dramatic effect. The teacher elicits student thinking about the circuit, and in particular the time it might take for a lamp some considerable distance from a switch or battery to light. The activity is designed around two conceptual tools referred to as 'learning demand' and the 'communicative approach' (Ruthven, Laborde, Leach, & Tiberghien, 2009). Learning demand (Leach & Scott, 2002) concerns analysing the 'gap' between students' current thinking and the canonical

*Figure 4. Questions highlighting the mapping of an analogy to electric circuits*

**Making sense of series circuits (1)**

*Using the supermarket delivery vans model, make complete sentences by matching the phrases on the left with those on the right. Use arrows.*

1) The bakery stores the goods in a similar way to...
2) The roads have a similar role to...
3) The supermarket allows goods from the van to be widely spread...
4) The vans that move around the roads are like...
5) Each individual van with its load of goods is like...
6) Traffic lights are like...

A) ...a single electron carrying some energy around the circuit.
B) ...like a lamp spreads out energy brought by the current.
C) ...a switch which can stop the current.
D) ...the wires that provide a pathway for the current.
E) ...the cell is a store of energy.
F) ...the current that flows around the circuit.

© epiSTEMe 2009/10

**Making sense of series circuits (2)**

*Using the supermarket delivery vans model, make complete sentences by matching the phrases on the left with those on the right. Use arrows.*

However, the supermarket delivery vans model is only a model, so...

1) Vans come in different shapes and sizes, ***whereas...***
2) Traffic lights only stop the vans in one place, ***whereas...***
3) Vans stop when drivers need rest or meals, ***whereas...***

A) ...a switch will immediately stop current flowing all the way around the loop.
B) ...the current does not stop as long as there is an energy source and a complete circuit.
C) ...all the electrons moving around a circuit are identical.

© epiSTEMe 2009/10

account presented in the curriculum - that is, it is a constructivist model stressing the importance of diagnostic assessment in classroom teaching (Taber, 2014). The communicative approach (Mortimer & Scott, 2003) refers to the kind of dialogic teaching referred to earlier where the teacher moves between exploring different ideas suggested by learners and presenting and advocating the scientific account set out as target knowledge in the curriculum.

## The Use of Teaching Analogies and Models

Another feature of existing good practice built into the epiSTEMe module was the use of teaching analogies for thinking about what is going on in circuits. In the Big Circuit activity, for example, as presented in the original Leeds teaching scheme, a 'teaching story' is introduced to compare the circuit with an everyday situation that would be accessible to learners: the delivery of bread from bakeries to keep supermarkets stocked by fleets of delivery vans (Hind, Leach, Lewis, & Scott, Not dated). A key feature of this analogy is that although it is the vans flowing around the distribution network, the number of vans is conserved as they act as carriers of something else - loaves of bread (see Figure 4). This is analogous to how electrons in circuits act as 'carriers of energy' allowing energy to be transferred from the store in the battery to the lamp (or other transducer) by a current that is constant around the circuit (as current reflects the amount of charge flowing at a point, not the energy associated with it).

The use of teaching analogies of this kind is ubiquitous in science teaching across a wide range of topics (Harrison & Coll, 2008; Harrison & Treagust, 2006). The principle here is simple enough: teaching is about making the unfamiliar familiar, and one way we can do this (especially where there is not the option of directly demonstrating a teaching point) is to make comparisons with what is already familiar. Teachers use explicit analogies as well as metaphors and similes to help learners anchor new ideas within existing propositional knowledge, and so to ensure teaching is perceived meaningfully and more likely to lead to learning (Ausubel, 2000). So it might be said that the nucleus is the control centre for a cell, that enzymes fits into substrates like a lock and key, and so forth.

Such devices are very common in teaching, although it is recognised that there are potential problems. Students at secondary level often display a relatively limited appreciation of the epistemological role and nature of models and analogies (Treagust, Chittleborough, & Mamiala, 2002) - for example treating comparisons more 'literally' or realistically than is intended. In the case of analogies, students may transfer inappropriate attributes from the analogue to the target (Nakiboglu & Taber, 2013; Taber, 2001) unless teaching is clear about the positive and negative aspects of the analogy (Gentner, 1983). Despite these limitations, previous work with trainee teachers teaching about the nature of ideas and evidence in science had suggested that there was considerable potential to support learning about electricity by working with analogies, models and creative writing (Taber, de Trafford, & Quail, 2006).

## MAKING MODELS AND ANALOGIES A CENTRAL FEATURE OF THE MODULE

The incorporation of teaching models and analogies in a module on electric circuits was not in itself novel, however the epiSTEMe module went beyond this. The module was designed to be in the spirit of the recent curriculum changes (discussed above) in that it foregrounded learning about the role of models and analogies in science alongside the learning of the specific topic of electric circuits. That is, the inclusion of models and analogies was not intended just to support learning about circuits, but also to support learning about a key feature of the nature of science (or 'how science works') that was highlighted in the new curriculum (QCA, 2007).

The intention then was to build synergy into the design (see Figure 5). The use of teaching models and analogies would help learners make the unfamiliar world of electrons and potential difference meaningful by comparison with familiar situations and experiences. However, the topic of electric circuits would also provide an authentic context for exploring how scientists use such devices as thinking tools in their work - for example in making predictions to test through empirical investigation.

This aim also had the advantage of offering a response to the minority of students (sometimes including some of those who already have a relatively strong concep-

*Figure 5. Synergy between learning about scientific ideas, and learning about the nature of science*

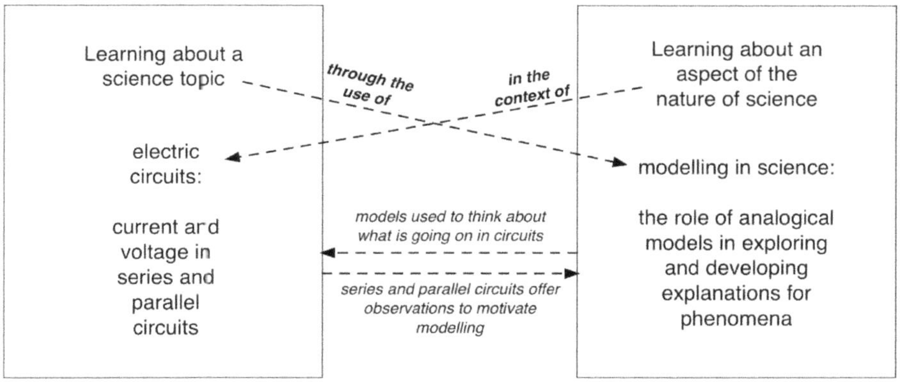


tual understanding of a topic) who consider the use of teaching analogies and some other models as 'silly' and feel the teacher is either being condescending in using them or intends them only for the low attaining students in the class.

## THE USE OF MULTIPLE MODELS

An important feature of this approach was the use of multiple models, in keeping with the principle that learning abstract scientific ideas is supported by the use of multiple representations (Tsui & Treagust, 2009; Tytler, Prain, Hubber, & Waldrip, 2013). Simply offering one model that generally 'worked' might have supported learning about electric circuits but without teaching about the role of models, and with the danger of inappropriate transfer of associations of the model, or the expectation that the model would always 'work' (apply) even though models and analogies generally have limited ranges of application.

Teachers were encouraged to elicit learners' own suggestions and develop those, but built into the teaching materials were three models that student were explicitly asked to consider and seek to apply. One of these was a version of the supermarkets/bread van model discussed above. A second was based on a physical model students could try in class using a loop of rope that was held in the hands of a series of people around the 'circuit' to represent current flow. The third model was a role-play (Dorion, 2009) where students take on the role of electrons moving packets of energy from a source (battery) to another circuit component (lamp).

*Figure 6. The module included opportunities to work with representations to model aspects of circuit phenomena*

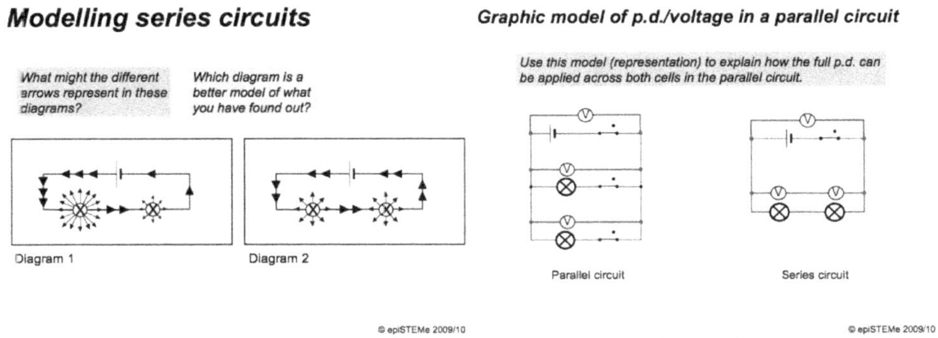


There is now an increasing awareness in science teaching that learning is often supported by both multi-modal teaching (Jewitt, Kress, Ogborn, & Tsatsarelis, 2001), and through asking students to find alternative ways of representing the same information (Tytler et al., 2013). The first analogical model was taught primarily through diagrams: the other two involved embodied learning - through students interacting with a physical model of current in a circuit, and playing a part in a physical simulation of current.

Offering three models motivated genuine questions about the extent to which the different models 'worked' in supporting thinking about different aspects of the actual circuits students could build, and how predictions informed by thinking about the different models were or were not supported by observations of actual circuits. In addition to the analogical models explored through the module, explicit opportunities were built into the module to consider the affordances of different kinds of representations of circuit phenomena (e.g., see Figure 6).

## Teaching about Models and Analogies in Science

The intention that analogies and models should take a central role in the module was reflected in the inclusion of explicit teaching about this theme early in the module. Slides to introduce the use of analogy in science were included in the teaching materials provided, along with related activities. These included asking learners to suggest their own analogies - an activity that had been successfully used in an earlier project (Taber, 2007).

The three different analogical models built into the module were introduced and explored through teacher led discussion. The students were then asked to work

*Figure 7. Students were asked to explicitly evaluate the models they had used throughout the module*

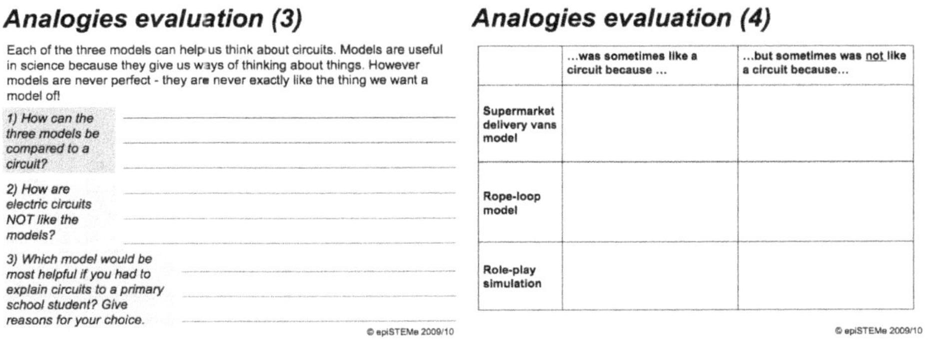

**Analogies evaluation (3)**

Each of the three models can help us think about circuits. Models are useful in science because they give us ways of thinking about things. However models are never perfect - they are never exactly like the thing we want a model of!

*1) How can the three models be compared to a circuit?*

*2) How are electric circuits NOT like the models?*

*3) Which model would be most helpful if you had to explain circuits to a primary school student? Give reasons for your choice.*

© epiSTEMe 2009/10

**Analogies evaluation (4)**

| | ...was sometimes like a circuit because ... | ...but sometimes was not like a circuit because... |
|---|---|---|
| Supermarket delivery vans model | | |
| Rope-loop model | | |
| Role-play simulation | | |

© epiSTEMe 2009/10

with the models when undertaking the 'P-O-E' based investigations of a sequence of circuits. During the teacher-led classroom discussions the teachers were asked to explore and work with student thinking about both the circuits themselves and the models. Later in the module students were asked to critique and evaluate the three analogical models they had used through the unit (e.g., see Figure 7).

## Extensive Use of Circuit Diagrams

As suggested above, circuit diagrams offer an additional challenge for students in circuit work. This was a concern for some of the teachers we worked with, as they rightly recognised how presenting formal circuit diagrams to students added to the cognitive demand of the work. On the advice of the teachers we included hybrid diagrams (showing pictorial representations of components in circuits) in the earliest activities of the module. However it was felt to be important to ask students to engage with formal circuit diagrams for much of the work as this is a core form of representation used in science that allows ready tracing of the key topological features of circuits (in particular where current splits in parallel branches).

Moreover, in a module with a strong focus on models and modelling in science, circuit diagrams offered an example of a commonly used representational model. It was also considered that, as with using the practical apparatus, asking students to undertake an extended sequence of activities using the representations would support developing familiarity to the point where this ceased to make a major demand upon student working memory.

We incorporated an initial diagnostic activity into the module asking students to match circuits from the two types of diagrams, thus giving teachers an opportunity

*Figure 8. Building familiarity with circuit diagrams is considered an important prerequisite to working effectively with such diagrams in circuit building*

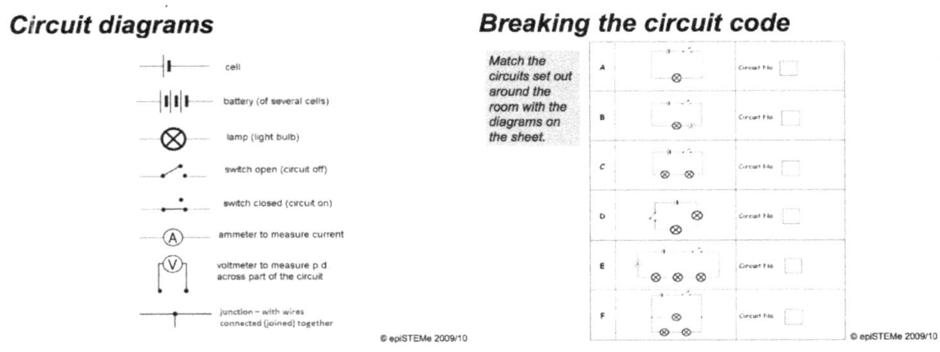


to see whether students could readily cope with the representations, for example perhaps based on earlier primary school work on electricity. Early in the module students were introduced to a small selection of circuit symbols to be used in the lessons, along the lines that "circuit diagrams are a special kind of model that is useful to *represent* circuits in science. Circuit symbols are like a special (graphical/ diagrammatic) language or code". The students then undertook an activity on 'breaking the circuit code' (see Figure 8) that asked groups to visit 6 different circuits set up at stations around the teaching room and work out which circuit matched each of six circuit diagrams on their worksheet.

This introductory activity preceded the group practical work where students were asked to think about circuits represented as diagrams in terms of the three analogical models, and then to build the circuits represented. At the end of module, one of the review activities provided was a game of circuit dominoes - which required students to recognise where differently drawn circuit diagrams represented substantially the same circuit (see Figure 9). This was provided with different levels of complexity, to allow differentiation in the challenge of the task.

## The epiSTEMe Module

After various drafts, and piloting by teachers in our partner schools, a version of the module was produced that the project team felt was ready for making available to teachers more widely. The module materials comprise a series of slides for teacher presentation to support discussion; a workbook for students; teachers' notes (see Figure 10) and technician notes. These are all available to any educator or researcher who contacts the authors.

*Figure 9. Review activities reinforced working with, and thinking about, the circuit diagram formalism*

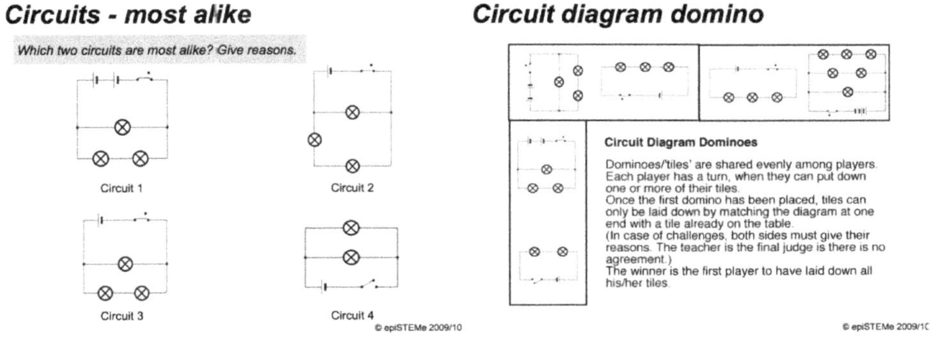


Although we recognised that teachers would need to organise material according to their school timetable structures (as length of lessons - classroom periods - vary between schools) and to meet the needs of particular teaching groups, we were encouraged by teacher partners to present the activities within nominal coherent lessons (see Figure 10). We expected teachers to retain the sequence of the module, but not to feel bound by the suggestions for how much material was to be included in a particular lesson.

## CURRENT CHALLENGES

One comment received in feedback on the electricity module was that the work on building the different circuits was time-consuming and involved students undertaking a number of similar activities (that is, building a sequence of circuits embedded within group-work structured around P-O-E). The implication was that school science should not be repetitive - even though arguably much professional science is precisely of this nature. This may reflect an apparent obsession within the school inspection system in the UK on 'pace': that students should be seen to be making progression in moving forward in their learning. Some teachers felt that school inspectors (or senior staff from their own schools conducting lesson observations) would expect to see obvious progression between clearly discrete activities - each with its own closure within the lesson. Teachers in England feel they are expected to demonstrate new learning at the end of each lesson, even though educational research shows that substantive conceptual change is a slow process that requires integration across sequences of learning activities (Vosniadou, 2008). Clearly there

*Figure 10. Teaching and learning activities were organised into possible lessons that could each be undertaken in classroom period of about an hour*

## Overview of the Teaching Notes



is a danger here of teachers focusing on achievable short-term objectives to the detriment of longer-term aims.

This can be a real concern if teachers are worried about spending extended periods developing ideas because they feel they should be seen to be moving on to something that is clearly (to students, and any visiting inspectors) 'different'.

Common criticisms of the English science curriculum have been the lack of depth which limits engagement with concepts - something that is of particular importance to the most gifted learners in science (Taber, 2010) - and the tendency for teachers to limit practical work to that considered to be clearly linked to formal assessment (Hacker & Rowe, 1997).

A serious concern then is that making our materials available unconditionally, without for example requiring attendance at related professional development sessions, risks our activities being used without being informed by the research-based design principles. Many teachers practise a form of professional bricolage, acquiring teaching materials to be 'mixed-and-matched' and adapted to fit existing teaching habits. Yet teaching with the epiSTEMe materials may not reflect the epiSTEMe approach unless teachers adopt something of the philosophy behind the project and incorporate the pedagogy we have put together rather than just use the materials. A key feature is the dialogic aspect, which requires both that teachers prepare students for effective group work, and that teachers orchestrate the shifts between inviting and exploring different views, and presenting the case for the scientific account.

Within the electricity module itself, our specific additional concerns are that teachers will not give students sufficient time to work carefully through the sequence of activities as intended, or may fail to maintain the exploration of the analogical models through the different circuit contexts that allows learners to appreciate how models are used and evaluated as thinking tools. In particular, unless teachers insist that learners take time to work through the P-O-E activities as instructed, shortcuts will be taken in building circuits before carefully thinking through what is expected to happen. The limited observational work we were able to carry out in the epiSTEMe project with teachers who had not been involved in the development process suggests these are real concerns, at least in the UK context.

## SOLUTIONS AND RECOMMENDATIONS

Our experience in piloting the materials with partner project teachers was that students certainly demonstrated learning gains in relation to understanding electrical circuits through the module. Pre- and post-tests were developed using assessment questions based on existing assessment materials for this topic to ensure content validity, as we intended to undertake a randomised field trial of the modules by comparing students in classes of teachers having attended two days of teacher development and using the materials, with students in (as far as possible) matched schools working with teachers teaching according to their usual schemes and approaches. (These teachers of 'control' classes were offered teacher development and access to all the materials at the end of this process. The trial has now been completed, although analysis of data is not yet complete.)

The epiSTEMe electricity module integrated teaching and learning about a science topic, electric circuits, with teaching and learning about a key feature of the nature of science, the role of models and modelling. Any learning gains in relation to this key curriculum aim would be in addition to the learning that took place about circuits themselves. As it would have been unfair to test students on this aspect of learning in classes where teachers were not following the epiSTEMe module, we did not collect data about this during the field trials.

The epiSTEMe project reinforced the possibility of designing teaching modules in science and mathematics according to what are now well-established pedagogic principles. The project also reminded us of the barriers to working in partnership with schools in such projects - personnel changes and constraints due to other school priorities limited the continuity of the wider development team and restricted the opportunities for effective piloting of materials. Two schools that worked with us throughout the development process have since worked towards embedding the pedagogy exemplified through epiSTEMe more widely into departmental teaching - but have to date succeeded to different degrees.

Our observations of classes using epiSTEMe materials taught by teachers who had attended our teacher development days reminded us of the difficulties of bringing about changes in teacher behaviour in their classrooms. Expecting teachers to shift towards more dialogic teaching approaches without extensive support and opportunities for feedback and review may be overly optimistic. Whilst this should remain an important aim, it is clear many teachers find it difficult to make substantial changes from familiar classroom approaches and this might reinforce the importance of research-informed *initial* teacher education programmes in setting up effective pedagogic habits from the start of a teaching career.

The materials from epiSTEMe are now available, and the authors would welcome approaches from those who wish to either critique them to inform their work in research-based instructional design, or even to test them out in teaching in their own local educational contexts. The electricity module might be of particular interest to those exploring how to embed learning about nature of science objectives into teaching of mainstream science topics. There has been debate about the best ways to teach nature of science objectives in relation to science 'content' objectives (Hodson, 2009), and the adoption of the electricity module design would benefit from careful examination in this regard. We would welcome evaluation of the module in diverse classroom contexts, especially where it is possible to (a) explore classroom processes (e.g. the nature of student group work; the extent of dialogicity in teaching); and (b) to simultaneously investigate learning gains across both the domains of physics subject knowledge (electric circuits) and the nature of science (the role of models and modelling in science).

## ACKNOWLEDGMENT

The research reported here was only made possible by the support of teachers from partner schools, and funding from the Economic and Social Research Council through grant RES-179-25-0003: 'Effecting Principled Improvement in STEM Education: Student Engagement and Learning in Early Secondary-School Physical Science and Mathematics'.

## REFERENCES

Abrahams, I. (2011). *Practical work in school science: A minds-on approach*. London: Continuum.

Abrahams, I., & Millar, R. (2008). Does Practical Work Really Work? A study of the effectiveness of practical work as a teaching and learning method in school science. *International Journal of Science Education*, *30*(14), 1945–1969. doi:10.1080/09500690701749305

Arnold, M., & Millar, R. (1987). Being constructive: An alternative approach to the teaching of introductory ideas in electricity. *International Journal of Science Education*, *9*(5), 553–563. doi:10.1080/0950069870090505

Ausubel, D. P. (2000). *The Acquisition and Retention of Knowledge: a cognitive view*. Dordrecht, The Netherlands: Kluwer Academic Publishers. doi:10.1007/978-94-015-9454-7

Baddeley, A. D. (2003). Working memory: looking back and looking forward. *Nature Reviews Neuroscience, 4*(10), 829-839.

Brook, A., & Driver, R. (1984). *Aspects of Secondary Students' Understanding of Energy: Full Report*. Leeds, UK: Centre for Studies in Science and Mathematics Education, University of Leeds.

Budde, M., Niedderer, H., Scott, P., & Leach, J. (2002). 'Electronium': A quantum atomic teaching model. *Physics Education*, *37*(3), 197–203. doi:10.1088/0031-9120/37/3/303

Chi, M. T. H. (2008). Three types of conceptual change: belief revision, mental model transformation, and categorical shift. In S. Vosniadou (Ed.), *International Handbook of Research on Conceptual Change* (pp. 61–82). New York: Routledge.

Chi, M. T. H., Slotta, J. D., & de Leeuw, N. (1994). From things to processes; a theory of conceptual change for learning science concepts. *Learning and Instruction*, *4*(1), 27–43. doi:10.1016/0959-4752(94)90017-5

Collins, H. (2010). *Tacit and Explicit Knowledge*. Chicago: The University of Chicago Press. doi:10.7208/chicago/9780226113821.001.0001

Cooper, J. (2007). *Cognitive Dissonance: Fifty years of a classic theory*. London: Sage.

DfEE/QCA. (1999). Science: The National Curriculum for England, key stages 1-4. London: Department for Education and Employment/Qualifications and Curriculum Authority.

Donnelly, J. (2001). Contested terrain or unified project? 'The nature of science' in the National Curriculum for England and Wales. *International Journal of Science Education*, *23*(2), 181–195. doi:10.1080/09500690120412

Dorion, K. R. (2009). Science through Drama: A multiple case exploration of the characteristics of drama activities used in secondary science lessons. *International Journal of Science Education*, *31*(16), 2247–2270. doi:10.1080/09500690802712699

Driver, R. (1983). *The Pupil as Scientist?* Milton Keynes, UK: Open University Press.

Driver, R., & Oldham, V. (1986). A constructivist approach to curriculum development in science. *Studies in Science Education*, *13*(1), 105–122. doi:10.1080/03057268608559933

Driver, R., Squires, A., Rushworth, P., & Wood-Robinson, V. (1994). *Making Sense of Secondary Science: research into children's ideas*. London: Routledge.

Duit, R. (2009). *Bibliography - Students' and Teachers' Conceptions and Science Education*. Retrieved from http://www.ipn.uni-kiel.de/aktuell/stcse/stcse.html

Duschl, R. A. (2000). Making the nature of science explicit. In R. Millar, J. Leach, & J. Osborne (Eds.), *Improving Science Education: the contribution of research* (pp. 187–206). Buckingham: Open University Press.

Feynman, R. (1965). *The Character of Physical Law*. Cambridge, MA: MIT Press.

Gauld, C. (1986). Models, meters and memory. *Research in Science Education*, *16*(1), 49–54. doi:10.1007/BF02356817

Gauld, C. (1989). A study of pupils' responses to empirical evidence. In R. Millar (Ed.), *Doing Science: images of science in science education* (pp. 62–82). London: The Falmer Press.

Gentner, D. (1983). Structure-Mapping: A Theoretical Framework for Analogy. *Cognitive Science*, *7*(2), 155–170. doi:10.1207/s15516709cog0702_3

Hacker, R. G., & Rowe, M. J. (1997). The impact of a National Curriculum development on teaching and learning behaviours. *International Journal of Science Education*, *19*(9), 997–1004. doi:10.1080/0950069970190901

Harrison, A. G., & Coll, R. K. (Eds.). (2008). *Using analogies in middle and secondary science classrooms*. Thousand Oaks, CA: Corwin Press.

Harrison, A. G., & Treagust, D. F. (2006). Teaching and learning with analogies: friend or foe? In P. J. Aubusson, A. G. Harrison & S. M. Ritchie (Eds.), Metaphor and Analogy in Science Education (pp. 11-24). Dordecht, The Netherlands: Springer.

Hind, A., Leach, J., Lewis, J., & Scott, P. (n.d.). *Teaching for understanding: Electic circuits: A teahcing scheme developed from research evdience on students' learning about electric circuits*. Academic Press.

Hodson, D. (2009). *Teaching and learning about science: Language, theories, methods, history, traditions and values*. Rotterdam, The Netherlands: Sense Publishers.

Howe, C., Tolmie, A., Thurston, A., Topping, K., Christie, D., & Livingston, K. et al. (2007). Group work in elementary science: Towards organisational principles for supporting pupil learning. *Learning and Instruction*, *17*(5), 549–563. doi:10.1016/j.learninstruc.2007.09.004

Instrument, S. (1989). The Education (National Curriculum) (Attainment Targets and Programmes of Study in Science). *Order*, 1989.

Jenkins, E. W. (2000). The impact of the national curriculum on secondary school science teaching in England and Wales. *International Journal of Science Education*, *22*(3), 325–336. doi:10.1080/095006900289903

Jewitt, C., Kress, G., Ogborn, J., & Tsatsarelis, C. (2001). Exploring Learning Through Visual, Actional and Linguistic Communication: The multimodal environment of a science classroom. *Educational Review*, *53*(1), 5–18. doi:10.1080/00131910123753

Johnstone, A. H. (1982). Macro- and microchemsitry. [Notes and correspondence]. *The School Science Review*, *64*(227), 377–379.

Johnstone, A. H. (1991). Why is science difficult to learn? Things are seldom what they seem. *Journal of Computer Assisted Learning*, *7*(2), 75–83. doi:10.1111/j.1365-2729.1991.tb00230.x

Jolliffe, T., & Baron-Cohen, S. (1999). A test of central coherence theory: linguistic processing in high-functioning adults with autism or Asperger syndrome: is local coherence impaired? *Cognition*, *71*(2), 149–185. doi:10.1016/S0010-0277(99)00022-0 PMID:10444907

Key Stage 3 National Strategy. (2002). *Framework for teaching science: years 7, 8 and 9*. London: Department for Education and Skills.

Kind, V., & Taber, K. S. (2005). Science: Teaching School Subjects 11-19. London: RoutledgeFalmer.

Kleine-Staarman, J., & Mercer, N. (2010). The guided construction of knowledge: Talk between teachers and students. In K. Littleton, C. Wood & J. K. Kleine-Staarman (Eds.), International Handbook of Research of Psychology in Education (pp. 75-104). Bingley, UK: Emerald.

Koffka, K. (1967). Principles of Gestalt Psychology. In J. A. Dyal (Ed.), *Readings in Psychology: Understanding human behavior* (2nd ed., pp. 9–13). New York: McGraw-Hill Book Company.

Kuhn, T. S. (1996). *The Structure of Scientific Revolutions* (3rd ed.). Chicago: University of Chicago. doi:10.7208/chicago/9780226458106.001.0001

Lakatos, I. (1999). Lectures on Scientific Method. In M. Motterlini (Ed.), *For and Against Method* (pp. 19–109). Chicago: University of Chicago Press. doi:10.7208/chicago/9780226467030.001.0001

Leach, J., & Scott, P. (2002). Designing and evaluating science teaching sequences: An approach drawing upon the concept of learning demand and a social constructivist perspective on learning. *Studies in Science Education*, *38*(1), 115–142. doi:10.1080/03057260208560189

Matthews, M. R. (1994). *Science Teaching: The role of history and philosophy of science*. London: Routledge.

McConnell, M. C. (1982). Teaching about Science, Technology and Society at the secondary school level in the United States. An educational dilemma for the 1980s. *Studies in Science Education*, *9*(1), 1–32. doi:10.1080/03057268208559893

Millar, R. (2004, 3-4 June 2004). *The role of practical work in the teaching and learning of science.* Paper presented at the High School Science Laboratories: Role and Vision, National Academy of Sciences. Washington, DC.

Mortimer, E. F., & Scott, P. H. (2003). *Meaning Making in Secondary Science Classrooms*. Maidenhead, UK: Open University Press.

Nickerson, R. S. (1998). Confirmation bias: A ubiquitous phenomenon in many guises. *Review of General Psychology*, *2*(2), 175–220. doi:10.1037/1089-2680.2.2.175

Parkin, A. J. (1993). *Memory: Phenomena, experiment and theory*. Hove, UK: Psychology Press.

Polanyi, M. (1962). Personal knowledge: Towards a post-critical philosophy (Corrected version ed.). Chicago: University of Chicago Press.

QCA. (2000). *Key stage 3 schemes of work*. Qualification and Curriculum Authority.

QCA. (2005). *Science: 2004/5 annual report on curriculum and assessment*. London: Qualifications and Curriculum Authority.

QCA. (2007). *Science: Programme of study for key stage 3 and attainment targets*. London: Qualifications and Curriculum Authority.

Renström, L., Andersson, B., & Marton, F. (1990). Students' conceptions of matter. *Journal of Educational Psychology*, *82*(3), 555–569. doi:10.1037/0022-0663.82.3.555

Ruthven, K., Howe, C., Mercer, N., Taber, K. S., Luthman, S., Hofmann, R., & Riga, F. (2010). Effecting Principled Improvement in STEM Education: Research-based pedagogical development for student engagement and learning in early secondary-school physical science and mathematics. In M. Joubert & P. Andrews (Eds.), *British Congress of Mathematics Education* (pp. 191-198). British Society for Research into Learning Mathematics.

Ruthven, K., Laborde, C., Leach, J., & Tiberghien, A. (2009). Design Tools in Didactical Research: Instrumenting the Epistemological and Cognitive Aspects of the Design of Teaching Sequences. *Educational Researcher*, *38*(5), 329–342. doi:10.3102/0013189X09338513

Shipstone‡, D. M., Rhöneck, C., Jung, W., Kärrqvist, C., Dupin, J.-J., Johsua, S., & Licht, P. (1988, July). A study of students' understanding of electricity in five European countries. *International Journal of Science Education*, *10*(3), 303–316. doi:10.1080/0950069880100306

Solomon, J. (1992). *Getting to Know about Energy - in School and Society*. London: Falmer Press.

Nakiboğlu, C., & Taber, K. S. (2013). The atom as a tiny solar system: Turkish high school students' understanding of the atom in relation to a common teaching analogy. In G. Tsaparlis, & H. Sevian (Eds.), *Concepts of Matter in Science Education* (pp. 169–198). Dordrecht, The Netherlands: Springer. doi:10.1007/978-94-007-5914-5_8

Taber, K. S. (2001). When the analogy breaks down: Modelling the atom on the solar system. *Physics Education*, *36*(3), 222–226. doi:10.1088/0031-9120/36/3/308

Taber, K. S. (2003). Lost without trace or not brought to mind? - a case study of remembering and forgetting of college science. *Chemistry Education: Research and Practice*, *4*(3), 249–277.

Taber, K. S. (2007). *Enriching School Science for the Gifted Learner*. London: Gatsby Science Enhancement Programme.

Taber, K. S. (2008). Towards a curricular model of the nature of science. *Science & Education*, *17*(2-3), 179–218. doi:10.1007/s11191-006-9056-4

Taber, K. S. (Ed.). (2009). *Progressing Science Education: Constructing the scientific research programme into the contingent nature of learning science*. Dordrecht: Springer. doi:10.1007/978-90-481-2431-2

Taber, K. S. (2010). Challenging gifted learners: General principles for science educators; and exemplification in the context of teaching chemistry. *Science Education International*, *21*(1), 5–30.

Taber, K. S. (2013). Revisiting the chemistry triplet: Drawing upon the nature of chemical knowledge and the psychology of learning to inform chemistry education. *Chemistry Education Research and Practice*, *14*(2), 156–168. doi:10.1039/c3rp00012e

Taber, K. S. (2014). *Student Thinking and Learning in Science: Perspectives on the Nature and Development of Learners' Ideas*. New York: *Routledge*.

Taber, K. S., de Trafford, T., & Quail, T. (2006). Conceptual resources for constructing the concepts of electricity: The role of models, analogies and imagination. *Physics Education*, *41*(155-160).

The National Strategies Secondary. (2008). *Explaining how electric circuits work: Science teaching unit*. Department for Children, Schools and Families.

Treagust, D. F., Chittleborough, G., & Mamiala, T. L. (2002). Students' understanding of the role of scientific models in learning science. *International Journal of Science Education, 24*(4), 357–368. doi:10.1080/09500690110066485

Tsui, C.-Y., & Treagust, D. (2010). Evaluating Secondary Students' Scientific Reasoning in Genetics Using a Two-Tier Diagnostic Instrument. *International Journal of Science Education, 32*(8), 1073–1098. doi:10.1080/09500690902951429

Tytler, R., Prain, V., Hubber, P., & Waldrip, B. G. (Eds.). (2013). *Constructing Representations to Learn in Science*. Rotterdam: Sense Publishers. doi:10.1007/978-94-6209-203-7

Vertes, R. P. (2004). Memory Consolidation in Sleep. *Neuron, 44*(1), 135–148. doi:10.1016/j.neuron.2004.08.034 PMID:15450166

Vosniadou, S. (Ed.). (2008). *International Handbook of Research on Conceptual Change*. London: Routledge.

Watts, M. (1983). Some alternative views of energy. *Physics Education, 18*(5), 213–217. doi:10.1088/0031-9120/18/5/307

White, R. T., & Gunstone, R. F. (1992). *Probing Understanding*. London: Falmer Press.

Whitehouse, M. (2002). Supporting Physics Teaching (11-14). *Physics Education, 37*(5), 363.

Wilensky, U., & Resnick, M. (1999). Thinking in Levels: A Dynamic Systems Approach to Making Sense of the World. *Journal of Science Education and Technology, 8*(1), 3–19. doi:10.1023/A:1009421303064

## KEY TERMS AND DEFINITIONS

**Constructivist Perspectives on Learning:** Constructivist perspectives on learning consider that knowledge is not 'out there', waiting to be found, but is constructed by people as they make sense of the world, and that there are constraints on this process (for example, limitations characteristic of human perception and cognition). Personal constructivism sees the key processes of learning occurring within the minds (and so the brains) of individual learners, whereas social constructivists put more emphasis on the ways culture and social interaction shape learning and the development of understanding. Whichever emphasis is adopted, it is recognised that what is learned is highly contingent on a range of factors that include elements of the learner's existing conceptual structure (e.g. prior knowledge and beliefs) and how these interact with specific features of teaching and the social context of learning. In the context of the reported project one example would be how the electrical circuits module was designed to give opportunities for the elicitation of common alternative conceptions (such as the idea that current values must change around a simple series circuit) and their consideration in relation to empirical evidence collected by students.

**Design of STEM Teaching Modules:** Teaching of formal curriculum is often organised into sections (often referred to as modules or units) based around a particular topic or concept area - such as electrical circuits. STEM teaching modules are these units of planned teaching in the relatively cognate subject areas of science, technology, engineering and mathematics. The design of STEM teaching modules in the reported project included considerations about the selection and sequencing of content, but also considerations about how features of effective and research-informed pedagogy are adopted when planning the teaching and learning activities and supporting curriculum and assessment materials. For example, the adoption of a constructivist perspective on learning informs the way teaching is designed to acknowledge and respond to students' existing ideas, and a commitment to dialogic teaching informs how canonical ideas are introduced and developed in the classroom and related to students' existing ideas.

**Dialogic Teaching:** Teaching is understood as behaviour which is intended to bring about learning. Dialogic teaching is that in which both teachers and learners make substantial and significant contributions to classroom talk. The teacher encourages learners to participate actively and so enables them to articulate, reflect upon and modify their own understanding, while also providing them with clear guidance, feedback and authoritative accounts of relevant knowledge when appropriate. It normally involves both teacher-led, whole-class sessions and group-based activities where learners can learn collaboratively. An important basis for dialogic teaching is that both the teacher and the learners appreciate the potential value of talk for learning, and of how that potential can best be realised.

**Learning about Electrical Circuits:** 'Electrical circuits' is here understood as a focus of the topic of 'electricity' which is set out as part of the lower secondary school curriculum in England. In particular, in the context of the reported project, this concerns learning about how electrical current flowing in a circuit relates to the configuration of the circuit (e.g. the number of resistive components and how they are arranged).

**Learning Science:** Learning is understood in this chapter as a change in the potential for behaviour. In the context of the reported project this could mean that after learning a student is able to offer an explanation of what electrical current is that they could not have offered before learning, or that after learning a student could offer reasons why using analogies to model circuits reflects scientific practice that they would not have been able to suggest prior to learning.

**School Science Practical Work:** The term 'practical work' within school science is usually intended to refer to laboratory or field work carried out by students. Such activity is often described by students as 'experiments' although much school practical work has involved practising of laboratory techniques, or carrying out procedures to demonstrate accepted (rather than to test conjectured) ideas. Practical work includes enquiry (or inquiry) work where students undertake authentic investigations as well as more routine activities. Sometimes such activities as the secondary analysis of existing data sets have been considered to fall under the heading 'practical work' although this does not involve students themselves in the 'practical' activities of collecting data through observations and measurements. Arguably, it is useful to distinguish learning activities that do have a practical (laboratory or field) component from the broader notion of 'active' learning where students are engaged in activities (group discussions, data analysis, model building) that do not involve specialised locations or apparatus. Collection and analysis of data by remote use of apparatus is becoming a more common type of practical work, and the collection and analysis of data produced by computer simulations may be seen as a borderline case of 'practical' work. In the context of the reported project the practical work undertaken was primarily the construction of electrical circuits by small groups of students to test their predictions and provide empirical evidence to inform discussion of their ideas.

**Teaching about Models and Analogies:** A model is a representation of something in another form (e.g. a mathematical representation of a pattern observed in measurements of some physical quantity) which is considered to be able to stand for some aspect of what is being modelled. Scientific knowledge is often formulated as models, and the development of scientific knowledge often involves the construction and testing of various kinds of models. Analogies are comparisons of structural similarity between different systems (such as comparing nucleus-electron interactions in an atom with sun-planet interactions in a solar system). The creative aspect of scientific work, which generates ideas to critique and test, often draws upon analogies as novel ways of thinking about a target phenomenon or concept. The topic of electricity is often taught at school level using teaching models and analogies, but teaching about models and analogies involves making explicit the roles of the models and analogies and acknowledging how this reflects aspects of authentic scientific practice.

**Teaching about the Nature of Science:** Teaching about the nature of science complements teaching about the output of the scientific process (i.e. consensus models and theories that are considered the 'content' to be taught) and is widely considered to be important both for future scientists and as part of the education of any scientifically literate citizen. Teaching about the nature of science includes consideration of both the fundamental commitments of science that inform what might be called the scientific attitude, or scientific values, and the processes of science. The latter goes beyond scientific method to appreciate both the way scientific knowledge may be robust yet always open to reconsideration, and how scientific knowledge develops from the mediation of creative human thinking through social/institutional processes. In the context of the reported project the main focus of teaching about the nature of science concerned how models are used as tools for developing explanations and for making predictions that can then be tested empirically.

# Chapter 7
# Presenting Physics Content and Fostering Creativity in Physics among Less-Academically Inclined Students through a Simple Design-Based Toy Project

**Nazir Amir**
*Greenview Secondary School, Singapore*

**R. Subramaniam**
*Nanyang Technological University, Singapore*

## EXECUTIVE SUMMARY

*One of the emphases of 21st century science education is in producing students who are creative and who can contribute to the economy. Physics affords immense scope in this regard. This study illustrates an instructional teaching approach to present the physics concepts of density and forces in liquids to kinesthetic students and, at the same time, offers an avenue to foster creativity among them through the fabrication of variants of a popular physics toy: the Cartesian diver. It was conducted*

DOI: 10.4018/978-1-4666-6375-6.ch007

*during curriculum time in a physics laboratory. Results showed that the students were able to showcase their creative abilities through knowledge from physics in this design-based toy project. Students found the pedagogical approach suitable for learning physics content and also a fun way to showcase their creative abilities through knowledge from physics. They also developed positive attitudes towards studying physics after going through this project.*

## 1. INTRODUCTION

In Sir Ken Robinson's lecture on the need to nurture creativity among students in order to meet the demands of the 21st century economy, he reminded educators that approaches to foster creativity among students is as equally necessary as teaching the subject content to them.

*My contention is that creativity now is as important in education as literacy and we should treat it with the same status. Robinson (2006)*

Examining the definitions of creativity in the works of Barlex (2007), Christensen, (1988), Guilford (1959), Robinson (2006), Spendlove (2005) and Torrance (1966; 1974), it is clear that it has to do with coming up with something original or novel and of value. The studies of Amabile (1982; 1988; 1996), Besemer (2010), Craft (2001), Cropley & Cropley (2010), Cziksentmihalyi (1998), Dacey & Lennon (2000), Feldman, Cziksentmihalyi & Gardner (1994), Rhyammer & Brolin (1999), and Vernon (1989) suggest that a way for physics teachers to promote creativity in the classroom is by guiding pupils through problem-solving contexts that are embedded in everyday life and which leverage on subject knowledge. The process of fostering creativity in physics amongst students can sharpen their skills in problem-solving, get them to be more inquisitive about how physics can be used to improve daily activities, and build up their confidence into thinking about how its use may value-add to the economy (Fisher, 2004, and Raviv, 2003). It could also instill the spirit of innovation among students and pave a path for them to be young inventors of our future.

In Singapore, students who do not perform academically well in the national Primary School Leaving Examinations (PSLE) are placed in the Normal Technical (NT) stream in secondary schools. While physics experiments in secondary school activity books published for NT students allow them to be engaged in learning physics through a hands-on approach, it is observed that many of these experiments lack instructional elements that would allow teachers to guide their students to showcase their creative abilities through knowledge from physics. A challenge is in crafting teaching approaches that are suitable to present physics content to these students while, at the same time, providing them with avenues to demonstrate their creative abilities through knowledge from the subject. The need to craft feasible teaching approaches to foster creativity among students during science curricula time in school cannot be at the expense of prescribed content, that is, it should be dovetailed with it. Teachers would buy-in to such teaching approaches if the activities are made appealing for students and carried out with the use of simple and inexpensive materials - in other words, it has to be a pedagogical approach that is feasible enough to be carried out in the classroom and one that NT students would find exciting and enjoyable.

Examining the literature, we find that little has been discussed about how teachers may adopt feasible classroom teaching approaches that foster creativity in physics amongst the less-academically inclined students during curriculum time. Teaching approaches to foster creativity in physics among these students, such as those in the NT stream in Singapore, may need to be different than those crafted for the more academically inclined students. This is primarily due to NT students' low levels of interest in studying physics when topics are not presented in ways that appeal to them and made relevant to their personal experiences (Amir & Subramaniam, 2009). Being predominantly kinesthetic learners, it is likely that such students can respond better to visual-spatial modes of learning rather than to visual-linguistic modes of learning (Ramadas, 2009). Findings from the works of Balchin (2005), Heacox (2002), Lee, Goh, Chia, et al. (2006), and Nunley (2006) highlight that from a teacher's viewpoint, approaches to foster creativity cannot be at the expense of presenting academic content to students.

Our observations show that apart from several schools sending only a handful of NT students to a few national science competitions, such as design-based science project competitions that are often carried out after school hours, little has been discussed on how avenues can be created for teachers to foster creativity in physics amongst NT students during science curriculum time. We also find that while physics experiments in the secondary school activity books published for NT students allow teachers to engage them in learning physics through a hands-on approach, several of these experiments lack specific instructional elements that would allow teachers to guide these students to make creative use of physics principles in coming up with novel ideas in the course of solving problems (Amir & Subramaniam, 2012). Many of the physics experiments in the secondary school activity books also seem to lack elements that can excite NT students in the learning of physics. Activities that are not appealing to these students risk the high likelihood of creating low levels of interest in the studying of physics. This may, in turn, affect their attitudes towards studying physics. In doing these experiments, students have been observed to go through a sequence of steps, as instructed by their teachers, with hardly much opportunities for them to exercise their creative skills. It has also been noted that students doing several of these experiments are not clear about the relevance of learning some of the skills and concepts taught through the physics experiment books. For example, several NT students have questioned how the skills learnt to measure the internal and external diameter of a test tube and compact disc (as described in their activity books) using a Vernier caliper would be of use in their everyday lives. It is indeed a challenge for teachers to convince these students to learn concepts and skills in physics when the content is not presented in ways that appeal to them and made relevant to their personal lives. This could be a factor that leads NT students to being somewhat unmotivated to study the subject and to teachers being somewhat frustrated in teaching them.

Based on the discussion points mentioned, we argue that while there has been emphasis in finding ways to promote inquiry in physics to NT students since the inception of the NT stream in Singapore in 1994, what seems to be lacking is the availability of feasible teaching approaches to present physics concepts and fostering creativity amongst these students through approaches that can appeal to them.

This chapter illustrates an instructional teaching approach that we crafted to present the physics concepts of density and forces in liquids and, at the same time, offering an avenue to foster creativity amongst NT students through the fabrication of variants of a simple and popular physics toy – the Cartesian diver. This toy was chosen as it can appeal to students and makes use of simple and inexpensive materials for learning physics content.

## 2. LITERATURE REVIEW

A number of articles have described how teachers can adopt feasible classroom teaching approaches to present physics concepts in contexts that appeal to students. Greenslade (2010) introduced his students to the concept of structural stability through discussing how the 'Leaning Tower of Pisa' in Italy is able to stand; he also made them think about how people are able to balance themselves when they walk on stilts. Chapman & Lewis (2001) made use of fast food (which is very relevant to the personal experiences of teenagers) to mimic an analogy on the concept of resistance in an electric circuit. Dishaw (2010) tapped on students' interest in military vehicles in presenting the concept of buoyancy of battleships. Ju (2005) made his students think about the concept of resonance through swings in a playground. Pendrill & Williams (2005), and Pendrill (2005) got students to think about how forces and acceleration are affected by the different shapes of slides and roller coasters. The concept of refraction of light to produce a rainbow was presented by Cockman, (2002), and Petterson & Williams (2004) through the common LCD projector in the classroom. The use of cartoons and movies can be a useful way to promote inquiry in physics to students (Rogers, 2007). Keogh & Naylor (1999) developed 'concept cartoons' aimed at getting students to discuss the use of science in a non-threatening context. Doherty, Rembert, Boice, et al. (1998) aimed to show that certain concepts in the popular 'Star Wars' trilogy seems incorrect - for example, concepts such as sound being able to travel in vacuum were highlighted in the movie. Rather than showing the physics concepts in certain segments of the movies, Daley (2004) encouraged them to search for movie clips to present their understanding of the physics behind these clips. Using a role-play approach to get students discover the physics that is within an activity can also lead to student engagement as well as injecting enthusiasm in the learning of the subject (Bonner, 2010; and Kofoed, 2006). Bonner (2010) describes how students learnt the concept of projectiles through a crime scene he created for them. Games also serve as a fun context to allow students to learn science –for example, Lowry (2008) described how the use of a simple, yet fun, vector game helped present the laws of gravitation to students.

Physics content can also be presented in ways that appeal to students though demonstrations. Barrett (2000) made use of two layers of liquid soap, together with toy dolphins, and placed these in a soap dispenser. The toy dolphins sink in the top colorless soap layer but float above the bottom layer (colored blue), and seem to be 'standing on their tails'. Pressing the dispenser handle causes the dolphins to 'dance'. dePino (2001) made use of marshmallow toy figurines and shaving cream in a vacuum jar for students to observe the effects of pressure. Gluck (2005) made use of a two-meter long aluminium rod to demonstrate principles of acoustics, such as pitch and loudness, to students. In one demonstration, he grasped his hand around different intervals of the aluminium rod to demonstrate the different types of sounds produced. Graf (2008) described how the use of a simple meter-rule, clamped to the edge of a laboratory bench, can be used to demonstrate the concept of projectiles – wooden blocks placed at different locations on the meter-rule can be 'launched' by bending the meter rule backwards. Upon release, students are able to see the flight of the wooden blocks to the floor and deduce the relationship between the various starting positions of take-off and distance travelled during flight. Gardner (1999) showed how a simple plastic bottle, balloon and pencil can be transformed to a pencil launcher through the concept of pressure. Shamsipour (2006) created a variation of Gardner's (1999) demonstration to foster critical thinking in students on the concept of pressure. Browne & Jackson (2007) showed how magnetized paper clips can 'react' in water to spur students' excitement in science. Froehle (2008) later came up with a slightly improved version that uses a second paper clip. Coffey (2008) allowed students to deepen their understanding of pressure through experimentation with the 'Diet Coke and Mentos' reaction in a soda bottle. Schlichting and Suhr (2010) came up with a simple variation of an old toy by using button and wire loops to make a buzzer. Dindorf (2001) showed how the concept of resonance can be exhibited through the simple use of chains (of various lengths) in the classroom.

Physics lessons can also be made appealing by infusing an 'element of surprise' in the demonstrations. Students often get captivated when teachers demonstrate concepts by presenting these as discrepant events. This is especially so when teachers are able to show how physics has been put to creative use in ways that defy natural phenomena and appear mysterious. This could trigger students into asking questions on the physics concepts that make these demonstrations work. Costa and Kallick (2000) refer to this process as one that gets students to '*respond with wonderment and awe*'. This encompasses an emotional element, which can be a powerful tool to stir curiosity and interest in learning about a particular concept. For example, Ruiz (2010) illustrates how Lenz's law can be demonstrated by dropping a brass piece and powerful magnet down a non-magnetic metal pipe. Students are able to see that the brass piece takes a very short time to drop out of the metal pipe while the magnet 'magically' takes a longer time. Examples of how concepts in refraction have been presented in interesting ways can be seen from a number of studies (Corrao, 2010; Gore, 2010; Ellenstein, 1982). Corrao (2010) explains how the archer fish is able to capture its prey by overcoming the refraction problem between water and air. Gore (2010) and Ellenstein (1982) describe how jelly marbles can 'hide' words that are written underneath an empty beaker and these words get 'unscrambled' only when water is added. Subramaniam and Riley (2008) did a demonstration with water using a glass bottle and a special plastic cap. The bottle is initially filled with water. The water does not flow out mysteriously when the bottle is inverted. To intrigue students further, matchsticks are inserted into the bottle when it is inverted (students are not aware of the plastic cap with the hole on the rim of the bottle). This demonstration gets students curious about concepts of surface tension and pressure. Featonby (2010) showed a variety of ways to introduce physics concepts through 'magical' demonstrations. In one demonstration that makes use of magnets, he inserted individual silk scarves into a circular tube. Upon pulling out the last scarf from the tube, the audience finds that all the scarves have been joined. The secret behind this demonstration is a small hidden magnet attached to the edges of the scarves.

Another way in which teachers have made physics appealing to students is by infusing a play element in lessons using toys. 'Play' is a critical issue to consider when introducing concepts as a means to make the learning of physics fun (Stables, 1997). The use of toys excites students and builds up their enthusiasm to learn physics (Güémez, Fiolhais & Fiolhais, 2009; Featonby, 2005; and Ucke, 2002). Toys are also not limited by a certain language, and tap on the kinesthetic learning style of students to promote inquiry. Books have been published to show how teachers can inject fun into their lessons by teaching science through the use of toys (McCullough & McCullough, 2000; Sarquis, Sarquis & Williams, 1995; Sarquis, Hogue, Sarquis,

Woodward, 1997; Sills, 1999; Sumners, 1997; Taylor, Poth & Portman, 1995). Rather than just getting NT students to play with toys, the experiences of educators (Kangas, 2010; McGarvey, 1995, Resnick, Berg & Eisenberg, 2000, and Thompson & Mathieson, 2001) suggest that it is possible to get these students to make toys through simple and cheap materials. Subramanaiam and Ning (2004) described a way for the concept of resonance to be taught through a toy that is made with a 40 cm x 1cm wooden rod, strings and pendulum bobs. The pendulum bob that is 'selected' mysteriously swings higher and higher than the other two pendulum bobs. The works of other educators (Featonby, 2005; Güémez et al., 2009; Planinsic, Kos & Jerman, 2004; Turner, 1983) have also shown how teachers can guide students to use a soft drink bottle (500ml or 1.5l) and a ketchup packet to come up with a simple Cartesian diver in teaching the concept of density.

The works of Meyer (2012) and Rowett (2010) highlight that a way to showcase creativity in physics would be to come up with ways to improve the functionality of a physics-based toy (a demonstration model) through contexts that are embedded in the use of physics principles. Studies of Austin & Shore (1995), Balchin (2005), Barak & Doppelt, (2000), Cross (2007), Doppelt (2005), Mackin, (1996), Slater (1996), Trumbo (2006) and Wiebe, Clark & Hasse (2001) suggest that NT students can be guided to record their learning of physics through the of use of sketches and annotations on design sheets rather than getting writing in full sentences.

Synthesizing ideas from the literature, it seems that a possible way to present physics concepts and foster creativity in physics at the same time among NT students is to use a design-based approach. It is essential to first show students a demonstration model of a physics-based toy (made from simple and cheap materials) that would appeal to them (such as the ones that infuse the 'element of surprise' in them), and then guide them to come up with variations of the toy through contexts that are embedded in the use of physics principles.

## 3. METHODOLOGY

This study was conducted with a class of 37 secondary three NT students (14 girls and 23 boys, averaging 15 years of age) through the use of a simple and popular physics toy – the Cartesian diver, which can be variously fabricated The physics toy was chosen as it makes use of simple and inexpensive materials, besides providing scope for students to be guided in making use of fundamental physics concepts to come up with variations in its context and functionality with respect to the demonstration version shown to them. The study was conducted during the physics curricula hours in a typical physics laboratory and took approximately 18 hours spread over six weeks. During this time, students were exposed to knowledge and skills across a number of topics that are within three chapters of the physics component of the NT science syllabus (measurement of mass and volume, density and forces) and one topic in the chemistry component (solute, solvent and solutions). Experience tells us that we would have saved only a little time if we had taught these topics through the use of didactic approaches, and yet not being able to generate as much interest and foster creativity in physics amongst these students.

## Procedure

There are two phases in the study. The first phase aimed at presenting a number of physics concepts and skills that are involved in making a Cartesian diver. The second phase aimed to equip students with problem-solving skills, such as brainstorming, and the introduction of the design process that would allow them to be familiar with several design principles in coming up with their own versions of the Cartesian diver. The latter included guiding them to merge fundamental physics concepts in the design and fabrication of the Cartesian diver.

Students were required to record their learning of physics concepts and skills gathered through this activity and to provide descriptions on the functionality of their designs through sketches and annotations in their design sheets (blank A4 papers) that would be compiled to form their portfolios.

## Phase One

In this phase, students were shown a simple Cartesian diver (made from plastic bottle and ketchup packet in tap water) (Figure 1) and given a demonstration of it in action.

The students were first made to think of how the diver (ketchup packet) in this demonstration version works. They were shown a video clip of two oranges – one that had its skin peeled and the other un-peeled. The students were made to think why the orange with the peeled skin sinks while that which is un-peeled floats. Many of them were able to reason that the un-peeled orange can float because there is a layer of air trapped between the fruit and its skin. They were then made to understand that air in the ketchup packet gets compressed when the bottle is squeezed. Reducing the amount of air would affect the buoyancy of the ketchup packet and cause it to sink. In order to convince them that air, being a gas, can be compressed while ketchup, being a liquid, cannot be compressed, students were told to work in pairs to separately explore the compressibility of air and liquid in a syringe. Students were then shown, with the aid of a video clip, how a submarine works. This was done to facilitate the linking of the concept to a real world context.

*Figure1. Cartesian diver demonstration model shown to students*

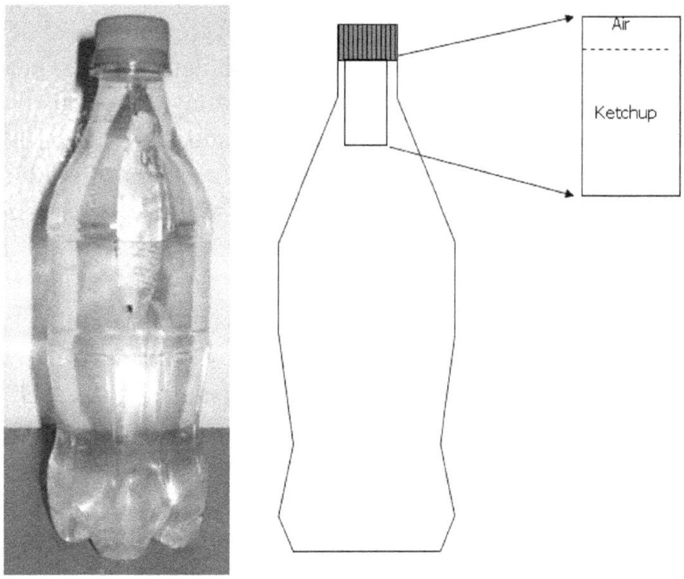


*Figure 2. Difference in buoyancy between two ketchup packets*

Another type of ketchup packet was then distributed to them – the type that has very little air in it. These packets are not able to float in tap water. The students were shown the difference in buoyancy between the ketchup packet that was used in our demonstration version and the one they received, through two beakers filled with water (Figure 2).

A brainstorming session was carried out to elicit answers from the students on how they would make the ketchup packets given to them float. As a build up to their thinking, video clips of people floating in the Dead Sea were shown. A discussion then ensued on why people were able to float in the Dead Sea but not easily in sea water in Singapore. It became apparent to the students that the Dead Sea has a high concentration of salt and this contributed to the high density of the sea water in it. Leveraging on this knowledge, salt packets and spatulas were distributed to them. They were told to keep adding salt to the water in the plastic beaker (containing their ketchup packet) and to keep stirring until all the salt has dissolved. They repeated this process and observed the buoyancy of the ketchup packet in the salt solution. The process was stopped only when their ketchup packet was able to float in the salt solution. They then transferred the salt solution and ketchup packet into a 500 ml soft drink bottle and observed the effect of their ketchup packets working as Cartesian divers in the salt solution. Along the way, students were introduced to the formula (Density = Mass ÷ Volume) as well as taught skills such as using an electronic balance and measuring cylinder. It became apparent to them that they would require these skills in order to gather the necessary data to calculate the densities of tap water and salt solution.

Two souvenirs (Figure 3) that had layers of liquids of various densities also served as teaching aids to deepen students' understanding of the concept of density. These souvenirs were used to get them to think about how the toy dolphins were

*Figure 3. Teaching aids on the concept of density*

able to float on the blue liquid layer and yet sink in the colorless liquid layer. When asked which of the two liquids had water and oil respectively, it was observed that several students were somewhat reluctant to mention that the colorless liquid is oil while the colored liquid is water. A possible reason is that students often associate water as a colorless liquid and oil as a colored liquid. In order to convince them that the colorless liquid in the souvenir is oil and that the colored liquid is water, a few drops of blue food coloring was dropped into a beaker half-full with tap water. The solution was stirred for a while. Johnson's baby oil (which is colorless) was then poured into the beaker. Again, the solution was stirred. It was not long before the layer of oil floated above the layer of the colored water. It was at this point that the students were convinced that the colorless liquid in the souvenir was indeed oil and that the blue solution is a more dense liquid, such as water.

The students were also made to realize that the metal piece attached to the base of each figurine in the toy (dolphins and Merlions) was necessary to make these stay upright. To reinforce their understanding of this concept, each student was given a piece of straw of about 5 cm in length. They were told to place the straw in a beaker filled with water and observe the way it floats on its side. Small lumps of plasticine were then distributed to each student. Students experienced how the amount of plasticine played a part in keeping the straw upright and how it contrib-

*Figure 4. Self-made 'hydrometer' floating upright in tap water*

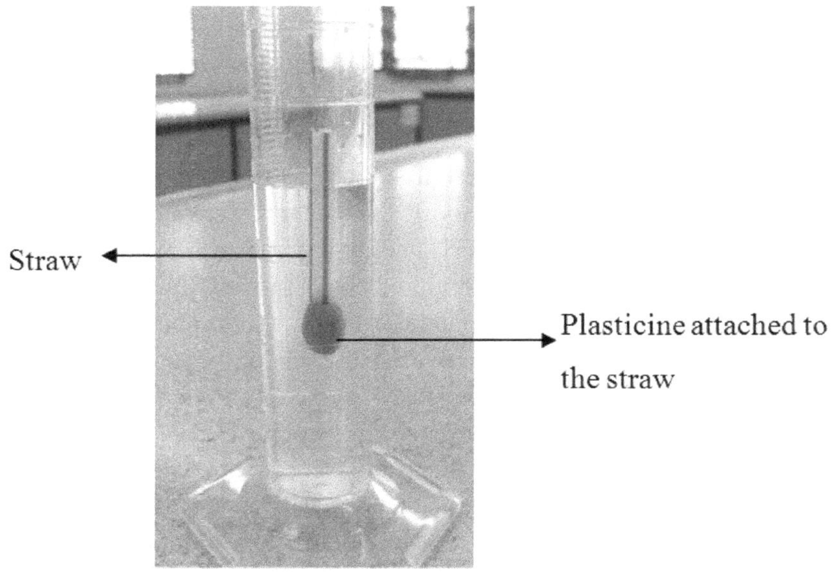

uted to the depth of the straw floating at the surface of the liquid. Most students were able to make connections that the plasticine depicted the metal piece in the toy dolphins that was required to keep the toy upright. Figure 4 shows a 'hydrometer' made by a student and which floats upright in a measuring cylinder containing tap water.

The students were then asked to observe the differences in depths of their 'hydrometer' in tap water and in various concentrations of salt solutions. This part of the activity made them realize that the depth of the 'hydrometer' depends on the concentration of salt that is dissolved in the liquid. Students were made to reuse the water in the measuring cylinder by pouring it into a plastic beaker each time they needed to remove their 'hydrometer'. They were also shown how a hydrometer played a part in checking the acid level in a car battery. This was done to show how this device is being put to use in the real world.

## Phase Two

In this phase, students were formed up in groups of threes. Each group was challenged to come up with an appealing version of the Cartesian diver using contexts embedded in the application of physics principles. A brainstorming session was carried out to elicit answers from students on various objects, other than ketchup packets, that could trap air and float in a soft drink bottle filled with water. To aid the brainstorming session, they were given a few recycled items as well as other materials commonly available in the physics laboratory. These included 'bendy' straws (straws with frills that allow a portion of it to bend, and commonly attached to drink packets), pen caps, eye droppers, balloons, etc. The students were guided to make use of the knowledge gained from the 'hydrometer' activity to think about how they could make these objects float upright in water and how they would provide a context that would appeal to the desired users of the toy.

As a trigger for this part of the activity, the students were shown the design of a hook Cartesian diver (Figure 5) fabricated by a group of NT students (Amir & Subramaniam, 2007). A brief description of how the hook Cartesian diver functions is now given. To bring the pen-cap hook down, a player needs to squeeze the bottle and maneuver the pen-cap hook to grab as many of the loops as possible. By controlled squeezing of the bottle, it is possible to control the motion of the hook in order to grab a number of loops within a stipulated time.

The design shows how students in that study were able to value-add to the demonstration version through the application of the physics concepts of forces (upthrust and downthrust), knowledge from materials science (plasticine weights wrapped with aluminium foil so that these do not stick to the pebbles), and through the use of a game context – one that appeals to children. This design served as an inspira-

*Figure 5. Infusing a game element into the hook Cartesian diver design*

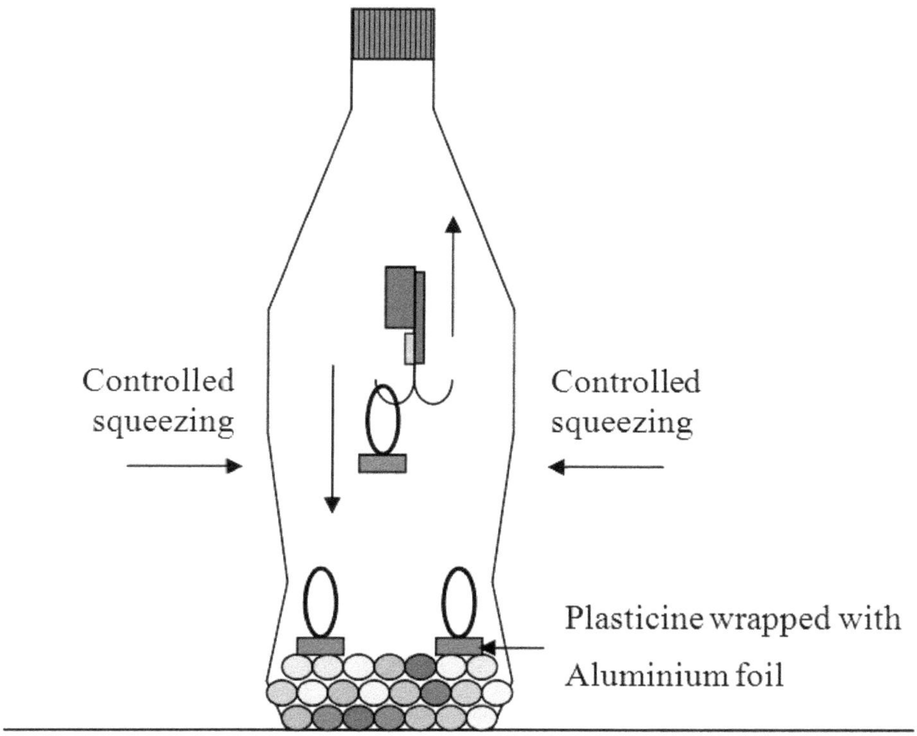


tion for the class to manipulate other simple materials in trying to come up with their own designs. It was clear that the students were discussing concepts of density in more depth as they brainstormed and discussed their ideas through sketches in their design sheets and by testing the buoyancy of simple materials. While facilitating the lesson, the students were seen discussing about the need for an object to 'trap' air so as to be able to float and also the need to compress air in order to reduce the buoyancy of the object so that it could sink.

## SURVEY INSTRUMENT

To survey the views of students who participated in this study, a survey instrument comprising 20 statements, placed on a 6-point Likert Scale (Strongly Disagree =1 and Strongly Agree = 6), was developed. There were four categories in the survey instrument:

1. Connections created between physics and design through the activity
2. Use of design sheets as a way to articulate creativity in physics
3. Fabrication of toy as a way to articulate creativity in physics
4. Attitudes towards learning physics after the design activity

Twenty statements were framed and distributed equally into the four categories. The first version of the instrument was face-validated by a few teachers, heads of departments, specialists from the Ministry of Education, and university staff. It was based on an earlier study involving the use of Design & Technology (D&T) projects (Amir & Subramaniam, 2012) but revisions were done to some of the statements in the instrument for use in this study.

## RESULTS

A number of creative versions that showed NT students' abilities to demonstrate their creativity in physics were evident through their design sheets and prototypes. We describe a few of them through Figures 6-18, along with a commentary.

*Figure 6. Cartesian diver with transparent pen cap*

## 1. Applying Knowledge from Optics to Show Compressibility of Air in the Diver

A group of students made use of their knowledge of compressibility of air (taught in the first phase) and linked it to principles in optics (transparency) in order to value-add to the design of the hook Cartesian diver that was shown earlier. Figure 6 illustrates how this group made use of a transparent pen cap instead of an opaque one. Users of this toy will be able to see the air being compressed each time the bottle is squeezed. Such a design can also be used as a teaching aid to show the compression of air when the bottle is squeezed.

A student in this group was also able to describe in one of his design sheets that '*only a little water can be seen*' before squeezing the bottle and that *'air compresses'* when the bottle is squeezed. An extract from the design sheet is shown in Figure 7.

## 2. Applying Knowledge from Fluid Movement to Produce a Spinning Effect on the Diver

Another group of students came up with an idea, which led to the development of a pen-cap hook diver that can spin as it dives and surfaces. Figures 8-9 show sketches

*Figure 7. Description of compression of air in a student's design sheet*

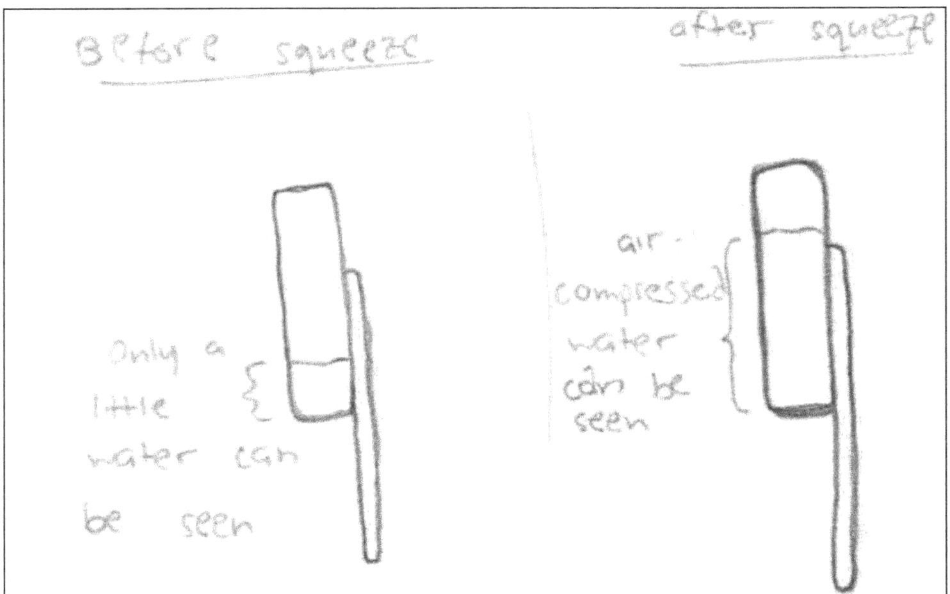

*Figure 8. Explanation on how the design was conceptualised from a propeller*

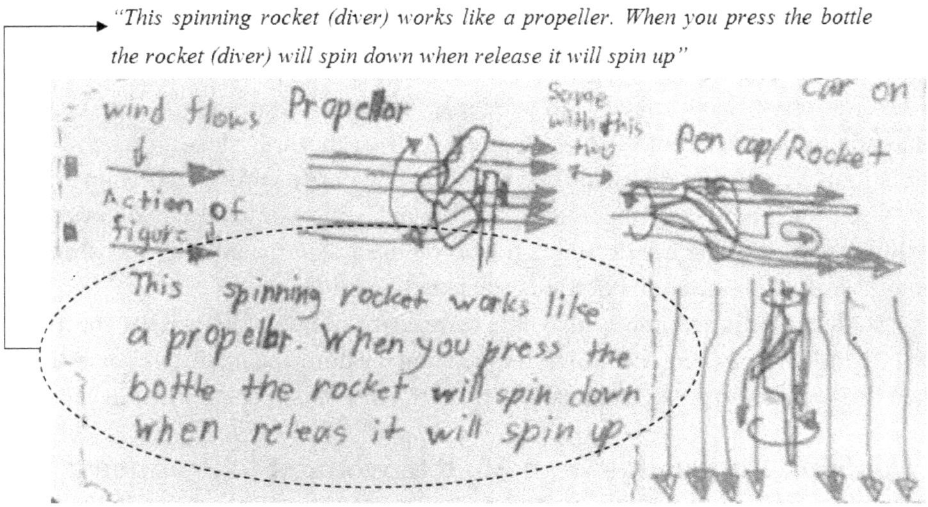


*Figure 9. Replicating the shape of the propeller onto their pen cap in the form of plasticine*

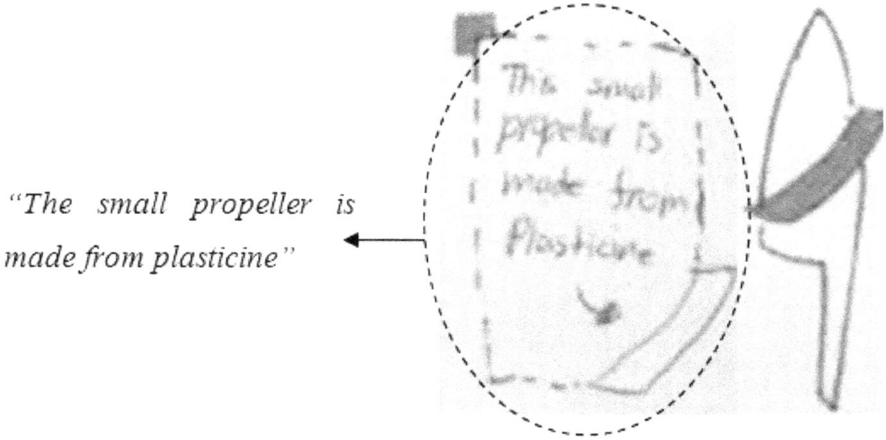


of how the group conceptualized this idea through the shape of a propeller attached to a boat. Sketches and annotations from a student's design sheet show how the shape of the propeller was formed in a simple way through the use of plasticine and attached to a pen-cap in order to make it spin. A sketch of how the pen cap spins, after the plasticine is attached to it, is shown in Figure 10.

*Figure 10. Sketch of how the 'spinning' diver moves in bottle*

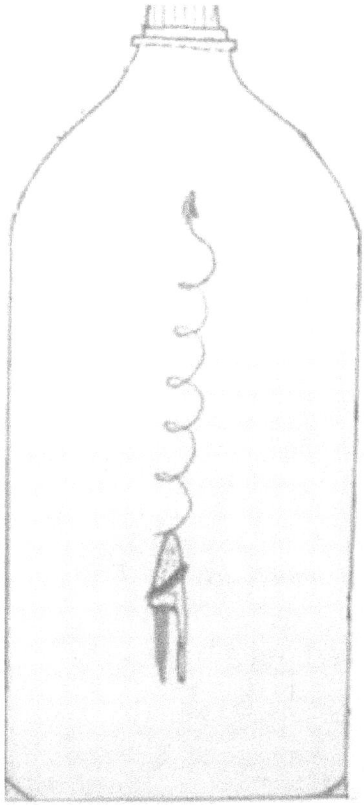

The students also mentioned that their pen cap had a hole at the top. This led to a risk of air not being able to be trapped in the pen cap. Water could flow in through this hole and cause the pen cap to sink. The group overcame this problem by placing a little plasticine to seal the hole at the top of the pen cap. Figure 11 illustrates this.

The above design inspired another group of students to come up with a similar design but making use of a larger amount of plasticine attached to the pen-cap-hook in a twirling fashion (Figure 12).

Another group of students searched for ideas from YouTube and was inspired to make the diver spin without the use of plasticine. They used a 'bendy' straw instead to make their version of a spinning diver. The 'bendy' straw was first trimmed to a shorter length such that parts that are not 'bendy' are of equal lengths. A washer (acting as weight) is inserted and made to rest in the bendy part of the straw. The two ends of the straw are sealed through the use of a flame. Two small holes

*Figure 11. Sketches and notes showing how the group ensured that air was trapped in their pen cap*

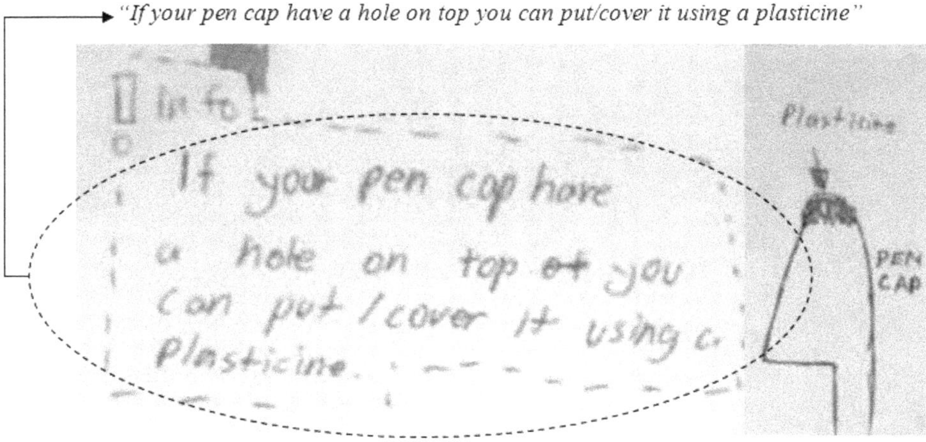


*Figure 12. Twirling pen-cap-hook Cartesian diver*

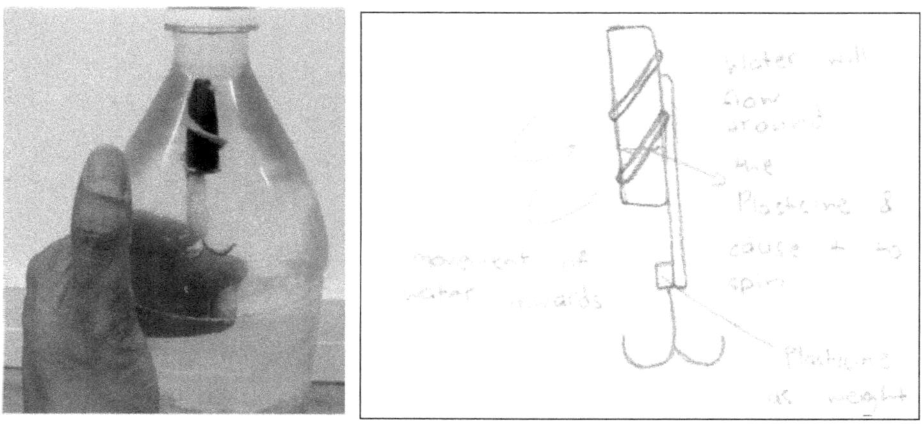

are made diagonally opposite each other on each end of the bent part of the straw. The students then tested the performance of this diver in a bottle filled with tap water. When squeezed, the air in the straw compresses. Upon release, the water that comes out from the small holes at the diagonal ends of the straw causes it to spin.

The students were able to describe this design through their design sheets. One of the student's design sheets (Figure 13) shows his ability to elaborate on a number of science concepts that are involved in this design. An extract from this design sheet

*Figure 13. Description of a student's version of spinning diver through the movement of fluids*

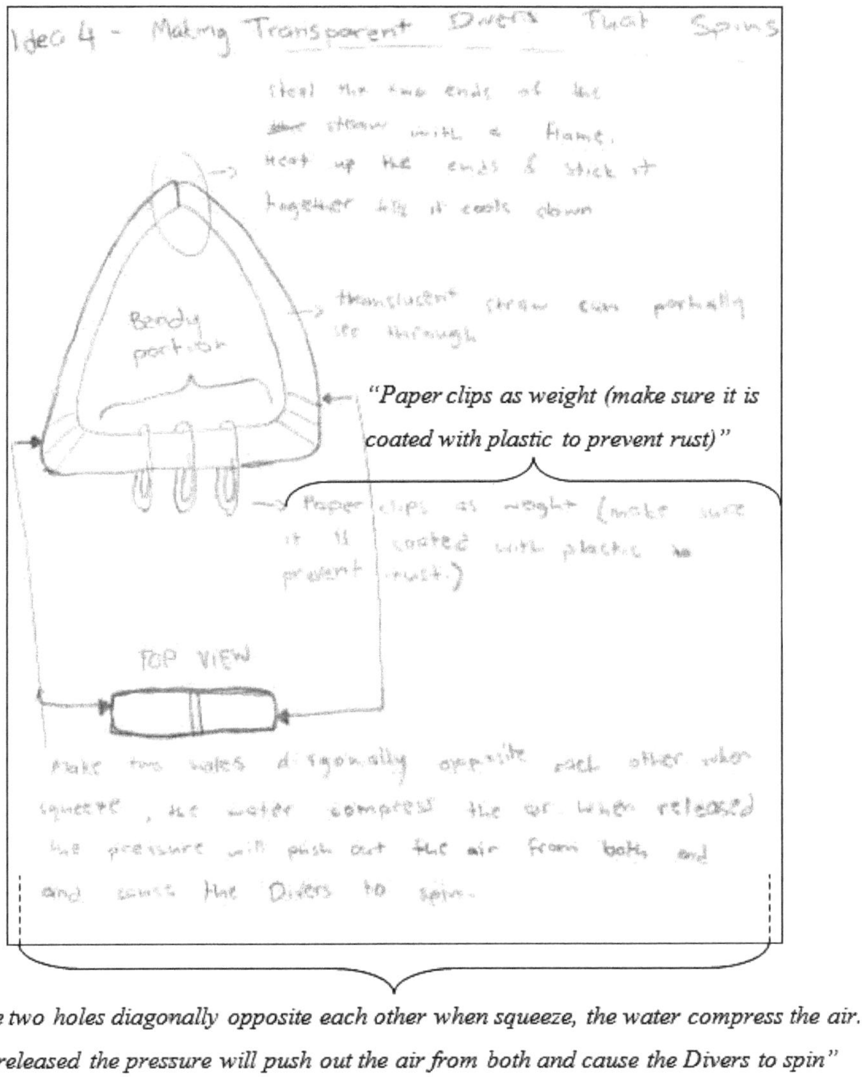


reads *'Paper clips as weight (make sure it is coated with plastic to prevent rust)'*. The student has shown evidence of his group's knowledge of physical properties of metals in terms of weight and type of surface. The student explains how the use of a few paper clips can be used to substitute for a single metal washer. He was also able to explain the need for such paper clips to be *'coated with plastic'* so as to prevent the formation of rust. The functionality of his group's design was clearly described in the design page. He mentions that the two holes that are created *'diagonally opposite each other'* on the bendy portion of the straw will cause the diver to spin.

Seeing the effect of several of the spinning divers spurred another two groups of students to come up with an entertaining version of yet another spinning Cartesian diver. By tapping on their skills picked up from making the simple 'hydrometer' from straws and plasticine in the first phase and linking it to the skills in making a spinning diver that was shown in Figure 13, these groups were able to come up with two other designs of Cartesian divers that illustrate the effect of two dolls dancing hand-in-hand with each other. One group's 'dancing dolls' design is shown in Figure 14. In both 'dancing dolls' designs, two straws of equal length are sealed with plasticine on one of their ends while the other ends are sealed through the use of a flame and hot glue. Each doll is linked to another doll through paper clips, which depict the 'hands' of the dolls. The functionality of the toy is practically the same as the one shown in Figure 13 except that a hole is now made only

*Figure 14. 'Dancing dolls' Cartesian diver*

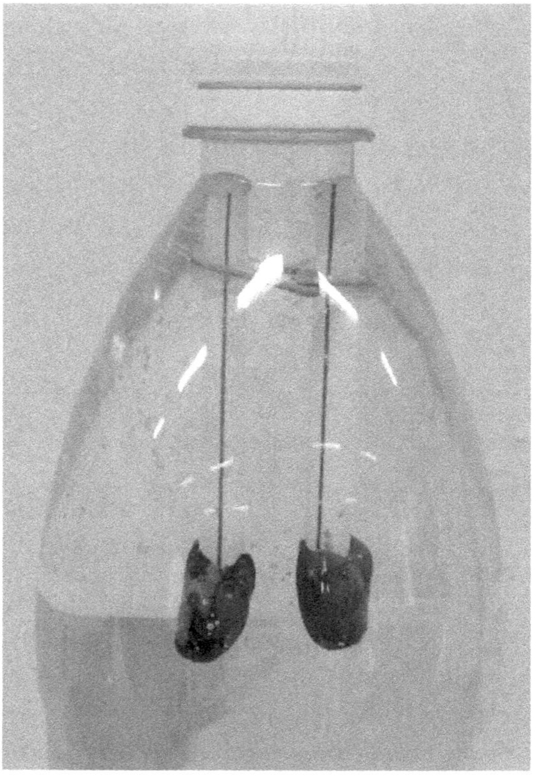

on one straw but still diagonal to each other on the opposite ends. The effect of the dolls dancing in these two designs brought a smile on those who viewed the version!

A student in one of these groups was able to elaborate on the fabrication techniques of her 'dancing dolls' design. It is clear that the student was able to explain the skills required to keep each doll afloat and to make the 'dancing doll' pair spin. She was able to explain these skills by making use of physics principles, such as bending the paper clips to make them resemble the hands of the dolls (concept of material properties), adding the right amount of plasticine to keep the doll afloat (concept of density) and getting the 'dancing doll' pair to spin (concept of movement of fluids). She also described her experimentation with a straw-and-plasticine hydrometer in a plastic beaker filled with water as part of getting each 'doll' to float upright. An extract of her design page (Figure 15) shows her reflections on the importance of having the right amount of weight added to the straw to keep it upright.

She also showed that a key consideration to make the dolls spin is that there should be only one hole on each straw (she mentioned how this could be achieved through the use of an optical pin in another design page). The description of her design is shown in Figure 16.

## 3. Applying Knowledge from Magnetism and Placing It in the Context of a Popular TV Character: 'Paul the Octopus'

Another group of students was inspired by the popular 'Paul the Octopus' sea creature that was made famous during the 2010 FIFA™ World Cup football tournament. This group made an entertaining version of the Cartesian diver that depicts the octopus predicting the outcome of a game. The group started off by conducting an experiment to investigate the possibility of using a balloon-and-paper-clips combination as a diver. When placed in a 1.5 litre bottle filled with water, the octopus head traps air inside it and causes it to float. The plastic-coated metallic paper clips depict the tentacles of the octopus while, at the same time, provide weight for the octopus to sink. Simple decorations (with a marker) were added to the balloon to make it depict the head of an octopus. This design is illustrated in Figure 17.

The group then made use of corrugated boards, cardboards, bamboo sticks and Velcro to fabricate the rest of their design (Figure 18). A small bamboo stick is inserted through a slot on a piece of square corrugated board (about 5cm x 5cm) and made to protrude at both ends. The 1.5 liter bottle containing the octopus diver is then placed onto the square corrugated board. Two other corrugated boards, each folded into a rectangular box (large enough to store a bar magnet) and which has part of its surface layered with a small piece of Velcro, is linked to the square corrugated board through small bamboo sticks. A bar magnet is hidden in one of these

*Figure 15. Description showing how the hydrometer is kept upright*

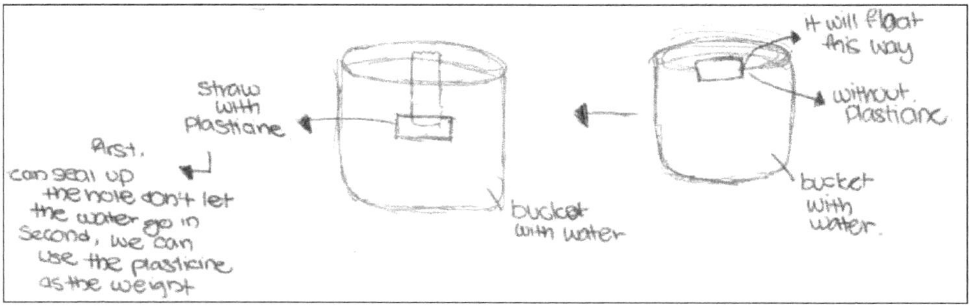


*Figure 16. Description of the 'dancing dolls' Cartesian diver*

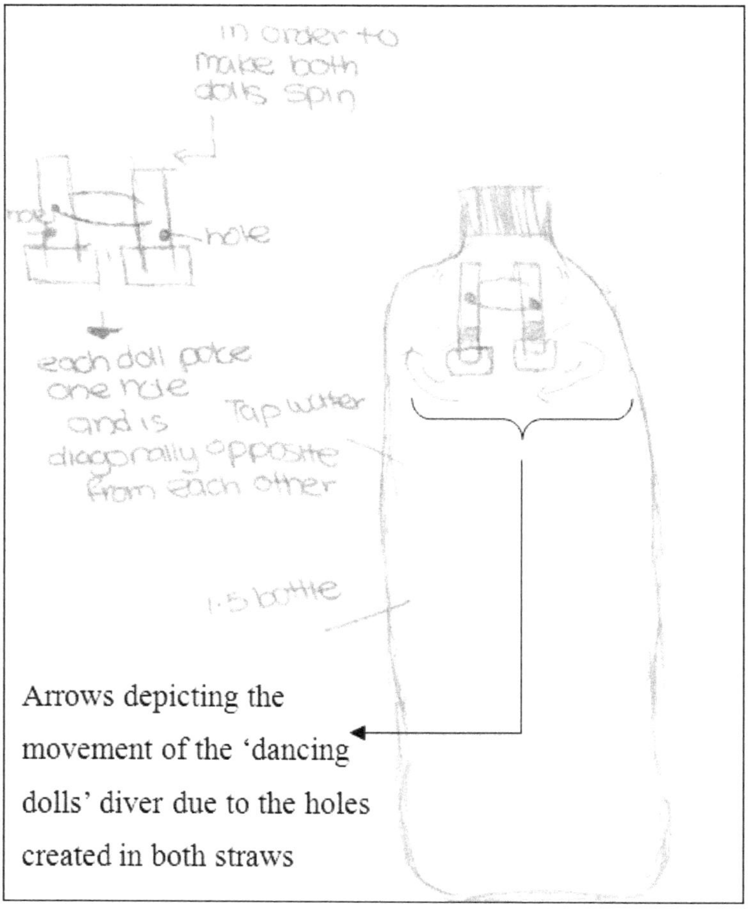

*Figure 17. Octopus diver made from balloon and paper-clips*

*Figure 18. Design of 'Paul the Octopus' Cartesian diver*

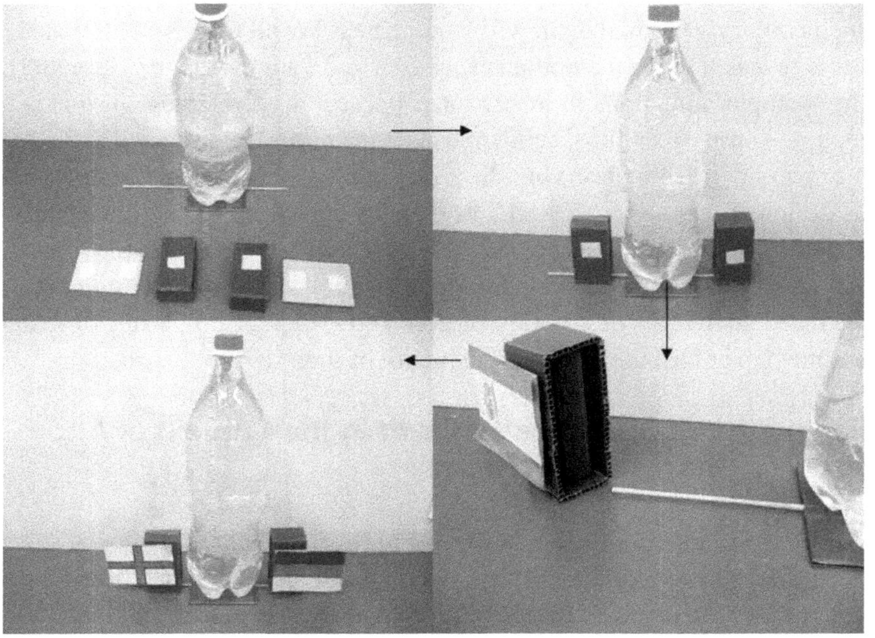

*Figure 19. Functionality of 'Paul the Octopus' Cartesian diver*

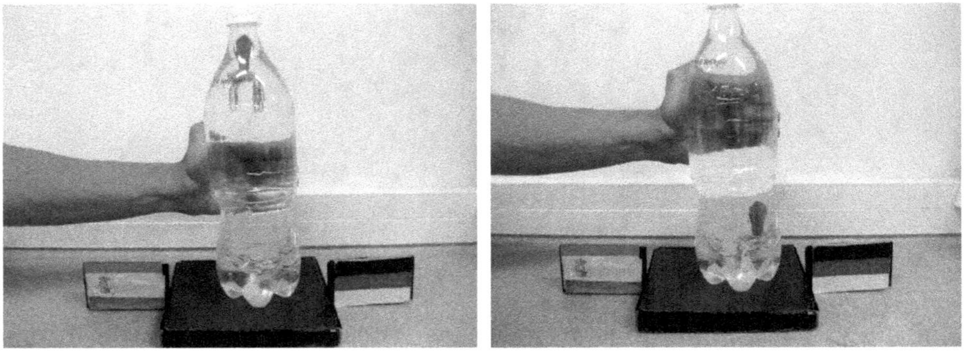

boxes, which is not obvious to the user of the game. Two rectangular pieces of paper, decorated as flags of different countries through the use of colored markers and stickers, are pasted onto similar sizes of rectangular cardboards. Small strips of Velcro are then attached behind these cardboards. These cardboards are mounted on to the rectangular boxes through Velcro strips. The final design depicts 'Paul the Octopus', ready to make a prediction on the country that will win a football game.

In Figure 19, the magnet is hidden very close to the German flag. To demonstrate the functionality of their prototype, the students in this group placed their toy in front of the class and asked their classmates to make a guess on Paul's prediction as to which country's football team will win the next World Cup football match. The answer is revealed when the bottle is squeezed and Paul moves the German flag.

The 'octopus' functions in this manner because its 'tentacles' are made from paper clips. These paper clips, being magnetic materials, are attracted to the hidden bar magnet. Apart from observing the amazement of the class at the functionality of this design, it was also observed how this design was able to stimulate curiosity in the students to think about how the use of magnetism has been placed in a context that appeals to children and adults. Figure 20 shows an extract from a student's design sheet, showing a description of how knowledge of properties of materials and magnetism contribute to the functionality of their design.

## 4. Placing the Hook Cartesian Diver in the Context of Another Popular TV Character: 'Mr. Bean'

Another group of students made a hook diver design similar to the version shown in Figure 5. They aimed to make their design appeal to users not just through a game context but also in a context familiar to users. They rode upon the popular TV char-

*Figure 20. Description related to 'Paul the Octopus' Cartesian diver*

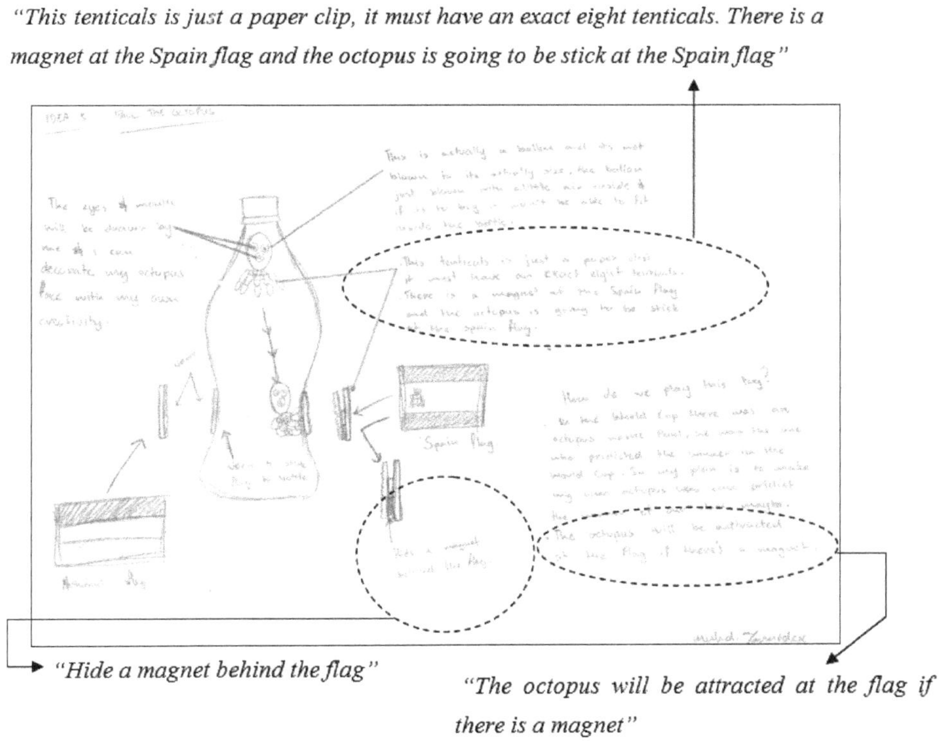


acter – 'Mr. Bean' to come up with their design. One of the group members had a toy figurine of Mr. Bean dressed up in only swimming trunks. The group attached this figurine to the top of a bottle cap through the use of hot glue. A small bamboo stick with an end attached to a string is secured to the top of the bottle cap in an inclined position through the use of hot glue. A string is attached from the top of the pen-cap-hook diver to the base of the bottle cap. This gives a realistic resemblance of a fishing line that goes down with the fishing hook each time the bottle is squeezed. The Mr. Bean-bamboo-stick-string combination depicts Mr. Bean standing in front of a fishing rod with the fishing line in a pond. The bottle cap is screwed onto a 500 ml bottle filled with water and Mr. Bean's 'clothes', which comprises trousers, shirt, shoes and even his favorite teddy bear, were made from clipart images downloaded from the Internet and printed on paper. The group laminated the printed papers to ensure that the images are waterproof. The challenge given to users of this toy is to try and 'fish out' Mr. Bean's clothes within a stipulated time frame, just like the hook diver version shown in Figure 5. Users of the toy were able to appreciate this

*Figure 21. Description on the functionality of 'Mr. Bean Finding His Clothes' Cartesian diver*

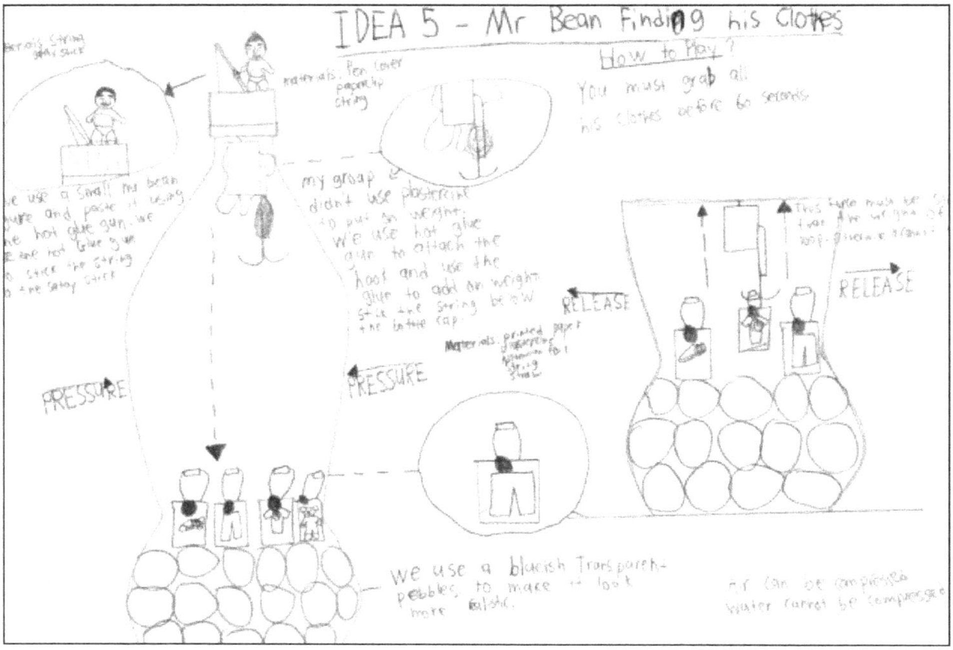


since they were able to see the movement of the fishing line through the transparent walls of the bottle.

The group entitled their design 'Mr. Bean Finding His Clothes'. An extract from a student's design page is shown in Figure 21. He was able to elaborate on the functionality of his group's design.

## VIEWS OF STUDENTS ABOUT THIS ACTIVITY

The survey instrument, which was administered to the students after the activity, had a Cronbach alpha of 0.82, which means that the reliability is above the recommended norm of 0.70 (Cronbach, 1970). Table 1 presents simple descriptive statistics for the survey.

Written feedback from student's after the activity are now presented.

*I enjoy making things through hands-on instead of reading the textbook. The textbook is a boring way of learning. It makes me sleepy while the portfolio is easier to let*

*Table 1. Students' responses through the instrument*

| | Categories | Mean | SD |
|---|---|---|---|
| | **Connections created between physics and design through this activity** | | |
| 1. | I learnt a physics concept through this design activity. | 5.82 | 0.39 |
| 2. | This design activity has made the learning of physics fun. | 5.94 | 0.24 |
| 3 | This activity has allowed me to learn skills to design a physics-based toy. | 5.91 | 0.29 |
| 4. | This activity has allowed me to learn skills to construct a physics-based toy. | 5.97 | 0.17 |
| 5. | This activity has helped me increase my understanding of several physics concepts. | 5.91 | 0.29 |
| | **Use of design sheets as a way to articulate creativity in physics** | | |
| 6. | I am able to describe the design context for my toy in my design sheets. | 5.85 | 0.44 |
| 7. | I am able to describe at least three different ideas in my design sheets before choosing my final design. | 5.74 | 0.90 |
| 8. | I am able to come up with at least one new idea that works on physics concept(s). | 5.94 | 0.24 |
| 9. | I am able to describe how to make my chosen idea using simple materials in my design sheets. | 5.85 | 0.44 |
| 10. | I am able to describe how physics concepts make my chosen idea work in my design sheets. | 5.91 | 0.29 |
| | **Fabrication of toy as a way to articulate creativity in physics** | | |
| 11. | I find that I get new ideas when my teacher exposes me to more physics concepts. | 5.97 | 0.17 |
| 12. | I find that I am able to apply more than one physics concept into my toy during the stages of constructing my toy. | 5.91 | 0.29 |
| 13. | I find that I am able to continuously improve my design using physics concepts during stages of constructing my toy. | 5.88 | 0.41 |
| 14. | I get my toy to work by combining what I have learnt in my physics lessons to what my teacher has taught me about design. | 6.00 | 0.00 |
| 15. | I made my toy in such a way that it will make other students curious to want to find out how physics concepts make it work. | 5.91 | 0.29 |
| | **Attitudes towards learning physics after the design activity** | | |
| 16. | I am interested to do more design projects that involve physics concepts. | 5.50 | 0.51 |
| 17. | I am more confident of learning physics after doing this design project. | 5.91 | 0.29 |
| 18 | I find that design is useful in helping me do well in physics. | 5.85 | 0.36 |
| 19. | I am able to show how creative I am by going through this physics-based design project. | 5.82 | 0.46 |
| 20. | I am further interested to show how creative I can be by doing similar physics-based design projects in the future. | 5.88 | 0.33 |

*our ideas to be drawn out. This way shows that NT students like me are creative and not lazy! - Female student*

*I enjoy learning physics this way because it is interesting. This project is fun and helps me learn more physics.- Female student*

*I think this way of teaching physics is more fun than learning through the normal way. We can also prove that we are able to learn. Think that all NT students should be learning physics through this way rather than from the textbook. - Male student*

*I am happy about physics. We want to prove that Normal Technical students are not stupid. If we continue learning like this, our physics can be very fun and we will do well! Textbooks are boring. Doing projects in school is totally fun!!! - Male student*

*I find physics amazing when we are given a chance to conduct our own experiments to make our chosen idea. Thank you cher (teacher)! You have made me enjoy physics! - Male student*

*I like the part when we do the Cartesian diver in our groups. I hope all the NT students in Singapore can learn this way. I am so proud to be an NT student! - Male student*

*Look at how creative we can be through this project! Tell them (the more academically inclined students) please don't look down on us NT! - Male student*

Results from Table 1 and comments from the students suggest that they enjoyed the activity very much. Most students are in strong agreement that they were able to make their toys work by linking their knowledge of fundamental physics principles to design principles (statement 14). This approach has contributed to their being able to understand a variety of physics concepts (statement 5), which in turn also led them to be more confident in learning physics (statement 17). Many students gave appreciative comments such as *'Thank you teacher!'* after the activity. They seem to appreciate being given the space to express their ideas through sketches and annotations in their design sheets and being able to experiment with these ideas during their physics lessons, as evidenced by the high scores and low standard deviations in the second and third categories. Several students, who often have difficulty in expressing their understanding of physics through verbal means and writing, found this approach effective in incorporating physics into their projects, as evidenced by a maximum score of six (with zero standard deviation) for statement 6.

However, not all students were able to describe at least three different ideas in their design sheets before choosing their final designs, as evidenced by a standard deviation of 0.90 for statement 7. This is also seen through the number of creative prototypes produced by the students. From the 12 student groups, about seven or eight designs can be said to be creative – this is still an impressive figure. Based on Torrance's (1979) framework for creativity (*fluency*, *flexibility*, *originality* and *elaboration*) and indicators of *originality* trait of creativity in products from Besemer's CPAM (Besemer, 1998), it was found that only two prototypes were original in terms of the functionality of the toys through the use of physics principles (dancing dolls and Paul the octopus). The others seem to be variations of the demonstration version shown to them or variations of ideas picked up from the Internet. However, this should not detract from the more fundamental import of the activity, which is to present a platform for NT students to learn and do physics in contexts that are appealing to them.

Physics teachers in the school liked this approach as it shows how students can be engaged to incorporate physics principles into prototypes so as to value-add to their functionality. The students also felt proud when other teachers gave them compliments after viewing their design pages and prototypes. Their morale was boosted even further when a few physics teachers brought some of these prototypes and design pages to their own physics classes and used them as demonstration kits to showcase how physics has been put into action in these toys and also used them as teaching aids to present a few topics in physics (such as density, fluid mechanics, optics and magnetism). The positive performance of the students seems to debunk the common assumption that the less-academically inclined students are not able to perform well when given problem-solving activities in physics.

## DISCUSSION

This study has suggested that a way to tap on the kinesthetic learning styles of less-academically students as well as foster creativity in physics among them is through a design-based classroom activity that appeals to them. This was done over two phases.

Synthesizing ideas from the discussion points in the literature, a possible way to present physics concepts and foster creativity in physics at the same time amongst NT students is to first show the students a demonstration version of a physics toy (made from simple and cheap materials) that would appeal to them (such as the ones that infuses an 'element of surprise' in them), and then guide them to come up with variations of the toy through contexts that are embedded in the use of physics principles.

This study has also shown that guidance by the teacher is a key feature in fostering creativity in physics amongst these students. Hérold & Ginestié, (2009) emphasized that guidance remains an important aspect of teaching in order to get students to perform well in design-based activities. Kirschner, Sweller & Clark (2006) warns that the lack of guidance in problem-based learning activities that aim to promote inquiry in physics may lead to students not learning the underlying learning physics concepts as well as they would have if they had been exposed to direct instruction.

Materials that were used in the activity were simple everyday items that are familiar to the students. They were able to make use of such materials to improve the functionality and appeal of the demonstration model that was shown to them. The creative manipulation and use of these simple materials to come up with such toys is in itself a testimony of these students being able to be trained to be creative thinkers in physics.

In the activity, care was taken not to 'spoon-fed' interesting ideas to students. Rather, the emphasis is on capitalizing on brainstorming activities to get them generate ideas on their own. The use of one creative version of the Cartesian diver – the hook diver has been used to trigger ideas in the students to come up with more creative versions. Along the way, in presenting the hook diver, students were encouraged to think about how they could embed elements that would appeal to younger children and adults through the use of simple materials. It was crucial that students had an opportunity to make the first version of the Cartesian diver through the ketchup packet and salt solution as this allowed tapping on the simple skills that they are familiar with. Doing this also helped them to develop confidence in learning the concept of density further through the prototypes.

## CONCLUSION

Norman (1993) encouraged physics educators to think about how physics can be presented to students through contextualized design-based activities. Jones & Moreland (2003) further stressed on the need for such activities to be carried out from 'bottom-up', classroom-based approaches and for educational researchers to show how such approaches can be carried out during lesson hours for the average busy teacher through descriptive case studies. A number of articles in the literature have addressed these needs and have shown that design-based learning have led to students being more motivated towards the learning of physics (Doppelt, et al., 2008; Fortus, et al., 2004; 2005). Some articles have also described how the adoption of design-based teaching approaches could provide a platform for teachers to craft lessons that would allow content and skills that are within the physics and mathematics syllabus to be presented to their students in ways that would 'make sense' to them (McCormick, 1997; Silk, 2010). Doppelt et al. (2008) and West (1997) further mention that such approaches have proven to be most helpful for low achieving students to understand physics and mathematical concepts.

What remains lacking is the number of case studies to show how feasible teaching approaches can be used to present physics content while at the same time fostering creativity in physics amongst the less academically inclined students, such as NT students in Singapore. We feel that what teachers need are feasible classroom teaching approaches that make use of simple and inexpensive materials. This study has shown how such a need can be significantly addressed. Students who went through the activity exhibited positive attitudes towards the learning of physics and just as importantly - their ability to prove to their parents, friends, teachers and themselves, that they too can be creative in physics.

Approaches such as those shown in this study can be used by classroom physics teachers to present physics content in interesting ways to the less academically inclined students and, at the same time, develop their problem-solving skills, which can lead them to showcase their inventive abilities. These are necessary competencies to be instilled amongst students in preparing them to meet the challenges of the 21st century.

## REFERENCES

Amabile, T. M. (1982). Social psychology of creativity: A consensual assessment technique. *Journal of Personality and Social Psychology*, *43*(5), 997–1013. doi:10.1037/0022-3514.43.5.997

Amabile, T. M. (1988). A model of creativity and innovation in organizations. *Research in Organizational Behavior*, *10*, 123–167.

Amabile, T. M. (1996). Creativity and innovation in organizations. Harvard Business School.

Amir, N., & Subramaniam, R. (2007). Making a fun Cartesian diver: A simple project to engage kinaesthetic learners. *Physics Education*, *42*(5), 478–480. doi:10.1088/0031-9120/42/5/004

Amir, N., & Subramaniam, R. (2009). Making a low cost candy floss kit gets students excited about learning physics. *Physics Education*, *44*(4), 420–428. doi:10.1088/0031-9120/44/4/013

Amir, N., & Subramaniam, R. (2012). Fostering inquiry in science among kinaesthetic learners through design & technology. In L. C. Lennex & K. F. Nettleton (Eds.), Cases on Inquiry Through Instructional Technology in Math and Science: Systemic Approaches (pp 221 - 257). Hershey, PA: IGI Global Publishing.

Austin, L. B., & Shore, B. M. (1995). Using concept mapping for assessment in physics. *Physics Education*, *30*(1), 41–45. doi:10.1088/0031-9120/30/1/009

Balchin, T. (2005). A creativity feedback package for teachers and students of design and technology (in the UK). *Design and Technology Education: An International Journal.*, *10*(2), 31–43.

Barak, M., & Doppelt, Y. (2000). Using portfolios to enhance creative thinking. *Journal-of-Technology-Studies*, *26*(2), 16–25.

Barlex, D. (2007). Creativity in school design and technology in England: A discussion of Influences. *International Journal of Technology and Design Education*, *17*(2), 149–162. doi:10.1007/s10798-006-0006-x

Besemer, S. P. (2010). Review of fostering creativity by Arthur and David Cropley. *Creativity Research Journal*, *22*(2), 236–237. doi:10.1080/10400419.2010.481543

Bonner, D. (2010). Increasing student engagement and enthusiasm: A projectile motion crime scene. *The Physics Teacher*, *48*(5), 324–325. doi:10.1119/1.3393066

Browne, K., & Jackson, D. P. (2007). Simple experiments to help students understand magnetic phenomena. *The Physics Teacher*, *45*(7), 425. doi:10.1119/1.2783151

Chapman, S., & Lewis, M. (2001). The McOhm: Using fast food to explain resistance. *Physics Education*, *36*(3), 255–260.

Cockman, J. (2002). A bright color mixer. *The Physics Teacher*, *40*(9), 553. doi:10.1119/1.1534824

Coffey, T. (2008). Soda pop fizz-ics. *The Physics Teacher*, *46*(8), 473–476. doi:10.1119/1.2999062

Corrao, C. T. (2010). The physicist fish. *The Physics Teacher*, *48*(7), 491. doi:10.1119/1.3488204

Costa, A., & Kallick, B. (Eds.). (2000). *Discovering and Exploring Habits of Mind*. Alexandria, VA: ASCD.

Craft, A. (2001). *An analysis of research and literature on creativity in education*. Qualifications and Curriculum Authority. Retrieved September 1, 2013, from http://www.euvonal.hu/images/creativity_report.pdf

Cronbach, L. J. (1970). *Essentials of psychological testing* (3rd ed.). New York, NY: Harper & Row.

Cropley, D. H., & Cropley, A. J. (2000). Fostering creativity in engineering undergraduates. *High Ability Studies*, *11*(2), 207–219. doi:10.1080/13598130020001223

Cross, N. (2007). *Designerly ways of knowing*. London, U.K.: Springer- Verlag.

Csikszentmihalyi, M. (1998). Society, culture and person: a systems view of creativity. In R. J. Sternberg (Ed.), *The nature of creativity* (pp. 325–339). Cambridge: Cambridge University Press.

Dacey, J., & Lennon, K. (2000). *Understanding creativity: the interplay of biological, psychological and social factors*. Buffalo, NY: Creative Education Foundation.

Daley, B. (2004). A project-based approach: Students describe the physics in movies. *The Physics Teacher*, *42*(1), 41–44. doi:10.1119/1.1639969

DePino, A. Jr. (2001). "Peeps," cream, heads, and food coloring in a vacuum jar. *The Physics Teacher*, *39*(1), 56–57. doi:10.1119/1.1343436

Dindorf, W. (1999). Strings of pearls resonance. *The Physics Teacher*, *39*(4), 251. doi:10.1119/1.1367800

Dishaw, J. P. (2010). Battleship buoyancy. *The Physics Teacher*, *48*(4), 242. doi:10.1119/1.3361992

Doppelt, Y., Mehalik, M., Schunn, C. D., Silk, E., & Krysinski, D. (2008). Engagement and achievements: A case study of design-based learning in a science context. *Journal of Technology Education*, *19*(2), 22–39.

Ellenstein, M. (1982). Magic and physics. *The Physics Teacher*, *20*(2), 104–106. doi:10.1119/1.2340960

Featonby, D. (2005). Toys and physics. *Physics Education*, *40*(6), 537–543. doi:10.1088/0031-9120/40/6/005

Featonby, D. (2010). Magic physics? *Physics Education*, *45*(1), 24–31. doi:10.1088/0031-9120/45/1/001

Feldman, D., Csikszentmihalyi, M., & Gardner, H. (1994). *Changing the World: A Framework for the Study of Creativity. West Point, CN*. Praeger.

Fisher, N. (2004). Creativity breeds best physicists. *Physics Education*, *39*(6), 472.

Fortus, D., Dershimer, R. C., Krajcik, J., Marx, R. W., & Mamlok-Naaman, R. (2004). Design-based science and student learning. *Journal of Research in Science Teaching*, *41*(10), 1081–1110. doi:10.1002/tea.20040

Fortus, D., Krajcik, J., Dershimer, R. C., Marx, R. W., & Mamlok-Naaman, R. (2005). Design-based science and real-world problem-solving. *International Journal of Science Education*, *27*(7), 855–879. doi:10.1080/09500690500038165

Froehle, P. (2008). Quick way to float a paper clip on water. *The Physics Teacher*, *46*(2), 70. doi:10.1119/1.2834520

Gardner, M. (1999). The jumping pencil. *The Physics Teacher*, *37*(3), 178. doi:10.1119/1.1527660

Gluck, P. (2008, March 06). Experiments for a special day. *Physics Education*, *43*(2), 189–197. doi:10.1088/0031-9120/43/2/009

Gore, G. R. (2009). Physics fun with jelly marbles. *The Physics Teacher*, *47*(9), 606–607. doi:10.1119/1.3264598

Graf, E. H. (2008). Projectile motion demonstration. *The Physics Teacher*, *46*(9), 553. doi:10.1119/1.3023659

Greenslade, T. B. (2010). Physics is all around us. *The Physics Teacher*, *48*(5), 338–340. doi:10.1119/1.3393071

Güémez, J., Fiolhais, C., & Fiolhais, M. (2009). Toys in physics lectures and demonstrations - a brief review. *Physics Education*, *44*(1), 53–64. doi:10.1088/0031-9120/44/1/008

Heacox, D. (2002). *Differentiating instruction in the regular classroom: How to reach and teach all learners, grades 3-12*. Free Spirit Pub.

Hérold, J. F., & Ginestié, J. (2011). Help with solving technological problems in project activities. *International Journal of Technology and Design Education*, *21*(1), 55–70. doi:10.1007/s10798-009-9106-8

Jones, A., & Moreland, J. (2003). Developing classroom-focused research in technology education. *Canadian Journal of Science. Mathematics and Technology Education.*, *3*(1), 37–50. doi:10.1080/14926150309556551

Ju, Y. (2005). The physics of having a swing. *Physics Education*, *40*(6), 534–536. doi:10.1088/0031-9120/40/6/004

Keogh, B., & Naylor, S. (1999). Concept Cartoons, teaching and learning in science: An evaluation. *International Journal of Science Education*, *21*(4), 431–446. doi:10.1080/095006999290642

Kirschner, P. A., Sweller, J., & Clark, R. E. (2006). Why minimal guidance during instruction does not work: An analysis of the failure of constructivist, discovery, problem-based, experiential, and inquiry-based teaching. *Educational Psychologist*, *41*(2), 75–86. doi:10.1207/s15326985ep4102_1

Kofoed, M. H. (2006). The Hiroshima and Nagasaki bombs: Role-play and students' interest in physics. *Physics Education*, *41*(6), 502–507. doi:10.1088/0031-9120/41/6/002

Lee, K. W. L., Goh, N. K., Chia, L. S., & Wan,, Y. K. (2006). *Report on Creativity in Science Education*. National Institute of Education, Nanyang Technological University Q 183.4.555.

Lowry, M. (2008). Teaching universal gravitation with vector games. *The Physics Teacher*, *46*(9), 519–521. doi:10.1119/1.3023651

McCormick, R. (1997). Conceptual and procedural knowledge. *International Journal of Technology and Design Education*, *7*(1-2), 141–159. doi:10.1023/A:1008819912213

McCullough, J., & McCullough, R. (2000). The Role of Toys in Teaching Physics. College Park, MD: American Association of Physics Teachers.

Mcgarvey, B., Marriott, S., Morgan, V., & Abbott, L. (1997). Planning for differentiation: The experience of teachers in Northern Ireland primary schools. *Curriculum Studies*, *29*(3), 351–364. doi:10.1080/002202797184080

McGervey, J. (1995). Hands-on physics for less than a dollar per hand. *The Physics Teacher*, *33*(4), 238–241. doi:10.1119/1.2344206

Meyer, D. (2012). Designing Design Challenges: Getting the details right. Using engineering problems to enact inquiry learning. *Science Teacher (Normal, Ill.)*, *79*(2), 58–62.

Norman, E. (1993). Science for design. *Physics Education*, *28*(5), 301–306. doi:10.1088/0031-9120/28/5/010

Nunley, K. F. (2006). *Differentiating in the High School*. Thousand Oaks, CA: Corwin Press.

Pendrill, A. M. (2005). Rollercoaster loop shapes. *Physics Education*, *40*(6), 517–521. doi:10.1088/0031-9120/40/6/001

Pendrill, A. M., & Williams, G. (2005). Swings and slides. *Physics Education*, *40*(6), 527–533. doi:10.1088/0031-9120/40/6/003

Pettersen, I. A. H., & Williams, G. (2004). Overhead projector doubles as a classroom 'rainbow machine'. *Physics Education*, *39*(6), 463. doi:10.1088/0031-9120/39/6/F02

Planinsic, G., Kos, M., & Jerman, R. (2004). Two-liquid Cartesian diver. *Physics Education*, *39*(1), 58–64. doi:10.1088/0031-9120/39/1/003

Ramadas, J. (2009). Visual and spatial modes in science learning. *International Journal of Science Education*, *31*(3), 301–318. doi:10.1080/09500690802595763

Raviv, D. (2003). Do We Teach Them How to Think? In *Proceedings of the 2002 American Society for Engineering Education Annual Conference and Exposition*. American Society for Engineering Education Session.

Resnick, M., Berg, R., & Eisenberg, M. (2000). Beyond Black Boxes: Bringing Transparency and Aesthetics Back to Scientific Investigation. *Journal of the Learning Sciences*, *9*(1), 7–30. doi:10.1207/s15327809jls0901_3

Robinson, K. (2006). *Do schools kill creativity?* Paper presented at the Technology, Entertainment and Design (TED) Conference. Retrieved September 1, 2013, from http://www.youtube.com/watch?v=iG9CE55wbtY

Rogers, M. (2007). An Inquiry-based Course Using "Physics?" in Cartoons and Movies. *The Physics Teacher*, *45*(1), 38–41. doi:10.1119/1.2409508

Rowett, T. (2010). *Paper Clip Top*. Grand Illusions Videos. Retrieved September 1, 2013, from http://www.youtube.com/watch?v=qgSiMHyJcxc

Ruiz, M. J. (2006). Lenz's Law magic trick. *The Physics Teacher*, *44*(2), 96–98. doi:10.1119/1.2165439

Ryhammar, L., & Brolin, C. (1999). Creativity research: Historical considerations and main lines of development. *Scandinavian Journal of Educational Research*, *43*(3), 259–273. doi:10.1080/0031383990430303

Sarquis, J., Hogue, L., Sarquis, M., & Woodward, L. (1997). *Investigating Solids Liquids and Gases with Toys*. Middletown: McGraw-Hill. Miami University.

Sarquis, J., Sarquis, M., & Williams, J. P. (1995). *Teaching chemistry with TOYS: activities for grades K-9. McGraw-Hill*. Learning Triangle Press.

Sarquis, M. (1997). *Exploring Matter with TOYS - Using and understanding the senses*. McGraw-Hill, Inc.

Schlichting, H. J., & Suhr, W. (2010). The buzzer—a novel physical perspective on a classical toy. *European Journal of Physics*, *31*(3), 501–510. doi:10.1088/0143-0807/31/3/007

Silk, E., Higashi, R., Shoop, R., & Schunn, C. (2010). Designing technology activities that teach mathematics. *Technology Teacher*, *69*(4), 21–27.

Sills, T. W. (1999). *Science Fun with Toys*. Chicago: Dearborn Resources.

Slater, T. F. (1996). Portfolio assessment strategies for grading first-year university physics students in the USA. *Physics Education*, *31*(5), 329–333. doi:10.1088/0031-9120/31/5/024

Stables, K. (1997). Critical issues to consider when introducing technology education into the curriculum of young learners. *Journal of Technology Education*, *8*(2), 50–65.

Subramaniam, R., & Riley, J. P. II. (2008). Physics trick gets students interested. *Physics Education*, *43*(4), 355–356. doi:10.1088/0031-9120/43/4/F06

Subramaniam, R., & Tiang, N. H. (2004). Pendulums swing into resonance. *Physics Education*, *39*(5), 395. doi:10.1088/0031-9120/39/5/F11

Sumners, C. (1997). *Toys in Space: Exploring Science With The Astronauts*. McGraw-Hill, Blue Ridge Summit.

Taylor, B. A. P., Poth, J., & Portman, D. J. (1995). *Teaching physics with toys: activities for grades K-9*. Learning Triangle Press.

Thompson, G., & Mathieson, D. (2001). The Mirror Box. *The Physics Teacher*, *39*(8), 508–509. doi:10.1119/1.1424606

Torrance, E. P. (1979). *The search for satori and creativity*. Buffalo, NY: Creative Education Foundation and Creative Synergetic Associates.

Torrance, E. P. (1990). *Torrance Tests of Creative Thinking*. Beakonville, IL: Scholastic Testing Services.

Trumbo, J. (2006). Making science visible: Visual literacy in science communication. In L. Pauwels (Ed.), *Visual culture of science: Re-thinking representational practices in knowledge building and science communication Hanover* (pp. 266–283). NH: Dartmouth College Press, University Press of New England.

Turner, R. C. (1983). Toys in physics teaching: Cartesian diver. *American Journal of Physics*, *51*(5), 475–476. doi:10.1119/1.13482

Ucke, C. (2002). Professor Sakai's paper-clip tops. *Physics Education (India)*, *19*(2), 97–100.

Vernon, P. E. (1989). The nature-nurture problem in creativity. In J.A. Glover, R.R. Ronning, & C.R. Reynolds (Eds.), Handbook of creativity: perspectives on individual differences. New York: Plenum Press.

West, T. (1997). *In the Mind's Eye: Visual Thinkers, Gifted People with Dyslexia and Other Learning Difficulties, Computer Images and the Ironies of Creativity*. Amherst, NY: Prometheus Books.

Wiebe, E. N., Clark, A. C., & Hasse, E. V. (2001). Scientific Visualization: Linking science and technology education through graphic communications. *The Journal of Design and Technology Education*, *6*(1), 40–47.

## ADDITIONAL READING

Amir, N., & Subramaniam, R. (2006). Making physics toys fosters creativity. *Physics Education*, *41*(1), 18–20. doi:10.1088/0031-9120/41/1/F05

Amir, N., & Subramaniam, R. (2007). Making a fun Cartesian diver: A simple project to engage kinaesthetic learners. *Physics Education*, *42*(5), 478–480. doi:10.1088/0031-9120/42/5/004

Amir, N., & Subramaniam, R. (2009). Making a low cost candy floss kit gets students excited about learning physics. *Physics Education*, *44*(4), 420–428. doi:10.1088/0031-9120/44/4/013

Amir, N., & Subramaniam, R. (2012). Fostering inquiry in science among kinaesthetic learners through design & technology. In L. C. Lennex & K. F. Nettleton (Eds.), Cases on Inquiry Through Instructional Technology in Math and Science: Systemic Approaches (pp 221 - 257). Hershey, PA: IGI Global Publishing.

Barak, M., & Doppelt, Y. (2000). Using portfolios to enhance creative thinking. *Journal-of-Technology-Studies*, *26*(2), 16–25.

Daley, B. (2004). A project-based approach: Students describe the physics in movies. *The Physics Teacher*, *42*(1), 41–44. doi:10.1119/1.1639969

Doppelt, Y., Mehalik, M., Schunn, C. D., Silk, E., & Krysinski, D. (2008). Engagement and achievements: A case study of design-based learning in a science context. *Journal of Technology Education*, *19*(2), 22–39.

Featonby, D. (2005). Toys and physics. *Physics Education*, *40*(6), 537–543. doi:10.1088/0031-9120/40/6/005

Featonby, D. (2010). Magic physics? *Physics Education*, *45*(1), 24–31. doi:10.1088/0031-9120/45/1/001

Fisher, N. (2004). Creativity breeds best physicists. *Physics Education*, *39*(6), 472.

Fortus, D., Dershimer, R. C., Krajcik, J., Marx, R. W., & Mamlok-Naaman, R. (2004). Design-based science and student learning. *Journal of Research in Science Teaching*, *41*(10), 1081–1110. doi:10.1002/tea.20040

Fortus, D., Krajcik, J., Dershimer, R. C., Marx, R. W., & Mamlok-Naaman, R. (2005). Design-based science and real-world problem-solving. *International Journal of Science Education*, *27*(7), 855–879. doi:10.1080/09500690500038165

Meyer, D. (2012). Designing Design Challenges: Getting the details right. Using engineering problems to enact inquiry learning. *Science Teacher (Normal, Ill.)*, *79*(2), 58–62.

Norman, E. (1993). Science for design. *Physics Education*, *28*(5), 301–306. doi:10.1088/0031-9120/28/5/010

Robinson, K. (2006). *Do schools kill creativity?* Paper presented at the Technology, Entertainment and Design (TED) Conference. Retrieved September 1, 2013, from http://www.youtube.com/watch?v=iG9CE55wbtY

Rowett, T. (2010). *Paper Clip Top*. Grand Illusions Videos. Retrieved September 1, 2013, from http://www.youtube.com/watch?v=qgSiMHyJcxc

Stables, K. (1997). Critical issues to consider when introducing technology education into the curriculum of young learners. *Journal of Technology Education*, *8*(2), 50–65.

Wiebe, E. N., Clark, A. C., & Hasse, E. V. (2001). Scientific Visualization: Linking science and technology education through graphic communications. *The Journal of Design and Technology Education*, *6*(1), 40–47.

## KEY TERMS AND DEFINITIONS

**Creativity:** A term used to describe the process of coming up with original and valuable ideas in the course of solving problems.

**Demonstration Model:** A standard design shown to students at the start of a project.

**Design Sheet/Design Page:** A sheet showing a student's thinking process in linking science to the design of his/her prototype.

**Design-Based Learning:** It is a form of learning where a prototype emerges as the final product of learning.

**Normal Technical (NT) Stream:** It is a system of education offered in secondary schools in Singapore to students who did not perform well in their primary school leaving examinations to qualify for the more academic streams.

**Prototype:** A product created by students through the design-based learning approach.

# Section 3
# Research

# Chapter 8
# Using Multidisciplinary Research Experiences to Enhance STEM Learning through Undergraduate, Team-Based, Summer Research Projects for At-Risk Students

**Jennifer Yantz**
*Middle Tennessee State University, USA*

**Thomas Cheatham**
*Middle Tennessee State University, USA*

**Brittany D. Smith**
*Middle Tennessee State University, USA*

**Donald Nelson**
*Middle Tennessee State University, USA*

**Ginger Holmes Rowell**
*Middle Tennessee State University, USA*

**D. Christopher Stephens**
*Middle Tennessee State University, USA*

**Elaine Bouldin Tenpenny**
*Middle Tennessee State University, USA*

## EXECUTIVE SUMMARY

*Undergraduate research can be one of the most important and influential learning experiences during a student's college career (Light, 2001). Significant retention value is achieved both through one-on-one contact with a faculty mentor (Campbell, 1997; Jacobi, 1991) and by interaction with peers in a learning community (Johnson, 2001). Colleges and universities are using undergraduate research experiences to*

DOI: 10.4018/978-1-4666-6375-6.ch008

*help improve student retention, graduation, and success in Science, Technology, Engineering, and Mathematics (STEM). However, undergraduate research is frequently reserved for the best and brightest students who have achieved junior or senior class status. This case study describes a team-based research experience designed for first-year, at-risk undergraduate students. For this project, the term "at-risk" is defined to be first-time, full-time freshman declared STEM majors with a weak mathematics background as measured by having an ACT-Mathematics sub score of 19 to 23, inclusive. In particular, this case study focuses on the multidisciplinary nature of some of the research projects and the benefits for the students in terms of confidence, depth of learning in STEM, and progress in understanding the scientific process.*

## ORGANIZATION BACKGROUND

This case took place at a public university in the Southeastern United States with an enrollment of approximately 25,000 students in 2012. Full-time undergraduate students made up 72% of the student population. Of these full-time undergraduate students, 21% declared a major in a science, technology, engineering, or mathematics (STEM) discipline. The number of males and females was almost perfectly balanced across the institution, but only 38% of undergraduate STEM majors were female. Minorities made up 31% of both the overall university population and undergraduate STEM majors.

Like many institutions of higher education, this university had shifted its focus to the retention of students to adjust to the state funding formulas that moved from "head count" to "outcomes." In particular, the university administrators are aware that the retention of students majoring in STEM fields was critical to meet the future workforce demands (Carnevale, Smith, & Strohl, 2010). University data showed that from 1999 to 2003 an average of 55% of students majoring in a STEM discipline progressed to their second year, and on average 35% progressed to their third year. In this same time period, only 16% of first-time, full-time students who started with a major in a STEM field actually graduated with a degree in a STEM field within six years.

To address the low graduation rate of students majoring in STEM fields, in 2010 the university applied for, and received, a $2 million, five-year National Science Foundation STEP 1.b grant called FirstSTEP: Mathematics as a FirstSTEP to Success in STEM (Grant No. 0969571). Recognizing that mathematics was often a barrier to student success in STEM disciplines, two of the grant's three components addressed student success in mathematics courses. The third component, called Summer Immersion, engaged low-achieving students in authentic undergraduate research experiences with the hope that it would increase students' retention in STEM majors and bolster their performance in future science courses. Outcomes from three years of the Summer Immersion program will be the focus of this chapter.

## SETTING THE STAGE

Despite the efforts to reform K-12 education, many students who enroll in public colleges and universities are underprepared in mathematics and science. According to ACT data from 2010 – 2012 ACT-tested graduating class (n = 1,167,221), one in ten students expressed an interest in a career from a STEM field (ACT, 2013). Of those students interested in a STEM education, only 57% met the mathematics benchmark score of 22 and only 41% met the science benchmark score of 24 (the science benchmark as of 2013 is now 23). The ACT benchmarks are minimum scores that predict student success in first-year, credit-bearing college courses. Meeting the benchmarks indicates only a 75% chance of the student earning a grade of C or better in first-year courses corresponding to the tested subject area (ACT, 2013). This data would suggest that the solution for increasing the number of STEM graduates lies not in increasing recruitment of STEM majors but in supporting the academic success of underprepared students who are already interested in a STEM career.

The lack of students' academic preparation has been attributed in part to the breadth of coverage given to science and mathematics topics in secondary education curricula (Kay & Greenhill, 2011). The emphasis on test scores has led to broad curricula where students are barraged with facts and information but are rarely given the opportunity to explore a topic in depth or discover important relationships and ideas on their own. Schwartz et. al. (2008) studied 8,310 students who were enrolled in introductory biology, chemistry, and/or physics courses at 55 randomly selected colleges and universities. Students who reported having studied one science topic in depth, for at least one month, in their secondary education were found to earn higher grades in all introductory science classes in their post-secondary education. The researchers found that studying a breadth of topics in secondary education settings did not correlate with improved grades in introductory post-secondary science classes.

It was this research that motivated the faculty and administration at the university to create the Summer Immersion undergraduate research program. The idea was to extend the findings of Schwartz, et. al (2008) to academically underprepared students early in their post-secondary education career with the hope that it would bolster their success in science courses taken after their participation in the program, and thus increase retention of at-risk students majoring in STEM disciplines. The primary purpose of the Summer Immersion program was to provide these students with the opportunity for the in-depth study of a science topic through research in a way that could not be delivered via traditional instruction in science courses. The details and structure of the Summer Immersion program will be discussed in the next section. This chapter focuses on three examples of projects in the Summer Immersion which were multidisciplinary-based. For the purposes of this chapter, the researchers consider multidisciplinary to mean that multiple disciplines bring their expertise together to work on the solution of a problem. This is similar to Resnick's (2012) explanation of multidisciplinary research.

*Multidisciplinary research is bringing disciplines together to talk about issues from each of their perspectives. They may collaborate, but they maintain a separation of their disciplines in that process. When the project is done, those disciplines go back to where they came from to start other projects. Interdisciplinary is bringing those same folks together in the same way, but using that expertise to create new instruments, models, approaches that couldn't occur if they were separately handled. (page 1)*

## CASE DESCRIPTION

The Summer Immersion undergraduate research program was designed to engage low-achieving students in authentic STEM research with the hope that they would develop stronger interest and abilities in areas of STEM. The program's participants and structure will be described in the next paragraphs. Detailed descriptions of the multidisciplinary teams will be included in the discussion of the program's structure. This will be followed by a discussion of the methods of evaluation, the assessment data collected, and a discussion of the outcomes from the first three years of the program.

### Participants

In the past, faculties have traditionally chosen the best and brightest undergraduate or graduate students to participate in their research projects. The success of this program, however, hinged upon the patience and willingness of faculty mentors to work with students who were inexperienced and at times unmotivated. The participants for the grant were selected in the summer prior to their first semester at the University. Two of the three grant components supported success in mathematics, so ACT-mathematics scores were used as criteria for admission into the program. Students were required to have an ACT-mathematics score between 19 and 23, inclusive, and to have declared a major in a STEM field.

*Table 1. Summer Immersion Participants for Years 2011, 2012, and 2013*

| | Male | Female | Total |
|---|---|---|---|
| White | 16 | 24 | 40 |
| Black or African American | 11 | 31 | 42 |
| Asian | 2 | 4 | 6 |
| Hispanic | 5 | 1 | 6 |
| Tctal | 34 | 60 | 94 |

The FirstSTEP program had 35 participants in 2010, 41 in 2011, and 41 in 2012. The participants for the first three years of the grant had an average ACT-composite score of 21.7, an average ACT-mathematics score of 21.3, and an average ACT-science score of 22.4. Additionally, participants were required to be first-time, full-time students. Thirty-six percent (n=117) of the students in the FirstSTEP program were first-generation college students (21 females and 21 males). The distribution of students by major was 47% Chemistry, 18% Engineering Technology, 16% Biology, 10% Computer Science, 5% Mathematics, and 3% Physics.

Out of the 117 FirstSTEP participants for the three years of the project described here, 94 were able to participate in the Summer Immersion, which occurred at the end of the participants' first academic year of post-secondary education. At this point, some students had exited the program due to attrition, academic suspension, or they had changed to a non-STEM major. The demographics for the Summer Immersion participants (Table 1) are similar to those of the overall participants. Students in good academic standing at the end of their first year might not have participated if they had job, military, or family obligations. The first-year students expressed a strong desire to participate again in the second year. Since funding was available in 2012 and 2013, second-year participants who demonstrated excellence as freshman were allowed to participate in a leadership role on a research team.

## Structure

The Summer Immersion program began with a solicitation for STEM faculty to submit research projects that were suitable for small teams of four or five STEM majors who had completed one year of college. The faculty were encouraged to describe their project at a level freshmen could understand and to develop projects at a level to which freshmen could contribute. This was a challenge for faculty who routinely conducted research with their best undergraduate and graduate students. The FirstSTEP team reviewed the proposals and selected those that engaged the

*Table 2. Examples of Some of the Summer Immersion Research Team Topics for the 2011 – 2013 Period*

| Summer | Discipline | Research Title |
|---|---|---|
| 2011 | Geosciences | Spatial Analysis of the Tilt and Tilt Direction of Carbonate Rock Layers in Rutherford County, Central Tennessee |
| 2011 | Biology | Quantification and Analysis of Chloroplast mRNA Processing |
| 2011 | Physics | Computer Simulation of Wave Propagation in Metamaterials |
| 2012 | Physics | Response to Optical Trapping by Red Blood Cells from a Transfused Sickle Cell Patient |
| 2012 | Engineering Technology | Wheel Hub Motor Development and Testing |
| 2012 | Chemistry | Characterization of Pharmaceutical Products via Raman and Infrared Spectroscopy Techniques |
| 2012 | Physics | Sounds Good To Me—Experimental Research in Digital Audio Correction |
| 2013 | Biology and Physics | Utility of Optical Laser Trapping to Determine the Malignant Potential of Cancer Cells |
| 2013 | Chemistry | Isolation of Compounds from Plant Sources, Including Traditional Medicines |
| 2013 | Math | What's the Point? A Geometric Exploration of Campus-based Applications |
| 2013 | Engineering Technology | Intelligent Mobile Robotics |
| 2013 | Biology | Laboratory-Directed Evolution of Salt-tolerant Luciferase |

students in hands-on activity, exposed them to multiple disciplines, could be completed within the timeframe (originally a month), and had reasonable budgets. A sample of selected projects is presented in Table 2. Faculty received a small stipend for their work, but for many their participation was motivated by a desire to fund one of their undergraduate or graduate research students, to further advance their own research agenda, and to contribute to the increased retention and success of at-risk students majoring in STEM fields

The faculty mentors were important elements of the success of the Summer Immersion program. They set the standards for group participation, behavior, safety, collaboration, discussions, expectations, and attitudes. With funds from the grant, each faculty mentor hired a capable upper-division or graduate research student to assist in leading the team. The best faculty mentors were those who were good organizers, good teachers, good researchers, able to engage students, able to express expectations clearly, and perhaps most importantly, adaptable. The faculty were not expected to spend every minute of the day with the research team. They were expected to be available, provide guidance and leadership for the research effort, be firm but patient as students learned to work in a team and perhaps in a discipline they did not know, and encourage students to exceed their expectations. At times during the research experience, faculty served as research leaders, teachers, parents, advisors, counselors, and friends. Recruiting the right faculty was critical to the success of the Summer Immersion.

Prior to the beginning of the Summer Immersion program, the FirstSTEP leadership met with the faculty mentors to ensure they were prepared for the maturity level and academic preparation of the Summer Immersion students. The first day of the program included an orientation to introduce all of the projects, mentors and research teams and to set clear expectations about attendance, participation, safety, and outcomes. The FirstSTEP student liaison and the Summer Immersion leader remained available throughout the program to address questions or non-compliance with the expectations set forth. After about three days, students and mentors usually settled into a routine. It was expected that student participants would work eight hours per day, five days per week for three or four weeks, depending on the University's summer schedule. This rigorous schedule was new to some students. Each faculty mentor was given the flexibility to arrange their research schedule within these parameters. The faculty were not expected to be with the students for eight hours each day, but it was expected that the research assistant would remain with the students at all times. Some bench research ran late into the night or on weekends and while it was expected that the students on these teams would remain fully engaged, accommodations to the schedules were made.

Near the midpoint of the research, all of the participants attended a working lunch where each team presented their research hypothesis, methods, and any preliminary results to the entire group. Each team member was required to speak at the presentation. At the end of the project, an abstract and a presentation were required. Although it was time consuming, this provided an opportunity for the students to summarize their work and to develop presentations and posters that could serve as a base for submitting to conferences held by the university and outside organizations. Initially, we expected students to be introduced to scientific research and learn some of the tools and techniques needed to do research. The students surpassed all expectations and in some cases completed publishable research which made important contributions to the scientific community.

## Multidisciplinary Research Teams

Several Summer Immersion projects were multidisciplinary in nature. The details of three multidisciplinary projects are given in the following paragraphs. Each description includes a project overview, the team composition, and a brief description of the multidisciplinary nature of the project.

## Project Example 1: Utility of Optical Laser Trapping to Determine the Malignant Potential of Cancer Cells

### *Project Overview*

Laser tweezers are formed by using lenses to focus laser light into a microscopic dot inside a liquid. Such focused lasers are capable of trapping and manipulating microscopic objects suspended in a liquid. In this study, two FirstSTEP research teams used laser tweezers to measure the response of cancer cells to deformations resulting from a viscous drag force. It has been well documented that cancers that originate from the same tissue (e.g. lung and breast cancers) exhibit different degrees of malignant properties (e.g., ability to induce blood vessel formation, metastatic potential) (Cross, et. al., 2008). Metastatic potential is the relative ability of the cancer cells to spread throughout the tissues of the body. The hypothesis was that the more metastatic/malignant cell lines would exhibit more elastic properties compared to the less metastatic/malignant cells. In this study the laser tweezers were used to measure the elasticity of two human breast cancer cell lines and two human lung cancer cell lines that have known metastatic properties (less aggressive/metastatic and more aggressive/metastatic). In this project students were expected to learn sterile tissue culture techniques; cell staining; enumeration and visible and fluorescent microscopy. They were also expected to conduct independent quantitative and qualitative analysis of the mechanical properties of various cell types using laser tweezers. At the end, the students were expected to analyze their results and draw conclusions.

### *Team Composition*

This research team was structured differently than the other Summer Immersion teams during the faculty application process in that two faculty from different disciplines, biology and physics, submitted a joint proposal. The proposal was accepted and students were selected to form two summer research teams for this project. A biology professor and a chemistry professor worked together to lead this project. The students were from a variety of disciplines: chemistry (two students), physics, science, and biology (three students). The science and biology majors each had declared the following concentrations: radiation theory, physiology, and microbiology, respectively.

### *Multidisciplinary Nature of the Project*

Students were required to interact with equipment in a physics laboratory, a biology laboratory, and a computer laboratory. In the physics setting they prepared samples on a microscopic slide, operated a high-power laser, collected hundreds of microscopic images, and analyzed these images using image analysis computer programing software. In the biology laboratory, they prepared growth media, solutions, and glassware. They learned how to macroscopically and microscopically observe cells, and how to determine viability of cells using cytometers. They also observed the differences in lung- and breast-cell lines using a fluorescent microscope. In the computer laboratory, the students conducted measurements on the size and shape of the cell images. They also carried out calculations and graphical analyses of the data.

## Project Example 2: Intelligent Mobile Robotics

### *Project Overview*

Small size mobile robots have been increasingly used in many applications such as industrial logistics, medical diagnostics, surgical operations, and search and rescue missions. Understanding robot-robot and robot-human interfaces are essential in successful design of such systems. Arduino boards have become an inexpensive and quite accessible tool in developing various platforms in the prototyping stages. Arduino boards provide motor controls and a myriad of Input/Output options with non-proprietary programming codes. In this project, students designed a many-robot system and investigated the various interfacing options that would best accomplish certain tasks which required the cooperation of many robots. Students utilized various sensors such as light, sound, proximity, and magnetic field to detect the robotic environment and help the robots make intelligent decisions. The project required research activities on error detection and analysis. Students designed and fabricated the robots used in the project. The faculty mentor provided instruction on robot design, programming, and sensor characterizations. Participating scholars were engaged in many diverse activities that required inquiry into the work of others in the field, including electronic component characterization and specifications, sensor selection, computer programming, behavior studies, and making modifications to the robots. To help achieve these steps, students utilized the resources of the Engineering Technology department and the Robotics, Electronics, and Machine Technology laboratories. Students gained valuable experience at several research levels. The participants were engaged in a literature search regarding the various concepts and techniques in multi-robot systems. They learned basic electronics and signal routing and processing. The participants programmed the robots in a procedural language using C++. Library resources were used for research into innovative techniques in designing multi-robot systems. Students worked in the robotics and electronics labs to design and fabricate the robots. They were also instructed in computer programming and robot-robot interfacing.

### *Team Composition*

The faculty mentor for this project was a Professor in the Department of Engineering Technology. This professor had experience leading undergraduate research teams and mentored a Summer Immersion team in 2012 as well. This experience with novice researchers was very helpful in student expectations with this multidisciplinary team. The project team included five students who had all recently completed their first two semesters of college. Their majors included two engineering technology majors, one science major, one physics major, and one computer science major. The computer science major was a specific request by the faculty mentor when the project proposals were submitted.

### Multidisciplinary Nature of the Project

The diverse backgrounds of the team members was very helpful in the successful completion of this project. Peer-to-peer teaching occurred in this project in the area of designing circuits as well as computer programming. The computer science major was the leader in terms of helping other students to learn programming. This project required the use of math, statistics, physics, computer programming as well as engineering fundamentals.

## Project Example 3: Sounds Good To Me – Experimental Research in Digital Audio Correction

### Project Overview

The goal of this project was to create a "virtual" anechoic chamber for acoustic measurements using computer processing to digitally remove room echoes from recorded sounds. High quality audio testing of items such as speakers, microphones, or the sound emission from industrial machinery, is typically conducted in an anechoic chamber. These specially built enclosures are designed to eliminate all sound reflections from the walls, floor, and ceiling by covering these surfaces with large pyramid-shaped sections of sound-absorbing material. Anechoic chambers are a very expensive and specialized research tool available at only a handful of universities and research labs around the world.

The objective of this project was to record sounds in a normal room—complete with reflections from the surroundings—and then use digital signal processing to remove the effect of reflections. The process of removing the effect of room reflections (reverberations) on a recorded signal involved creating a digital filter that reversed the effect of all of the reflections arriving at the microphone. Although this process may appear difficult, there was a fairly straightforward method to creating a suitable filter from a test signal recorded in the reverberant space. It has been shown in MATLAB simulations that it is possible to make a very good (but not perfect) echo removal filter.

A physics graduate student was pursuing the experimental side of the project to evaluate how effectively this technique can be implemented in practice. The three-week FirstSTEP summer team was involved in acquiring test signals, helping the graduate student and faculty mentor create filters in MATLAB, and collecting real audio data (pure sine waves, voice, and music) to evaluate how well these filters work in removing the reverberant effect of the room. The aspect of data collecting allowed students to have hands-on experience with data acquisition hardware and to identify the effect of the wall reflections of a room on a signal. The data acquisition helped students understand the problem at hand because they could graphically see the difference between the signal they were sending out and the one that they recorded in a real space. This understanding of the problem segued into the research aspect, which was to develop a signal manipulation process to restore the room recorded signal to its initial clean form. That process involved physics, mathematics, and computer skills. The product from the team was a large set of carefully acquired and well documented audio data taken in a variety of environments that allowed for the refinement and testing of new methods of echo removal filters that formed the graduate students' physics research thesis work.

### *Team Composition*

The faculty mentor for this project was a physics professor who had extensive experience with multidisciplinary and interdisciplinary research. He had conducted several projects working with biology and chemistry faculty and undergraduate students in these three disciplines. For this project, however, he specifically requested that he would need students with backgrounds in computer programming, physics, and mathematics to successfully complete the project. His team had students from each of those disciplines, bringing their expertise and ideas together to help solve the problem.

### *Multidisciplinary Nature of the Project*

In addition to being a physics project, this project involved aspects of computer science, engineering, and mathematics. In the realm of computer science, students learned to write code that attempted to digitally remove echoes from sound samples. For mathematics, students learned to model and analyze audio samples mathematically. In the realm of engineering, students learned about and constructed instruments for propagating echoes, reducing echoes, and transferring sounds. Students also constructed and modified simple circuit boards.

Participants' majors included computer science, mathematics, and physics. Students were encouraged to bring their various discipline-specific skill sets to the project. They utilized their own major-field skills in accomplishing various parts of the project, and they also shared these skills with one another and engaged in informal peer-teaching.

## Summer Immersion Evaluation

The Summer Immersion project included several evaluation components that used both formative and summative assessments. Survey instruments with closed and open -ended questions were used to collect the students' perception of the benefits and effectiveness of the program. A survey based on Bloom's Taxonomy of the cognitive domain was developed and given to team leaders and students to determine the Depth of Science Experience (DOSE) for the projects in which they were involved. The SURE III Survey and the corresponding Prereflection Survey helped assess student attitudes and gains made in science learning (Lapatto, 2004). Observations of teams were conducted as well as interviews with students and faculty mentors to corroborate evidence collected in the survey instruments. The lessons learned from evidence collected in each of these assessments was used to make improvements in subsequent years of the program. In this section we first present a summary of the assessment data followed by a discussion of the implications for the retention of at-risk students in the Summer Immersion program.

## Results

A typical measure of success of undergraduate research is the production of presentations and publications. However, the students participating in this introductory undergraduate research program had just finished their freshman year. Therefore, the expectation for presentations and publications was not the same for them as it would be for juniors and seniors completing undergraduate research. The FirstSTEP project team was pleased with the initial successes for these young students who found opportunities to present and publish their research. Students were involved in research that led to three presentations and four publications. Additionally, students participated in five poster presentations in 2012, and four more student poster presentations are forthcoming from the 2013 Summer Immersion projects.

## Summer Immersion Project Assessment Data (2011-2013)

The 20 Summer Immersion participants who completed surveys in 2011 were asked to rate their perception of the gains in learning that they had achieved on a scale of one to five, with one meaning no gain and five meaning great gains. Students reported an average of 4.2 for gains in enthusiasm for the subject and an average of 4.1 for gains in connecting key ideas with other knowledge. The students reported an average of 4.1 for gains in confidence that they understood the main concepts and an average of 4.3 for gains in confidence that they could perform work in their subject areas. Also noteworthy, the students were asked to rate the helpfulness of different aspects of the program using a scale of one to five, with one meaning no help and five meaning great help. The item which received the highest average rating of 4.6 was the category *participating in group work*.

The following list is a sample of comments provided by students to the open-ended survey items. While this is an abbreviated list, it is representative of the comments. The list below does not include any negative comments because all responses received from the students were positive.

## Summer Immersion Student Comments

- "I learned more knowledge in a four weeks span than I would in a whole semester."
- "(Summer Immersion) has strengthened my critical thinking skills."
- I will take away from this program "problem solving skills."
- "This Summer Immersion has opened my eyes to what a real team project will be like in a real job."
- I will take away from this program "more confidence and a [sic] open mind."
- "This experience has made me have [sic] the confidence to stick to my major and even maybe go on the research side of it."
- "It has made me want to focus more and stay dedicated to STEM."

In 2012, two existing survey instruments were used to assess how the students were impacted by the program. A questionnaire from Pacifici and Thomson (2011) was used to measure students' perceptions of their efficacy related to conducting research. This survey was completed both before and after the research experience. Evidence of the students' confidence was found in these categories:

1. Confidence in their ability to do science research,
2. Capability of conducting undergraduate research in science,
3. Knowledge and skills required to do research, and
4. Being well prepared to do undergraduate research in science.

The students reported an approximate average of three out of five in each of these categories at the beginning of the Summer Immersion. Their post scores increased an average of three-tenths for items one and two above and two-tenths for questions three and four.

In 2013, the SURE III survey was used to assess students' perception of their own abilities related to scientific research before and after participation in the Summer Immersion program. Of the 2013 Summer Immersion participants, 60% indicated on the SURE III survey that they were very satisfied with the research experience, 21%

were mildly satisfied, 15% were neutral and only one of the students reported being very dissatisfied. The SURE Survey was given throughout the nation which allowed a national comparison of the data collected in this study. However, the Summer Immersion is different from the majority of undergraduate research experiences in the nation. Summer Immersion is for at-risk students who have only completed the first year of college. Therefore, these national comparisons must be used carefully. For example, it is not surprising that the satisfaction results for our FirstSTEP students were slightly lower than the overall SURE III results in which 91% of participants were mildly or very satisfied with the research experience.

When asked if they would participate in a research experience again in the future, 44% of the Summer Immersion participants said it was very likely, 41% said it was likely, 12% said it was unlikely and only one student reported they would not participate again in a research experience. Again, this was slightly less than the overall SURE II participants who reported 92% were likely or very likely to participate again in undergraduate research as compared to 85% of the FirstSTEP participants.

The FirstSTEP students for Summer Immersion 2013 reported an average self-confidence gain after the summer immersion of 3.61 out of 5, which was higher than the SURE III survey respondents' average gain of 3.42. Among other items, the SURE III Survey again asked students to rate their perception of the benefit of working with other students in a group. Of the 34 Summer Immersion participants surveyed in 2013, twelve (35%) rated working with others as the best part of their experience and twelve (35%) indicated it moderately enhanced the experience. Six students thought working with others did not affect the experience and four reported that it was the worst part, or moderately detracted from the experience.

During the second year of Summer Immersion, a new instrument was developed to assess the extent to which the Summer Immersion program met one of the most important goals, which was to provide the students an "in-depth" research experience. The instrument, called "DOSE" (Depth of Science Experience), was designed to measure the frequency of the students' involvement in activities from the Bloom's Taxonomy cognitive domains of remembering, understanding, applying, analyzing, evaluation, and creating (Bloom, 1956). The instrument had five items in each of the six Bloom's taxonomy categories for a total of total 30 items. The students and mentors were asked to report the frequency of the students' involvement in each of the 30 items without the instructor or student mentor's guidance. The possible responses for each item were never, sometimes, and often with assigned scores of zero, one, and two, respectively. Thus, there were ten possible points for each Bloom's category. The questions were presented in a random order. Two science educators reviewed the content of the instrument. The instrument was administered to both students and faculty one day prior to the end of the experience. The internal reliability of the instrument was measured using Cronbach's alpha which has an

*Table 3. DOSE results: Sounds Good to Me project*

| Bloom's Taxonomy Level | Category | Mentor | Student |
|---|---|---|---|
| 1 | Remembering | 7.5 | 6.7 |
| 2 | Understanding | 5.0 | 6.8 |
| 3 | Applying | 6.0 | 6.2 |
| 4 | Analyzing | 6.0 | 6.0 |
| 5 | Evaluating | 5.0 | 5.3 |
| 6 | Creating | 5.5 | 7.2 |

acceptable level of 0.70 for validity (Stemler & Tsai, 2008). The overall Cronbach's alpha was found to be 0.930, meaning that the instrument had acceptable inter-rater reliability. The student and faculty responses were tabulated and compared.

The student and mentor evaluation results for the three projects that we have highlighted in this case study are shown in Tables 3 through 5. The mentors generally gave a lower rating to the frequency of engagement in each category of activities. Exceptions were found in data reported from the *Sounds Good to Me* project in the category of remembering (Table 3) and data reported from the *Robotics* project in the category of applying (Table 5). The most notable differences between the mentor and student ratings can be seen in the higher-order thinking categories of evaluation and creating where the mentor's ratings were consistently lower than the students' ratings. The students' perception of their involvement in the higher levels of Bloom's Cognitive domains without instructor or student mentor guidance occurred, on average, at least sometimes (a rating of 5 out of 10) for all three projects described here. The independent work at these higher level domains is a desired result of this undergraduate research experience for these students who have completed only two semesters of college.

*Table 4. DOSE Results: Biology/Physics Project*

| Bloom's Taxonomy Level | Category | Mentor | Student |
|---|---|---|---|
| 1 | Remembering | 7.0 | 7.1 |
| 2 | Understanding | 6.2 | 7.3 |
| 3 | Applying | 7.4 | 7.9 |
| 4 | Analyzing | 6.8 | 7.3 |
| 5 | Evaluating | 4.8 | 7.1 |
| 6 | Creating | 5.2 | 6.4 |

*Table 5. DOSE Results: Robotics Project*

| Bloom's Taxonomy Level | Category | Mentor | Student |
|---|---|---|---|
| 1 | Remembering | 8.0 | 9.0 |
| 2 | Understanding | 7.0 | 9.5 |
| 3 | Applying | 9.0 | 8.8 |
| 4 | Analyzing | 7.5 | 8.8 |
| 5 | Evaluating | 6.0 | 8.0 |
| 6 | Creating | 6.0 | 8.8 |

Another assessment that occurred within the Summer Immersion was project team observations. Teams were randomly selected and a graduate mathematics and science education doctoral student observed the team and took field notes. The Sounds Good to Me project was selected and observed several times throughout the three weeks of the project. The observations revealed that the students seemed to feel overwhelmed at the beginning of the research experience, which is most likely due to the fact that these students had never been exposed to research prior to their Summer Immersion experience and most were not physics majors. One student verbalized this when he stated, "It [the research project] is just a lot to take in at once." However, as students progressed through the research project, they appeared to develop a level of comfort with the material and exhibited great gains in the knowledge about the project. In an observation conducted towards the end of the project, it appeared as if each student had taken ownership of a role in the project based on their content background. For example, the computer science majors developed programs that produced various sounds, the mathematics major performed calculations and checked measurements, and the physics majors constructed the pipes that the sound was sent through and adjusted circuits. Each member used their unique content knowledge to work together on this research project.

Conversations, both formal and informal, during and after the projects that were completed, revealed that participating in multidisciplinary projects for younger students provided special learning opportunities. Students were able to observe and participate in content and topics that they will learn more about in future courses. For example, the students in the Robotics project learned about circuits. In the following semester, one of the students took a physics course in which they learned about circuits. The experiences from the summer research project gave the student an advantage in the knowledge of the application of circuits. The student reported that learning during the course was easier because of the prior exposure and the learning that had occurred in the Summer Immersion. The coursework further reinforced the learning from the summer, as well. Additionally, the students benefited from seeing how multiple disciplines were related. In the robotics project, students were able to see relationships between math, physics, engineering technology, and computer science. The robotics team and the cancer cell team reported an average self-confidence gain after the Summer Immersion of 4.0 and 3.3 out of 5, respectively, which was comparable to the SURE III survey respondents' average gain of 3.42.

## Discussion

The findings from the assessment of the Summer Immersion program were derived from several independent lines of evidence. These included surveys of students' self-assessed learning gains which include both numerical ratings and open-ended comments; interviews with faculty mentors and teaching assistants; and observations of student participants. Analysis of the data collected resulted in the following conclusions:

1. Motivation and attitude are keys to success in the Summer Immersion experience and STEM in general.
2. Students who participated in the Summer Immersion program reported that they experienced gains in content knowledge.
3. Student who participated in the Summer Immersion program reported that they experienced gains in research skills.
4. Students who participated in the Summer Immersion program reported that they have an increased desire to study STEM fields.
5. Students who participated in the Summer Immersion program are more self-confident in their ability to complete a STEM major.

The success of the project is due in large part to the willingness of faculty mentors and teaching assistants to engage students. Faculty mentors and teaching assistants were aware of the students' level of motivation and desire to learn throughout the project. Early in the research experience they reported that students "appeared apprehensive" with faculty, were "afraid to ask questions," and "did not provide answers to questions" with which they should be familiar. Students were quiet, hesitant, and did not show natural curiosity.

Faculty mentors and teaching assistants created additional activities or opportunities for every student to contribute to the research in a way that suited their interests and abilities. As the program progressed, students became more comfortable working with the faculty mentor and research assistant. Their interest in the project, and STEM in general, grew and many of the students reported considerable gains in enthusiasm for STEM and their confidence to succeed in their chosen major. Students reported that the research assistants were enthusiastic about their subject and offered encouragement.

The evidence points to three main goals accomplished by the Summer Immersion program. First, it built relationships between students and faculty, which research shows is critical to the student's intellectual development and confidence (Thiry, Laursen, & Hunter, 2011). It has been the experience of the project team that participants were reluctant to conduct even the most basic conversations with faculty members. Summer Immersion allowed the students to work full-time with faculty members in an informal setting for three weeks which resulted in the development of positive relationships with their mentors. Second, the Summer Immersion program exposed the participants to authentic STEM careers through experiencing what STEM professionals do on a day-to-day basis. Summer Immersion corrected the students' misconceptions of what real science is and how it is done. Finally, many students had incorrect views of scientific or academic adversity. For instance, they may have either avoided (or been protected from) adversity or they may have been exposed to an unhealthy kind of adversity. Summer Immersion gave the students an opportunity to experience scientific and academic adversity in a safe environment. In addition, students had a chance to witness how their faculty mentors handled scientific adversity, which both humanized the faculty and demystified adversity.

The FirstSTEP project also found that the integration of multiple disciplines into a research project provided the students with additional benefits. Students learned more about related STEM disciplines than if each project had only one discipline in the research. Being young in their college career (at the end of their first year of academic study), the students were still exploring their academic majors. Students were exposed to new academic disciplines when the projects were multidisciplinary in nature. Following the Summer Immersion experience, a couple of students changed their majors to one of the majors they were exposed to in their multidisciplinary summer research project. Even if the exposure was relatively minor to multiple disciplines, the students gained from the exposure because they will take courses which use the other discipline later in their college career. For some students this happens the semester after their summer research experience and for others it comes later. Finally, the experience of being the student mentor in the group for their discipline was very powerful both for learning and building confidence. The student leaders who had to teach their peers developed a deeper understanding of the material in their disciplines and gained confidence in their abilities when their peers understood their explanations. Multidisciplinary projects were a win-win learning experience for the student participants.

## CURRENT CHALLENGES FACING THE ORGANIZATION

### Changes along the Way

The first year of the project, the FirstSTEP leadership team selected the faculty projects and assigned students to research teams based on their major and what the leadership team knew about the students. While logical, this turned out to be a poor approach. Several students were "unhappy" with their assignments. In subsequent years, the team decided to let the students select the projects in which they had an interest by ranking their top 3 projects. If more projects were available than were needed, projects were eliminated based on the level of student interest. The team then tried to assign a student to one of their top two choices. This method yielded less unhappiness about assignments. Occasionally a student would get started with a project and learn that it was not what they expected. The FirstSTEP team accommodated a limited number of changes when it was possible.

After observing the first summer research session, the FirstSTEP project team was able to anticipate possible difficulties that the faculty mentors might face such as the need to scaffold student learning, to make learning active, and to avoid lecturing to students for extended time periods. Starting in the second year, the project team worked hard to ensure that the faculty mentors were aware of the complexities of conducting research with students who had only completed their first year of college as compared to the typical junior or senior research participants. The project team provided training to prepare the faculty mentors for possible issues that might arise due to the students' weaker academic backgrounds.

### Issues That Still Remain

The project team has not been able to determine before admittance to the program, which students are motivated to learn, are willing to be helped, and are willing to do their part to succeed. It is clear, however, that if any of these three conditions are not met, the summer research team will be weak and will not be an effective learning experience. Additionally, it was evident that overall efforts to help students persist as a STEM major are less likely to be successful with students who lack the characteristics described above.

Housing and scheduling continue to be challenges for the Summer Immersion program. A four-week program starting immediately after the spring semester ends and ending before the June summer session begins is preferred. This was problematic on both ends. At the start, housing problems existed because the dorms were not open the first week after the spring semester or maintenance projects scheduled for this time period limited the availability of resources such as hot water. If the program lasted four weeks, the risk was that the students would be prevented from enrolling in the June summer term. It would also be difficult to find faculty mentors who could devote time to the program during a regular summer session. Despite these challenges, the FirstSTEP project team still believes that the immersive format of the program – working intensely for a short period of time – offered the most benefits to the students.

## SOLUTIONS AND RECOMMENDATIONS

Throughout the first three years of management of the Summer Immersion program, best practices for the administration of a team undergraduate research program for underprepared STEM majors early in their college career became evident to the project team. First, it is important that program administrators carefully choose student assistants and faculty mentors who work well with students of all abilities. Working with inexperienced and underprepared students required a great deal of patience and personal attention as well as strong organization skills. The faculty needed to think about the research project from the eyes of a novice student. This enabled the faculty member to help the students build new skills in a manner that fostered students' successes. The students initially lacked confidence in their knowledge and abilities, and needed time to grow comfortable with the concepts, methods, and laboratory equipment so that they could work independently. They needed to learn to ask for help.

It was important to keep team sizes manageable so that the student assistants and faculty mentors could be effective in their roles. It was difficult for one person to simultaneously keep four or five students who are new to research engaged and working independently. At the same time, faculty mentors should avoid assigning five students to a job that could be done by two students. An alternative could be asking small groups to each do the same experiment or activity and then compare results, or even compete in some way. Finally, program administrators and mentors should be prepared for declining interest towards the end of the project. The students and faculty had been intensely focused on their project for several weeks, so an enthusiasm builder late in the project was helpful to keep momentum going. Having the students make midterm presentations with the statement of their research problem and their research method helped ensure that students were focused on the research and not lost in simply skill-building. This presentation also helped the students organize information in their minds and on paper before the project was too far underway. About seven working days into the project seemed to be the appropriate amount of time for the students to be able to verbally communicate about their research hypothesis and methods. Hearing what the other students were doing for their research in this manner was also valuable for the students. The final presentations on the last day of the Summer Immersion provided an important place for the students to practice communicating science, an important learning opportunity that none of these students had experienced before. All students were required to speak at some point during the approximately 10 minute presentations. The Summer Immersion project administrators continued to be amazed at the growth in the students' scientific knowledge about the research project and in the growth in the students' confidence in their ability to understand and present their research. From observing the gains for students engaged in multidisciplinary projects, more faculty were encouraged to propose multidisciplinary projects.

## ACKNOWLEDGMENT

This material is based upon work supported by the National Science Foundation under Grant Number 0969571. Any opinions, findings, and conclusions or recommendations expressed in this material are those of the author(s) and do not necessarily reflect the views of the National Science Foundation. Without this support we would not have been able to study innovations that support the success of under-prepared STEM majors. We gratefully acknowledge the support of our Provost, Dean, STEM department chairs and our Internal and External Advisory Boards who provided feedback and suggestions that helped us improve our program. We acknowledge the great work of our Coordinators Brad Rudnik and Teresa Wade.

## REFERENCES

ACT. (2013). *The condition of college & career readiness 2013 national.* Retrieved from ACT Improve Yourself website: http://www.act.org/research/policymakers/cccr13/pdf/CCCR13-NationalReadinessRpt.pdf

Bloom, B. S. (1956). *Taxonomy of educational objectives, Handbook I: The cognitive domain*. New York: David McKay Co Inc.

Campbell, T., & Campbell, D. (1997). Faculty/Student mentor program: Effects on academic performance and retention. *Research in Higher Education, 38*(6), 727–742. doi:10.1023/A:1024911904627

Carnevale, A. P., Smith, N., & Strohl, J. (2010). *Help wanted: Projections of jobs and education requirements through 2018*. Washington, DC: Georgetown University Center on Education and the Workforce.

Cross, S. E., Jin, Y., Tondre, J., Wong, R., Rao, J., & Gimzewski, J. K. (2008). AFM-based analysis of human metastatic cancer cells. *Nanotechnology*, *19*(38), 384003. doi:10.1088/0957-4484/19/38/384003 PMID:21832563

Interview, R. J. (n.d.). *Interdisciplinary and Multidisciplinary Research.* Retrieved from http://www.4researchers.org/articles/5213

Jacobi, M. (1991). Mentoring and undergraduate academic success: A literature review. *Review of Educational Research*, *61*(4), 505–532. doi:10.3102/00346543061004505

Johnson, J. (2000). Learning communities and special efforts in the retention of university students: What works, what doesn't, and is the return worth the investment? *Journal of College Student Retention*, *2*(3), 219–238. doi:10.2190/V0PA-BL4B-1X2L-W5VT

Kay, K., & Greenhill, V. (2011). Twenty-first century students need 21st century skills. In G. Wan, & D. Gut (Eds.), *Bringing schools into the 21st century* (pp. 41–65). New York: Springer. doi:10.1007/978-94-007-0268-4_3

Light, R. (2001). *Making the most of college: StudentssSpeak their minds*. Cambridge, MA: Harvard University Press.

Lopatto, D. (2004). Survey of undergraduate research experiences (SURE): First findings. *Cell Biology Education*, *3*(4), 270–277. doi:10.1187/cbe.04-07-0045 PMID:15592600

Pacifici, L. B., & Thomson, N. (2011). Undergraduate science research: A comparison of influences and experiences between premed and non–premed students. *CBE Life Sciences Education*, *10*(2), 199–208. doi:10.1187/cbe.11-01-0005 PMID:21633068

Pellizzaro, A., Welker, G., Scott, D., Solomon, R., Cooper, J., & Farone, A. et al. (2012). Direct laser trapping for measuring the behavior of transfused erythrocytes in a sickle cell anemia patient. *Biomedical Optics Express*, *3*(9), 2190–2199. doi:10.1364/BOE.3.002190 PMID:23024913

Schwartz, M. S., Sadler, P. M., Sonnert, G., & Tai, R. H. (2009). Depth versus breadth: How content coverage in high school science courses relates to later success in college science coursework. *Science Education*, *93*(5), 798–826. doi:10.1002/sce.20328

Stemler, S., & Tsai, J. (2008). Best practices in interrelater reliability: Three common approaches. In J. Osborne (Ed.), *Best practices in quantitative methods* (pp. 29–49). Thousand Oaks, CA: Sage Publications. doi:10.4135/9781412995627.d5

Thiry, H., Laursen, S. L., & Hunter, A. (2011). What experiences help students become scientists? A comparative study of research and other sources of personal and professional gains for STEM undergraduates. *The Journal of Higher Education*, *82*(4), 357–388. doi:10.1353/jhe.2011.0023

## ADDITIONAL READING

Adedokun, O. A., Zhang, D., Carleton Parker, L., Bessenbacher, A., Childress, A., & Daniels Burgess, W. (2012). Understanding how undergraduate research experiences influence student aspirations for research careers and graduate education. *Journal of College Science Teaching*, *42*, 82–90.

Harsh, J. A., Maltese, A. V., & Tai, R. H. (2011). Undergraduate research experiences from a longitudinal perspective. *Journal of College Science Teaching*, *41*, 84–91.

Kardash, C. M. (2000). Evaluation of an undergraduate research experience: Perceptions of undergraduate interns and their faculty mentors. *Journal of Educational Psychology*, *92*(1), 191–201. doi:10.1037/0022-0663.92.1.191

Lei, S. A., & Chuang, N. (2009). Undergraduate research assistantship: A comparison of benefits and costs from faculty and students' perspectives. *Education*, *130*, 232–240.

Seymour, E., Hunter, A., Laursen, S. L., & DeAntoni, T. (2004). Establishing the benefits of research experiences for undergraduates in the sciences: First findings from a three-year study. *Science Education*, *88*(4), 493–534. doi:10.1002/sce.10131

Singer, J., & Zimmerman, B. (2012). Evaluating a summer undergraduate research program: Measuring student outcomes and program impact. *Council on Undergraduate Research Quarterly*, *32*, 40–47.

## KEY TERMS AND DEFINITIONS

**Early Undergraduate Research:** Research within the first two years of college.

**Summer Immersion Program:** The three to four week early research program for FirstSTEP participants.

# Chapter 9
# Collaborative Teams as a Means of Constructing Knowledge in the Life Sciences:
## Theory and Practice

**Grant E. Gardner**
*Middle Tennessee State University, USA*

**Kristi L. Walters**
*East Carolina University, USA*

## EXECUTIVE SUMMARY

*The use of small collaborative learning teams in STEM classrooms is not new to the field of education. At the undergraduate level, evidence continues to accumulate that organizing students into groups in which they engage in knowledge construction by completing active learning tasks is an effective means to achieve student-learning objectives. However, this teaching method is rarely used by postsecondary faculty, especially in large-enrollment classes. An argument for the efficacy of this method is presented in three parts. This chapter first outlines the theoretical basis for collaborative group learning. Grounded in the literature, this theory is then translated into practice by discussing evidence-based advantages and challenges to creating collaborative learning environments. The chapter concludes with a discussion of a case study examining how the first author has implemented this method of collaborative instruction with a unique means of structuring groups within a large-enrollment non-majors biology classroom.*

DOI: 10.4018/978-1-4666-6375-6.ch009

## ORGANIZATION BACKGROUND

Recent science education policy documents recommend that students learning science at all levels should be modeling the process of scientific discovery in their classrooms through inquiry-driven learning experiences (National Research Council, 2000). In the context of undergraduate life sciences education the American Association for the Advancement of the Sciences' (2011) *Vision and Change in Undergraduate Biology Education* also highlights the benefits of modeling the process of science after and during formal instruction by adopting student-centered classrooms at all levels from K-16. "In practice, student-centered classrooms tend to be interactive, inquiry-driven, cooperative, collaborative, and relevant. Classes authentically mirror the scientific process, convey the wonder of the natural world and the passion and curiosity of scientists, and encourage thinking" (AAAS, 2011, p. 7). This pedagogy, often called *scientific teaching*, is based on the idea that both the teaching and learning of science should model the methodologies of science and worldviews of scientists (Handelsman, *et al.*, 2004).

One of the most common means through which inquiry-based, student-centered instruction is implemented is by organizing students in larger classroom environments into small learning teams that promote cooperation, collaboration, and interaction in a more targeted manner than attempting to promote student-learning at the whole class level. These teams are typically groups of four to five students that work together to achieve classroom learning objectives in conjunction with, or independent of, the instructor (depending on the particular instructional methodology being implemented). At the postsecondary level, organizing students into collaborative groups in which they engage in knowledge construction by completing active learning tasks (within these groups) has been shown through Discipline-Based Education Research (DBER) to be an effective means to achieve critical learning objectives in Science, Technology, Engineering, and Mathematics (STEM) fields (Bowen, 2000; Springer, Stanne, & Donovan, 1999). More specifically, collaborative group work in in STEM classrooms increases academic achievement, promotes positive attitudes, increases students' reasoning ability and promotes student retention (Armstrong, Chang, & Brickman, 2007; Bowen, 2000; Jenson & Lawson, 2011; Johnson, Johnson, & Smith, 1998; McKinney & Graham-Buxton, 1993) as well as numerous other cognitive and affective advantages.

Despite the voluminous evidence base for team learning as a means for structuring effective student learning environments, widespread implementation and sustainability of these types of classrooms models at the undergraduate levels remains a challenge. This is often due to STEM faculty being uncomfortable with, or outright resistant to these research-based methodologies. As Tanner (2009) states in her series on undergraduate biology teaching and learning, "(O)ften, we as instructors feel that we need to be an intimate part of each student's learning, when in fact it is more important that we construct opportunities for them to do the learning themselves" (p. 94). The question becomes, in a historical paradigm of a lecturing "sage on the stage" that delivers content to students through the power of prose, why and how should postsecondary instructors be assisting students in constructing their own knowledge? More specific to this discussion, what advantages are there to utilizing small groups as a classroom tool for knowledge construction? The following chapter presents a case of one instructors' implementation of a research-based instructional strategy in a large enrollment non-majors life sciences course.

## Context of the Case

The lead author and instructor during this case reported here has worked at several large public universities in the southeastern United States including one with the third largest undergraduate population in North Carolina (East Carolina University) and currently at one with the largest undergraduate population in Tennessee (Middle Tennessee State University). Both of these institutions consist of populations of students that are largely rural, low-to-middle-class socio-economic status, and many who are first-generation college students. Like many state-funded institutions, the last few years have seen a decrease in funding from the state governing bodies with a subsequent increase in faculty classroom responsibilities, time commitments, and class sizes. In addition, many state funding formulas are moving from being based purely on enrollment numbers to putting more weight on retention rates and time-to-completion.

As an instructor, the lead author has had the responsibility of teaching both majors and non-majors biology students with classrooms consisting of upward of 250 students in a single class section. In fact, as I write this, East Carolina University has recently opened a 500-seat section of introductory biology for non-majors students to be taught in the campus performing arts building which typically shows artistic performances and movies. What this serves to highlight is the challenges currently presented to public universities that are attempting to maximize opportunities for student success while subsequently making decisions based on limited funding (which often works against student success and retention).

The lead author has a terminal degree in science education and is committed to implementing active, student-centered pedagogies in the course. Active techniques utilized in the course were "flipped" teaching methods with students viewing an audio-recorded lecture (that aligned with their assigned reading) prior to coming to class; limited use of in-class lectures and being sure that lecture time incorporated dynamic visuals such as animations, etc.; having students work in-class in small groups on exercises, problems, and case studies; and the use of student response devices (i.e. "clickers") as a means of formative assessment. The instructor was also assisted by an undergraduate assistant who was available to rotate around the room during class time to facilitate student learning during group work.

For this particular case, we discuss some of the means through which the lead author implemented team learning in a non-majors biology class of 252 students enrolled at East Carolina University (enrolling approximately 26,000 undergraduate students) in the Fall of 2011. The majority of the students were freshmen and had an average age of 20.5 years old with a range of 19 to 23 (with three outlier continuing education students with ages of 38, 33, and 33). The class consisted of 74.6% females and 25.4% males. The majority of the students self-identified as Caucasian (58.8%) with the rest of the students identifying as African American (20.4%), Hispanic (6.4%) or self-identified as Other (14.4%). These demographics are similar to that of the university as a whole except for the larger distribution of females enrolled in the course (the University distribution is about 59% female, 41% male). Full university Institutional Review Board was exempt status obtained prior to collecting and reporting data on students. Students were required to read and sign an online permission letter prior to participation.

## SETTING THE STAGE

In this case we present an argument for the benefits of small group work in undergraduate life science classrooms in three parts. First, we briefly outline some of the theoretical bases for small group learning as it is embedded in the framework of social constructivism. Next, we review some of the advantages of group instruction at the undergraduate level as supported by the science education and DBER literature. Finally, as an instructive case study, we describe how we have implemented this method of instruction within the lead author's large non-majors biology classroom. The final section offers evidence on the areas of success in this approach, describing pilot data we have collected to evaluate the impacts of collaborative group structure, utilizing a unique team-building method, on achieving student learning objectives. The goal of this chapter is to assist in translating theory into practice for those interested in the idea of utilizing small collaborative teams to promote learning in undergraduate science classroom as well as reinvigorate the discourse in scholarship in this field.

### Learning in Groups: The Theory of Social Constructivism

Using small collaborative groups in order to achieve critical learning objectives is not a new concept in science education, having been initially established with the theoretical and empirical work of the social psychologist Lev Vygotsky (1978). One of his most influential hypotheses to the field of education was that higher cognitive processes were developed through, and could not be divorced from, social interactions. In other words, learning cannot be conceptualized as an individual process as learning environments are culturally-embedded and the student must filter verbal instruction through the medium of language (that is in turn defined by cultural norms) (Vygotsky & Cole, 1978). Vygotsky claimed that communication between individuals was one of the key factors that facilitates the acquisition of conceptual understanding through the process of shared learning in a social setting. All students experience cognitive growth despite their differential academic positions because of their social interactions in the defined learning environment. In fact, within this sociocultural framework, cognitive diversity is viewed as an asset to learning and not a hindrance (as it is sometimes viewed by instructors). How often have instructors lamented having to "teach down" to their "weaker" students while fantasizing of classrooms filled with their "brighter" students? From the perspective of social constructivism, building of new knowledge requires diverse learning environments as all students benefit from interacting with a range of knowledge and ability levels. Learning is viewed as a social activity facilitated by more capable peers.

While much of Vygotsky's work focused on young children, his observations of the learning process have been applied to all ranges of leaners. Through his empirical work, Vygotsky observed that the memory skills of a younger child could be improved by working with an adult or more-capable peer (*i.e.*, someone with more highly developed memory skills). To explain this phenomenon, he theorized the Zone of Proximal Development (ZPD) as a hypothetical range of potential cognitive development of an individual learner that is maximized when learning takes place in collaboration with more capable peers. This theory of social construction of knowledge within a ZPD have been extended to other areas of social learning such as social interdependence theory that asserts that there are certain learning objectives that can *only* be achieved when situated in social settings (Johnson & Johnson, 2009). This idea of knowledge construction as a social enterprise is not unusual to scientists and engineers who frequently view knowledge building as a process of "standing on the shoulders" of others to move the collective understanding of their field forward. For example, imagine the validity you would grant our argument provided here without the frequency of citations provided from those who have come before us?

Despite the theoretical arguments of Vygotsky and their direct application to learning in STEM fields, in our experience, undergraduate classrooms continue to frequently reward the success of the individual learner at the expense of the classroom as a whole (or smaller collaborative groups). Individual competition is either explicitly or implicitly encouraged at the expense of the social construction of knowledge. Much of this individualistic classroom culture, at least in our classrooms in the United States, is a reproduction of the larger western culture that praises individual achievement over collectivization (Johnson & Johnson, 2009). On a smaller scale, as anecdotally evidenced by numerous conversations the lead author has had in his office over the years, for many students, conceptual learning is a secondary (or perhaps even a lower priority) goal than ensuring future career success. Individual competition is seen as the only means to get into highly competitive graduate or professional schools. Often times, this is true and only reinforces individual competition. Students' culture outside the classroom and pressure to succeed tends to promote individually competitive classrooms that can stifle learning based on a social constructivist framework. Instructors are often equally to blame by reproducing this competitive culture by supporting this hyper-individualist discourse in both word and deed. For example, studies have shown that the classroom culture created by instructors can greatly influence the learning orientation of students (Patrick *et al*., 2001).

Lest we, as authors, be branded as anti-individualist, let us be clear that there are numerous academic and pragmatic benefits to creating an individually competitive classroom culture. However, our argument lies in the volumes of theoretical and empirical data that the desired learning outcomes in STEM fields such as engagement, motivation, utilizing evidence to make arguments, developing coherent explanations for natural phenomena, communicating, and justifying arguments are promoted by *collaborative* and *interactive* classroom norms (Bowen, 2000; Springer *et al.*, 1999). There are certainly sub-sets of learning objectives in the science and engineering fields that would be promoted by individual learning environments as well. However, since social constructivism encourages communication and interactions with other people to enable learning, team learning is an ideal choice for the classroom that wishes to promote these types of values and habits of mind in science students. It is, of course, not the only choice (see AAAS, 2011 for other teaching methodologies that support higher-order learning in undergraduate classrooms). Collaborative learning brings groups of students together and allows them to work as a team to discover a solution to a problem, promoting the type of cooperation we often value in our own scientific or engineering pursuits. There is also much empirical support for these claims. For example, in a meta-analysis of cooperative versus competitive student interactions on problem-solving tasks, the cooperative groups consistently outperformed individuals on all forms of problem-solving including linguistic, symbolic, operational and ill-defined problems (Qin, Johnson, & Johnson, 1995).

Having established theoretical support for small group learning in the STEM fields it is important to now address what explicit advantages small collaborative group instructions provides for not only undergraduate students but also for faculty in undergraduate classroom settings. These advantages to collaborative group learning are by no means comprehensive, but provide a general categorical framework for examining some of the more poignant aspects of group learning outcomes. The advantages we list are supported by evidence from discipline-based education research as well as empirical studies in other fields. This discussion is intended to point the reader to potential resources for understanding the advantages of team learning in undergraduate STEM classrooms.

## Advantages to Collaborative Group Learning

*Collaboration in academic settings increases student retention in STEM pipelienes.* In an age when the United States is losing its competitive advantage in STEM fields and student attrition in these fields is increasing at all levels (National Science Board, 2013), one of the biggest advantages to small group instruction is its impact on retention at postsecondary institutions. For example, in a recent National Science Board (2013) study, students in 4-year institutions beginning in the academic year 2003/04 with intentions to major in science and education had 67% retention when measured in spring 2009. Work in general academic experiences has shown that one of the most powerful predictors of student retention (as well as their involvement in other aspects of the institution) has to do with personal and social interactions with both faculty and other students in their field (Tinto, 1987). These social interactions seem to be especially vital to at-risk students such as women and under-represented minorities (Jonides, 1995). The importance of teaching styles that promote social interactions in the classroom seem to be particularly important in STEM fields where students often perceive the teaching faculty as distant and the classroom culture as individualistic and highly competitive. These negative perceptions of how the fields of science and engineering work often become a major impetus for students moving into other disciplinary fields or leaving the university all together (Seymour & Hewitt, 1997; Tobias, 1992).

However, the benefits (*i.e.,* lack of anonymity and decreased student passivity) of a more intimate classroom can be achieved through team-based work even in large-enrollment classes with several hundred students. By establishing small student groups, the class size is effectively reduced and students have readily-established peers. In the first authors' instruction with freshman, I have frequently had students return to me years later and say that they are still friends with their small groups from their initial class and that these groups were critical for emotional and academic support throughout their program. The use of team-based learning in the classroom often means that the majority of class time is spent on active-learning exercises which helps keep the students engaged, increases the interactions between peers, and allows the student to take greater responsibility for their own learning (Michaelsen, 2002). The increased accountability and autonomy felt by the students also often has the side-affect of improving attendance and class preparedness as well (Michaelsen, 2002).

*Collaborative group work is responsive to student diversity.* Retention of women and underrepresented minorities in the fields of science and engineering is beneficial to maintaining a diverse community of scholars and promoting differential means of looking at the natural world. One of the benefits of having students interact in collaborative teams is that it promotes diverse interactions and reduces racial, ethnic, and physical ability stereotypes, as well as improves the personal self-esteem of minority students (Michaelsen, 2002). Students are better able to complete the border crossing into the science and engineering community of practice when they view their voice as valued within that community. Small group learning often allows those individuals to have their voice heard. It has been demonstrated that cooperative learning systems appear to improve academic and attitudinal outcomes for a diversity of students including women and under-represented minority groups. The latter groups also seem to prefer collaborative learning opportunities in contrast to competitive ones (Cabrera *et al.*, 2002).

In addition, to diversity of cultural backgrounds, students enter the classroom with a diversity of learning styles and abilities. There is some debate in the literature as to the structure and empirical basis for learning styles, however, it does seem apparent that students maintain preferences for how they integrate new information into their current constructs as well as how they process that information (Pashler, McDaniel, Rohrer, & Bjork, 2009). If nothing else, collaborative group learning provides an additional venue for which students who thrive in this type of learning environment can promote their own understanding. In addition, as part of the liberal arts goals of many institutions of higher education, when students are allowed to work with individuals of differing background with different worldviews, this promotes their own experiential education.

*Collaborative group work helps students develop higher order thinking skills.* Increasing student participation in the learning process not only improves students' comprehension of the material through active discussion with their classmates, but can also encourage higher-order cognitive skills such as critical thinking skills (Crowe, Dirks & Wenderoth, 2008). Group work can encourage students to use higher-order cognitive skills (*i.e.,* the Application, Analysis, Synthesis, and Evaluation levels of Bloom's Taxonomy) a well as improve the students' comprehension of the material by discussion with their classmates (Crowe *et al.*, 2008). Many science and engineering faculty report that productive functioning in a group to complete a task is one of the most important learning objectives for their students during their undergraduate career. For example, simply encouraging students to talk in the classroom can be essential to learning, as it helps them integrate new information into their cognitive structure (Tanner, 2009). The interaction between the students as they discuss the assignment helps improve their memory, stimulates cognitive function, and also reinforces the students' societal connections through these social exchanges (Michaelsen, 2002). In a recent study Jenson & Lawson (2011) demonstrated that inquiry biology instruction (grounded in students completing activities in collaborative groups) significantly increased their reasoning ability (as measured by the Classroom Test of Scientific Reasoning). Practice presenting their ideas to the team can also help an anxious student feel more comfortable. When a student can successfully explain a concept in their own words to a peer, it helps them to better understand the information and is less intimidating than presenting to the entire class.

*Collaborative group work prepares students for the world outside of academia.* Collaborative group learning has been shown to help students achieve not only goals for future careers, but also to achieve some of the goals for a liberal education such as a commitment to social engagement, cultural interest, and community leadership (Sweet & Pelton-Sweet, 2008). Being part of a team can provide emotional and social support, as well as the development of vital interpersonal skills necessary for future employment. Some students may be uncomfortable working in teams, as they feel they know more than their peers and believe they are unlikely to benefit from a group experience (Crowe *et al.*, 2008). Others may be unwilling to discuss their thoughts in front of fellow students (or the instructor) for fear of being ridiculed. However, these are the students that often benefit the most from participating in the group, as it encourages them to form their own explanations and assists with the construction of new knowledge (Tanner, 2009). In addition, specific work skills such as interpersonal skills and teamwork learned during collaborative experiences are some of the most in demand skills of future employers.

*Collaborative group work promotes faculty professional development and innovation.* It is no surprise to many in the field of undergraduate education that many university courses are lecture-based with students passively receiving information (Cox, McIntosh, Reason, & Terenzini, 2011). As mentioned before, this passivity in the classroom frequently leads to student dissatisfaction with their science courses and eventual attrition to other fields (Seymour & Hewitt, 1997). This dissatisfaction becomes a two-way street, with faculty feeling the effects of student disappointment in their success in courses. However, the relative ease of enacting collaborative group work within science classrooms opens up opportunities for professional development of faculty. Perhaps the greatest appeal of using team-based learning in the classroom is that the majority of class time is spent on activities instead of lecture (Michaelsen, 2002). It increases the interactions between students and the instructor and allows the students to self-teach the material. This frees up time for the instructor to pursue other obligations and encourages autonomy in the students. The instructor can focus on the more challenging concepts, while students cover the basics on their own, in small groups, or outside of class time. In addition, during team-based learning activities the active instructor can more easily identify areas of confusion and student misconception as they are verbally working through the concepts.

## CASE DESCRIPTION

In the current environment of ubiquitous budget cuts for all levels of education, classroom sizes are continuing to grow as both monetary and time support for instructors is diminishing in parallel (Crowe *et al.*, 2008). Nowhere is this felt more than in undergraduate classrooms where class sizes in traditional lecture courses can sometimes reach up to 300 students. Most K-12 classrooms have at least some established student-teacher ratio maximum that apparently dissolves as a critical consideration in student learning the instant students matriculate from high school. It is no secret that achieving more complex learning objectives, maintaining positive student attitudes, and encouraging retention becomes more challenging as the class size swells. As we have established, student-centered teaching mechanisms such as the utilization of collaborative learning teams might mediate these challenges somewhat. Benefits of a more intimate classroom can be achieved through team-based work even in courses with large classrooms.

In the following we describe how the lead author has effectively "reduced" his class size by creating learning teams in his large introductory biology courses with the intent of maximizing student acquisition of critical learning objectives. We visualize our particular version of collaborative team learning as defined by Fink (2002) as an instructional method that is intended to support the development of high-performance learning teams and provide the chance for these groups to work together during the completion of significant learning tasks. It brings groups of students together (often with different abilities, interests, and levels of motivation) and allows them to work together to discover a solution to a problem or socially construct their knowledge in the course.

For many of the positive reasons mentioned above, we are committed to promoting small-group collaborative learning opportunities for students in these classes. The question that is frequently asked (even by departmental colleagues) is: "how?" How do you cover all the material expected of you in an introductory course and still have time for small group instruction? How do you convince students this is a worthwhile learning method? How do you ensure student interactions in a large lecture hall where seating does not promote these interactions? How do you form groups to ensure that they are productive? How can you manage such a large number of students in small groups and still ensure that they are achieving the learning goals you set for them? We would like to address each of these questions in turn in the following section from our experiences and aligned with this specific case. Within this case study we include specific examples of how the lead author has attempted to address these challenges as well as provide data from an empirical pilot study that has sought to examine the most productive means of structuring these learning teams within the context of non-majors undergraduate life science students.

## CURRENT CHALLENGES FACING THE ORGANIZATION

*How do you cover all the material expected of you in an introductory course and still have time for small group instruction?* This is one of the most frequent questions we get asked and also the one that requires the biggest paradigm shift in how an instructor thinks about teaching and learning. Instructors with this questions are essentially asking: "if I do not stand up during class and tell them what they need to know (*e.g.*, the "sage on the stage" philosophy), but instead provide them with higher level learning activities in small groups how will they ever learn all the material?" Our response refers largely back to the values of a social constructivist classroom that stress the importance of providing students with meaningful learning opportunities and an ability to address learning on their own as opposed to *delivering* material to students as if they were empty vessels (Tanner, 2009).

We as instructors need to realize that providing students with information by "delivering content" is no longer as critical a role for an instructor as it has been in the past. We live in the *google* age where information is largely at everyone's finger tips. In fact, I inform my students on the first day of class that I will not spend time providing them with a lot of facts or definitions during in-class time. To reinforce the reasoning behind this idea I tell them that I could likely ask them the most difficult factual question in biology I could think of, and they would be able to hop on their smart-phones, tablets, or laptop computers and deliver an answer to me. However, the key is that they may not understand the answer and also that they do not have any concept of its validity in the context of the question asked. This is where restructuring classroom philosophies comes into play. If the goal is to have our students know how to locate, validate, and manipulate scientific information we need to give them guided opportunities in class to practice these skills not talk at them like an audio-driven search engine of information.

As a mechanism of structuring the course to promote student factual acquisition of material outside of class while maintaining in-class time for working with this material, the lead author has implemented a "flipped classroom" strategy. I do this by posting Microsoft PowerPoint lecture notes on our online course management website. These PowerPoint lecture notes directly correspond to the assigned readings. I then create *Camtasia* files with these PowerPoints by voicing over a lecture that corresponds to the reading content, essentially providing the lecture the night prior to the class in a format that Millennial students understand; digital media. These lectures are typically 15 to 30 minutes in length and can be paused or re-wound so that students can use them as a structured reading guide. Students are expected to complete the review of the information before coming to class so that we can spend actual class time working on higher-level tasks utilizing the material presented the night before.

Although there are some critiques to this mode of instruction that have begun to arise in the literature. One of the most pointed is the assumption that students will engage with the material outside of class and actually comprehend it without the aid of an instructor. One interesting twist on this is combining a learning cycle philosophy with a flipped classroom methodology by having students encounter material first by *engaging* and *exploring* with new material in the classroom, utilizing out-of-class learning tools to *explain* new material and the classroom experiences, then finally bringing the *extension* and *evaluation* activities back into the classroom (Schneider, Wallace, Blikstein, & Pea, 2013). This requires envisioning the flipped classroom as spanning over multiple days but seems to hold some promise in promoting student learning in early comparative studies (see the Schneider *et al.*, study above) but little empirical work has been gathered.

How do you convince students this is a worthwhile learning method? Establishing these benefits is critically important from the *first day of class*. We find that in our large-enrollment undergraduate courses the first day is often spent with the faculty instructor going over the syllabus with the students in a teacher-centered method of instruction. The instructor often reads through the syllabus with the assistance of a PowerPoint outline, while highlighting course policies, rules, and procedures. As the old adage goes, "you only have one chance to make a first impression." We believe this is true in this situation as well. By reading over the syllabus the first day of class you establish that the course will be teacher-centered and will require students to abide by rules and procedures in order to be successful. Where are the student-centered goals of supporting learning in a classroom environment shaped in this way on the first day?

To address this issue in the first author's course, we begin learning in groups the moment the students walk in the classroom. Group composition is established before the first day of class (see below) and students find their groups when they walk in the door. Group lists are posted all around the classroom and the students are encouraged to check the list on their online course management website prior to class so they can look up their group number to which they have been assigned. A map of the classroom is displayed on the overhead so that students know the general region to congregate in order to locate their group. They are immediately set to a learning task after meeting and talking briefly with their group a short 5-10 minute ice breaker session. In the wrap-up session at the end of the class information is provided on the benefits of working in groups to the students including: 1) that working in groups will help promote higher level cognitive skills and critical thinking, 2) that working in groups will promote their acquisition of life skills in communication and working with a group to accomplish a task, 3) that working in groups will reduce their anxiety by effectively reducing class size, and 4) working in groups will provide immediate points of contact for support both in and outside of the classroom environment.

*How do you ensure student interactions in a large lecture hall where seating does not promote these interactions?* According to Michaelsen (2002) the first step to implementing team learning is to prepare the learning environment. Since most large classes are taught in lecture halls with set seating, it may be difficult to impossible to rearrange seats for the groups, however, simply having a permanent area designated for each group within the classroom can provide them with a sense of "home". This space needs to facilitate communication between the students (with eye contact) and allow access for everyone (*i.e.*, to expedite handing out papers and to allow the instructor to monitor student progress). Additionally, it is beneficial to ensure that spaces are easily accessible for students with physical challenges. Again, we believe the crux of this issue is for an instructor to release some of their own need for control over the classroom setting and allow students to explore in comfortable settings themselves. We typically encourage students not to remain in their seats (that are organized for a typical instructor-focused lecture classroom), but instead to get up, move around, sit on the floor, go outside, etc. We do encourage them to stay close by so that the instructor can continue to circulate around and address groups with questions or provide real-time feedback as they complete group work. In addition, we have had the benefit of receiving undergraduate assistants in some of our large courses that allow for multiple monitors to circulate and assess the progress of the group work.

In a way this is often the most difficult challenge to addressing small group learning as classrooms promoting small group interactions often tend to be limited (at least on our campus) and in high demand. In addition, classroom in which we teach rarely remain consistent from semester to semester. Manipulating the learning environment is often the variable under which we feel we have the least control.

*How do you form groups to ensure that they are productive?* Although the concept of using learning teams to promote science learning is not new, there is still debate in the literature as to the most effective means of structuring these groups to best promote interactions and to maximize learning. In general, this argument revolves around whether homogenously or heterogeneously structured groups are better to advance learning. This idea is further complicated by the fact that there is little consensus in the literature about exactly what variables are being referred to when one speaks of homogenous or heterogeneous. It is worth taking a portion of this manuscript to discuss this briefly below, as this is the basis for some of the pilot data we will share.

Frequently, team members are randomly assigned to their groups (Dolmans & Schmidt, 2006). In the 1970's Belbin noted group function could be improved by controlling the structure (McHarg, Kay, & Coombes, 2012). He also observed that teams composed of individuals with different characteristics were the best functioning because they had different strengths and weaknesses. He developed a self-perception inventory that grouped participants into one of eight categories. These labels were then used to compose the teams. This helped reduce problems for teams with a poor group dynamic. Michaelsen (2002) also encourages heterogeneous teams based on the distribution of student talents and liabilities evenly between the groups. He recommends using work experience, access to technology, and demographic data to create the teams.

Mello and Ruckes (2006) found that heterogeneous teams were better at dealing with changes and challenging situations, but that the members' different backgrounds and views could become a weakness since they made different choices. In this situation, homogeneous teams have an advantage, because they have similar inclinations and tend to work better together. Mello and Ruckes (2006) hypothesized that a heterogeneous team is better informed than a homogeneous team because of their diverse characteristics, however, a heterogeneous team may still find it difficult to work together. Mello and Ruckes (2006) further note that a homogeneous team may have similar background knowledge and will therefore have less information available to make decisions.

According to Wright and Drewery (2006) team diversity based on race, ethnicity, gender, and other factors can initially lead to division in the team, but this is often corrected by spending more time getting to know each other. Eventually, the differences perceived by the team members become insignificant and the group becomes a cohesive team. Wright and Drewery (2006) also noted that teams with members from different cultures might have diverse methods for dealing with conflicts within the team. Groups organized by gender have some interesting dynamics. Ro and Choi (2011) observed that groups were more successful when women outnumbered men and that male dominated group performance was worse than female dominated or mixed groups. Women also performed better in all female groups, while men performed better in mixed groups. Unfortunately, women tend to be more stressed when working in groups than their male counterparts.

In the review by Dolmans and Schmidt (2006), they reported the significance of motivation in teams, and noted that student motivation impacted group efficiency and communication. It is well known that motivation plays a key role in student learning (Koballa & Glynn, 2007). They observed that students with low levels of motivation interfered the most with the learning process. This may be connected to the students' lack of interest in the course and the low level of motivation is the manifestation of the decreased interest. In other words, because the less motivated students lack interest in the class, they were more disruptive and contributed less to the team. The motivational levels of the group can also effect the cognitive function of the students (Dolmans and Schmidt, 2006). This is likely connected with the effects of social constructivism. Since the whole team is not working together, the dysfunction frustrates the other team members and impacts their ability to learn. Also, they are not receiving the benefits of team communication and the social interactions that properly functioning teams enjoy. De Grave, Dolmans, & Van Der Vleuten (2002) directly observed the significance of motivation on group function, and noted its importance.

In thinking about the important variables that might be used to construct groups with non-majors biology students, we hypothesized that the construct of motivation to learn science might actually be most critical in structuring collaborative learning groups (Glynn, Brickman, Armstrong, & Taasoobshirazi, 2011). Prior to the beginning of class our students were asked to take a six-scale motivation instrument (Tuan, Chin, & Shieh, 2005) online through the Blackboard learning platform. Data was downloaded to spreadsheet software and then sorted along a continuum of high to low motivation students. Collaborative groups were structured heterogeneously based on a six-scale motivation instrument using a "snaking method". In other words, the individual with the highest motivation score was placed in group one, the next individual in group two and so on until there was a total of 50 collaborative groups with five individuals in each; consisting of students ranging along the motivational construct. Students remained in these groups for the entire semester.

*How can you manage such a large number of students in small groups and still ensure that they are achieving the learning goals you set for them?* Using a pre-post design, an assessment was made of students' changes in attitudes toward biology, perceptions of the science of biology, perceptions of biologists, attendance rates, and achievement in the course. Much to the surprise of the first author, it was discovered that initially, many students reported a moderate to high level of motivation (along the sum of the six-scale instrument) to learn science prior to the course with 0% of the students falling into a poor motivation quartile, 1% of the students falling into a low motivation quartile, 56% of the students falling into a moderate motivation quartile, and 43% of students falling into a high motivation quartile. These results should be viewed cautiously as they are self-report and recorded prior to when the students had close contact with me. In other words, it is entirely possible they were trying to make a good initial impression.

The first three outcomes were assessed using a pre-post instrument design: students' attitudes toward biology (Biology Attitude Assessment), students' perceptions of biology, and students' perceptions of biologists. The Biology Attitude Assessment was a newly designed Likert-type assessment instrument that was validated with the literature and analyzed for reliability and factor structure. A five factor solution was determined including student confidence in biology, student interest in biology, perceived social relevance of biology, perceived personal relevance of biology, foreignness of biology as a subject, and comfort in the laboratory (Bartlett's test of sphericity $\chi^2 = 3{,}235.34$, df = 528, $p < 0.0001$; Kaiser-Meyer-Olkin sampling adequacy KMO = 0.893; 59.37% of the variance was accounted for by the model). The two perceptual measures were items that began "biology is…" and "biologists are" with a list of dichotomous descriptors that students were asked to indicate the decree to which they perceived a particular descriptor was accurate. In addition we utilized student attendance scores as recorded by students' clicker response systems and students' achievement in the class as measured by final exam grades.

Initial results show that on the Biology Attitude Assessment students demonstrated a significant increase in confidence ($p = 0.004$), interest ($p = 0.012$), social relevance ($p = 0.01$), and confidence in the laboratory ($p = 0.017$) utilizing a Wilcoxon sign rank test. There was no change in students perceived personal relevance of biology or perceived foreignness of biology as a subject. On perceptual measures, students increasingly viewed biology as more enlightening ($p = 0.006$), more gratifying ($p = 0.006$), more patterned ($p = 0.033$), more unbiased ($p = 0.003$), and more ethical ($p = 0.021$). Students also perceived biologists as more hard-working ($p = 0.040$), more ethical ($p = 0.022$), and more likely to be female ($p = 0.048$) following the course. The first author felt that many of our learning objectives were achieved for this course.

## SOLUTIONS AND RECOMMENDATIONS

Within this chapter we have attempted to provide both theoretical and research-based arguments for the benefits of team-based and collaborative learning methods within the undergraduate science classroom. Social constructivism is a robust theory that lends credence to research and teaching in group based learning environments. We also acknowledge that educational theory might be foreign to many postsecondary faculties, but we encourage readers to examine these ideas and assess their validity as well as partner with educational theorists in your home institution to pursue these ideas further.

Research on team-based, collaborative, and cooperative learning has a long history, but it is we feel the field has begun to stagnate. Most importantly, there is copious evidence that small group learning works, but the literature has begun to wane on the evidence as to how to create the most effective learning ecology for these groups (learning environments, group structure, etc.) (Anderson & Nielson, 2013). We encourage other faculty to continue the dialogue on the efficacy of small collaborative group work and collect assessment data to evaluate and validate the discipline-based work that they are completing. Our work continues as we further examine the structure of groups utilizing motivation variables and we are currently conducting comparative and experimental studies to more closely examine how this type of group structure effect student learning along cognitive and affective variables (Gardner, under review).

Finally, we introduced anecdotal and empirical case study evidence for the effectiveness of utilizing team learning within our particular classroom context. In addition, we discussed some of the pragmatic hurdles that we have faced in our own classrooms and how we have dealt with these issues. In this sense, we hope we have provided some practical solutions and recommendations for classroom instructors to effectively and efficiently convert their classrooms to a small-group based learning environment. We encourage suggestions and future dialogue. There is a need for more work in the area as well as a need to translate maintain a bridge between the theoretical and practical so as to create learning environments that are not only theoretically sound but that can be implemented in a wide range of contexts (Daniel, 2012).

## REFERENCES

American Association for the Advancement of Science (AAAS). (2011). *Vision and Change in Undergraduate Biology Education*. Washington, DC: Author.

Andersen, H. M., & Nielsen, B. L. (2013). Video-based analyses of motivation and interaction in science classrooms. *International Journal of Science Education*, *35*(6), 906–928. doi:10.1080/09500693.2011.627954

Armstrong, N., Chang, S. M., & Brickman, M. (2007). Cooperative learning in industrial-sized biology classes. *CBE Life Sciences Education*, *6*(2), 163–171. doi:10.1187/cbe.06-11-0200 PMID:17548878

Bowen, C. W. (2000). A quantitative literature review of cooperative learning effects on high school and college chemistry achievement. *Journal of Chemical Education*, *77*(1), 116–119. doi:10.1021/ed077p116

Cabrera, A. F., Crissman, J. L., Bernal, E. M., Nora, A., Terenzini, P. T., & Pascarella, E. T. (2002). Collaborative learning: Its impact on college students' development and diversity. *Journal of College Student Development*, *43*, 20–34.

Cox, B. E., McIntosh, K. L., Reason, R. D., & Terenzini, P. T. (2011). A culture of teaching: Policy, perception, and practice in higher education. *Research in Higher Education*, *52*(8), 808–829. doi:10.1007/s11162-011-9223-6

Crowe, A., Dirks, C., & Wenderoth, M. P. (2008). Biology in bloom: Implementing Bloom's Taxonomy to enhance student learning in biology. *CBE Life Sciences Education*, *7*(4), 368–381. doi:10.1187/cbe.08-05-0024 PMID:19047424

Daniel, D. B. (2012). Promising principle: Translating the science of learning to educational practice. *Journal of Applied Research in Memory and Cognition*, *1*(4), 251–253. doi:10.1016/j.jarmac.2012.10.004

De Grave, W. S., Dolmans, D. H. J. M., & van der Vleuten, C. P. M. (2002). Student perspectives on critical incidents in the tutorial group. *Advances in Health Sciences Education: Theory and Practice*, *7*(3), 201–209. doi:10.1023/A:1021104201303 PMID:12510142

Dolmans, D. H. J. M., & Schmidt, H. G. (2006). What do we know about cognitive and motivational effects of small group tutorials in problem-based learning? *Advances in Health Sciences Education: Theory and Practice*, *11*(4), 321–336. doi:10.1007/s10459-006-9012-8 PMID:16953462

Fink, L. D. (2002). Beyond small groups: Harnessing the extraordinary power of learning teams. In L. Michaelsen, A. Knight, & D. Pink (Eds.), *Team based learning: A transformative use of small groups*. Sterling, VA: Stylus Publishing.

Gardner, G. E. (Manuscript submitted for publication). The effects of constructing collaborative groups using motivation on undergraduate student attitudes and perceptions of biology and biologists. *CBE Life Sciences Education*.

Glynn, S. M., Brickman, P., Armstrong, N., & Taasoobshirazi, G. (2011). Science motivation questionnaire II: Validation with science majors and nonscience majors. *Journal of Research in Science Teaching*, *48*(10), 1159–1176. doi:10.1002/tea.20442

Handelsman, J., Ebert-May, D., Beichner, R., Bruns, P., Chang, A., & DeHaan, R. et al. (2004). EDUCATION: Scientific Teaching. *Science*, *304*(5670), 521–522. doi:10.1126/science.1096022 PMID:15105480

Jensen, J. L., & Lawson, A. (2011). Effects of collaborative group composition and inquiry instruction on reasoning gains and achievement in undergraduate biology. *CBE Life Sciences Education*, *10*(1), 64–73. doi:10.1187/cbe.10-07-0089 PMID:21364101

Johnson, D. W., & Johnson, R. T. (2009). An educational psychology success story: Social interdependence theory and cooperative learning. *Educational Researcher*, *38*(5), 365–379. doi:10.3102/0013189X09339057

Johnson, D. W., Johnson, R. T., & Smith, K. A. (1998). *Active learning: Cooperation in the college classroom*. Edina, MN: Interaction Book Company.

Jonides, J. (1995). *Evaluation and dissemination of an undergraduate program to improve retention of at-risk students*. Ann Arbor, MI: University of Michigan College of Literature, Science, and Arts.

Koballa, T. R. J., & Glynn, S. M. (2007). Attitudinal and motivational constructs in science learning. In S. K. Abell, & N. G. Lederman (Eds.), *Handbook of research on science education* (pp. 75–102). Mahwah, NJ: Lawrence Erlbaum Association Publishers.

McHarg, J., Kay, E. J., & Coombes, L. R. (2012). Students' engagement with their group in a problem-based learning curriculum. *European Journal of Dental Education*, *16*(1), e106–e110. doi:10.1111/j.1600-0579.2011.00682.x PMID:22251332

McKinney, K., & Graham-Buxton, M. (1993). The use of collaborative learning groups in the large class: Is it possible? *Teaching Sociology*, *21*(4), 403–408. doi:10.2307/1319092

Mello, A. S., & Ruckes, M. E. (2006). Team composition. *The Journal of Business*, *79*(3), 1019–1039. doi:10.1086/500668

Michaelsen, L. K. (2002). Team-based learning in large classes. In L. K. Michaelsen, A. B. Knight, & L. D. Fink (Eds.), *Team-based learning: A transformative use of small groups* (pp. 157–171). Westport, CT: Praeger Publishers.

National Research Council. (2000). *Inquiry and the National Science Education Standards: A guide for teaching and learning*. Washington, DC: National Academies Press.

National Science Board. (2013). *STEM Education Data and Trends*. Retrieved October 1, 2013, from http://www.nsf.gov/nsb/sei/edTool/timeline.html#5

Pashler, H., McDaniel, M., Rohrer, D., & Bjork, R. (2009). Learning styles: Concepts and evidence. *Psychological Science in the Public Interest*, *9*, 105–119.

Patrick, H., Anderman, L. H., Ryan, R. M., Edelin, K. C., & Midgley, C. (2001). Teachers' communication of goal orientations in four fifth-grade classrooms. *The Elementary School Journal*, *102*(1), 35–58. doi:10.1086/499692

Qin, Z., Johnson, D. W., & Johnson, R. T. (1995). Cooperative versus competitive efforts and problem solving. *Review of Educational Research, 65*(2), 129–143. doi:10.3102/00346543065002129

Ro, H., & Choi, Y. (2011). Student Team Project: Gender Differences in Team Project Experience and Attitudes Toward Team-Based Work. *Journal of Teaching in Travel & Tourism, 11*(2), 149–163. doi:10.1080/15313220.2011.575022

Schneider, B., Wallace, J., Blikstein, P., & Pea, R. (2013). Preparing for future learning with a tangible user interface. The case of neuroscience. *IEEE Transactions of Learning Technologies*, *6*(2), 117–129. doi:10.1109/TLT.2013.15

Seymour, E., & Hewitt, N. M. (1997). *Talking about leaving: Why undergraduates leave the sciences.* Boulder, CO: Westview Press.

Springer, L., Stanne, M. E., & Donovan, S. S. (1999). Effects of small-group learning on undergraduates in science, mathematics, engineering, and technology: A meta-analysis. *Review of Educational Research, 69*(1), 21–51. doi:10.3102/00346543069001021

Sweet, M., & Pelton-Sweet, L. M. (2008). The social foundation of team-based learning: Students accountable to students. *New Directions for Teaching and Learning, 2008*(116), 29–40. doi:10.1002/tl.331

Tanner, K. D. (2009). Talking to learn: Why biology students should be talking in classrooms and how to make it happen. *CBE Life Sciences Education, 8*(2), 89–94. doi:10.1187/cbe.09-03-0021 PMID:19487494

Tinto, V. (1987). *Leaving college: Rethinking the causes and cures of student attrition.* Chicago, IL: University of Chicago Press.

Tobias, S. (1992). *Revitalizing undergraduate science: Why some things work and most don't.* Tucson, AZ: Research Corporation.

Tuan, H.-L., Chin, C.-C., & Shieh, S.-H. (2005, January). The development of a questionnaire to measure students' motivation towards science learning. *International Journal of Science Education, 27*(6), 639–654. doi:10.1080/0950069042000323737

Vygotsky, L. S., & Cole, M. (1978). *Mind in society: The development of higher psychological processes.* Cambridge, MA: Harvard University Press.

Wright, N. S., & Drewery, G. P. (2006). Forming cohesion in culturally heterogeneous teams: Differences in Japanese, Pacific Islander and Anglo experiences. *Cross Cultural Management: An International Journal, 13*(1), 43–53. doi:10.1108/13527600610643475

## ADDITIONAL READING


Allen, D., & Tanner, K. (2009). *Transformations: Approaches to college science teaching*. New York, NY: W. H. Freeman Scientific.

Barkley, E. F., Cross, K. P., & Major, C. H. (2004). *Collaborative learning techniques: A handbook for college faculty*. Academic Press.

Bruffee, K. A. (1998). *Collaborative learning: Higher education, interdependence, and the authority of knowledge*. Academic Press.

Gillies, R. M. (2007). *Cooperative learning: Integrating theory and practice*. Academic Press.

Hamada, M. (2013). *Active and collaborative learning: Practices, problems, and prospects*. Academic Press.

Handelsman, J., Miller, S., & Pfund, C. (2007). *Scientific teaching*. New York, NY: W. H. Freeman.

Hmelo-Silver, C. E., Chinn, C. A., Chan, C., & O'Donnell, A. M. (2013). The international handbook of collaborative learning. Academic Press.

Michaelson, L. K., Knight, A. B., & Fink, L. D. (2004). *Team-based learning: A transformative use of small groups in college teaching*. Academic Press.

Michaelson, L. K., Parmalee, D. X., McMahon, K. K., & Levine, R. E. (2007). *Team-based learning for health profession education: A guide to using small groups for improving learning*. Academic Press.

Millis, B., & Rhem, J. (2010). *Cooperative learning in higher education: Across the disciplines, across the academy*. Academic Press.

O'Donnell, A. M., Hmelo-Silver, C. E., & Erkens, G. (2012). *Collaborative learning, reasoning, and technology*. Academic Press.

Slater, S. J., Slater, T. F., & Bailey, J. M. (2010). *Discipline-based education research: A scientist's guide*. New York, NY: W.H. Freeman Scientific.


## KEY TERMS AND DEFINITIONS

**Collaborative Learning:** An instructional method in which students work together in small groups toward a common goal.

**Cooperative Learning:** A structured form of group work where students pursue common goals while being assessed individually.

**Discipline-Based Education Research (DBER):** A general form of education research that is embedded in and informed by the content discipline in which it contextualizes itself.

**Social Constructivism:** A theory of learning that posits that the 'mind' is located in the individual-in-social action and that learning is a process of enculturation into a community of practice.

# Chapter 10
# Interdisciplinary Problem-Based Learning Practices in Higher Education

**Despo Ktoridou**
*University of Nicosia, Cyprus*

## EXECUTIVE SUMMARY

*More and more students in higher education are enrolling on interdisciplinary programs. This phenomenon occurs since universities are breaking the boarders of a single subject area. At the university of Nicosia, the lecturer of two interdependent courses: MGT-372 Management of Innovation and Technology and MIS-151 Business Software Applications attempted to bring together students from different disciplines to explore the two topics. More specifically, through Interdisciplinary Problem-Based Learning (IPBL), the lecturer (author) aimed to eliminate the fragmentation and the learning of isolated skills and investigate students' motivation for learning and their level of active engagement through the use of technology (Google Apps). To address the above, the study employed a case study approach, collecting qualitative data through student focus groups, online/in-class observations, and lecturers' comments. The study showed that students seemed intrigued and satisfied working on interdisciplinary tasks, shared prior and newly researched knowledge, as well as acquired an integrated viewpoint and solution-focused strategies deriving from those disciplines.*

DOI: 10.4018/978-1-4666-6375-6.ch010

## ORGANIZATION BACKGROUND

More and more students in higher education are enrolling on interdisciplinary programs within multidisciplinary departments. This phenomenon occurs since universities are breaking the boarders of a single subject area. Interdisciplinary learning (IL) initiatives are multiplying throughout higher education at an extraordinary rate (Creamer& Lattuca 2005; DeZure 1999). At the university of Nicosia, the lecturer, of two interdependent courses: MGT-372 Management of Innovation and Technology and MIS-151 Business Software Applications attempted to bring together students from different disciplines to explore the two areas: management of technology innovations and software applications. The topics varied and focused on technology innovations, software applications, information communication technologies, social networking technologies and e-business.

The implementation of IL approach, for the purpose of the current study, aimed to seek meaningful connections between the two courses were students could complete a critical analysis of those connections, for example integrate Social media technologies in an enterprise.

With the current study the lecturer intended to provide students new opportunities to experience deep, challenging and relevant learning through stimulating activities using knowledge and learning from two different courses.

The two aforementioned courses were united on one relevant topic/issue/problem, through the analysis, comparison, and contrast of perspectives contributed by each discipline. Contextualized, complex, open-ended, authentic problem sets and innovative projects based on the union of the two courses, were presented to students individually and/or in groups for investigation. A class/virtual discussion followed as well as a group progress report on earlier/present learning issues and future plans (Ktoridou 2010).

More specifically, through Interdisciplinary Problem-Based Learning - IPBL the author aimed to eliminate the fragmentation and the learning of isolated skills allowing students through the use of technology (Google Apps) to increase their motivation for learning as well as their level of active engagement. To address the above, the study employed a case study approach, collecting qualitative data through focus groups with students, online/in-class observations and lecturers comments.

This section discusses the essential background information on Interdisciplinary Learning and Problem-Based Learning. It then proceeds with a brief introduction to Cloud Computing Applications and Services along with a description of Google Applications for education. Finally brief descriptions of the two courses used for the purposes of case study: MIS-151 Business Software Applications and MGT-372 Management of Innovation and Technology follow.

*Figure 1. Interdisciplinary learning diagram*

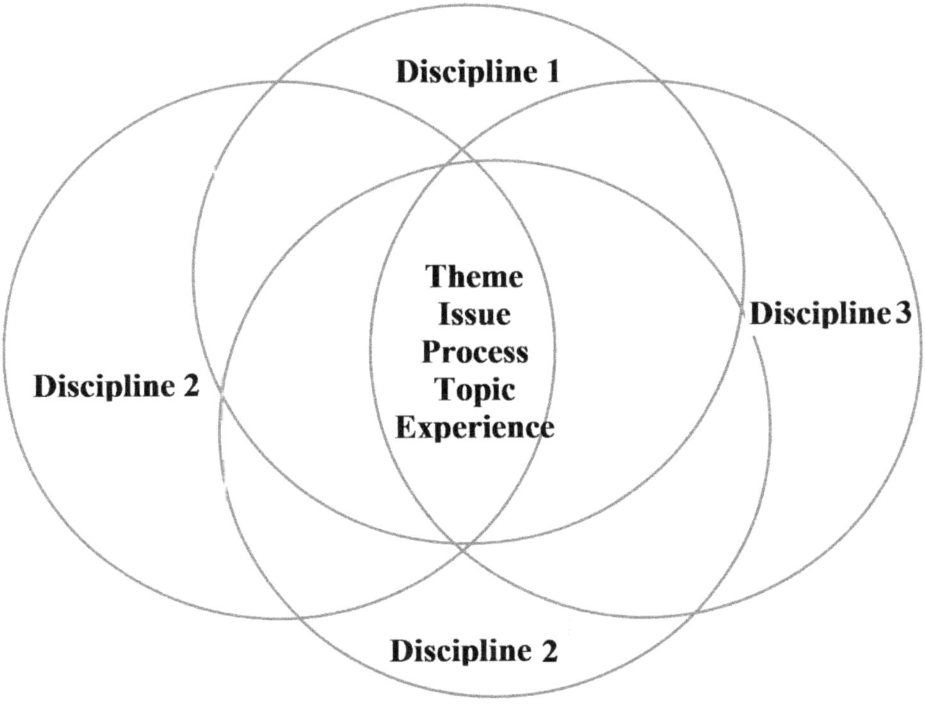


## Interdisciplinary Learning

Interdisciplinary education is defined as the process in which two or more subject areas are integrated with the goal of fostering enhanced learning in each subject area. According to Jacobs, the disciplines may be related through a specific theme, issue, problem, process, topic, or experience (See Figure 1) (Jacobs, 1989).

Implementing IL in the course curriculum offers educators opportunities to enhance and enrich their teaching processes and also create exciting learning experiences for their students. The concept of interdisciplinary education recognizes the interrelationships among distinct subjects. Educators arrange their curriculum around common themes, issues, processes, topics, experiences and skills across disciplines to facilitate learning.

According to Repko (2009), IL fosters advances in cognitive ability while other educational researchers (Kavaloski 1979 and Newell 1999), have identified a number of distinct educational benefits of interdisciplinary learning including gains in the ability to: recognize particularities, think critically, tolerate uncertainty, recognize and appreciate ethical concerns. IL recognizes the interrelationships of one subject to another by emphasizing connections between traditionally distinct disciplines (Repko 2009; Kavaloski 1979; Newell 1998).

IL was on the different topics of the aforementioned courses to acquire a deep and thorough understanding of complex issues related to the curriculum and help students: a) identify insights from the aforementioned subject areas that contribute to an understanding of the issue under consideration and b) develop the ability to integrate concepts and ideas from these subject areas into a broader conceptual framework of analysis (Starting Point Teaching and Learning Economics 2010). A comprehensive search of the educational literature (Ackerman, 1989; Ackerman & Perkins 1989; Field, Lee & Field 1994) describe the anticipated learning outcomes of IL, that cover the development or enhancement of improved thinking and learning skills (Ivanitskaya, Deborah, Montgomery & Primeau, 2002):

- **Ackerman (1989):** Flexible thinking; Understand of the strengths and limitations of disciplines; Ability to assess value to knowledge gained.
- **Ackerman & Perkins (1989):** Enhanced thinking and learning skills; Improved higher-order cognitive skills; Improved content retention; Capacity for proactive and autonomous thinking skills; Ability to devise connections between seemingly dissimilar contexts.
- **Field, Lee, & Field (1994):** Ability to tolerate ambiguity or paradox; Sensitivity to the ethical dimensions of issues; Enlarged perspectives and horizons; Ability to synthesize or integrate; Enhanced creativity, original insights or unconventional thinking; Enhanced critical thinking; Capacity to perceive a balance between; subjective and objective thinking; Humility, sensitivity to bias, and empowerment; Ability to demythologize experts (Ivanitskaya, Deborah, Montgomery & Primeau, 2002).

With this work the author attempted to unite two different undergraduate courses (MIS-151 Business Software Applications and MGT-372 Management of Innovation & Technology) on one relevant topic/issue/problem, through the analysis, comparison, and contrast of perspectives contributed by each discipline. More specifically, through IL the authors aimed to eliminate the fragmentation and the learning of isolated skills allowing students through the use of technology (Google Apps) to increase their motivation for learning as well as their level of active engagement.

## Problem-Based Learning

Problem-based learning (PBL) can be defined as an instructional method characterized by the use of "authentic" problem sets as a context for students to develop critical thinking and problem solving skills, and acquire the necessary course concepts. PBL is considered as an increasingly essential part of education reform around the world. Michel Bischoff, and Jakobs, 2002). Domin and Dutch defined PBL as the approach that challenges students to learn through engagement in a real problem or situation (Domin, 1999; Duch, 1995). The main principle of PBL is for students to play the role of problem-solvers and develop critical thinking abilities, knowledge acquisition, decision making, teamwork and productive collaboration skills, self-evaluation, and flexibility to accept the change (Ryan and Quinn, 1994). More specifically, the PBL process is as follows: Initially the educator presents the problem, then the students explore the involved learning issues and they define possible problems; in a group environment they investigate potential solutions by researching prior and new knowledge essential for solution finding; and finally they document their problem solution (See Figure 2).

Schmidt in his work defined PBL as the approach that is based on the following principles of cognitive psychology (Schmidt 1993):

1. Activation their prior knowledge;
2. Elaboration prior knowledge through joined discussions;
3. Reform of prior knowledge to adjust to the presented problem set;
4. Learn in a conceptual complex authentic context of a problem;
5. Foster curiosity due to relevance of the problem. (Schmidt, 1993)

PBL challenges students to learn through engagement in a real problem where problem solving strategies, context knowledge and skills develop by placing them playing the role of problem-solvers. More specifically the process used in PBL is separated in the following four stages:

**Stage 1:** *"Problem Presentation"* is when students are presented with a problem case or project. They organize and elaborate their ideas and prior knowledge through group discussions and attempt to identify the broad nature of the problem.

**Stage 2:** *"Concerns Posing"* students through group discussion a collaborative environment is developed and they define their concerns -what they know and what they do not know.

**Stage 3:** *"Concerns Ranking"* here all concerns that were posted during the previous stage are ranked by the students, in order of importance. More specifically, it is decided which concerns/questions will be followed up by all group members and which concerns/questions will be assigned to individual members so that later they will come and present their finding to the rest of the group. In addition discussions with the lecturer over the needed resources needed in order to research the learning issues take also place during this stage.

*Figure 2. Problem-based learning process*

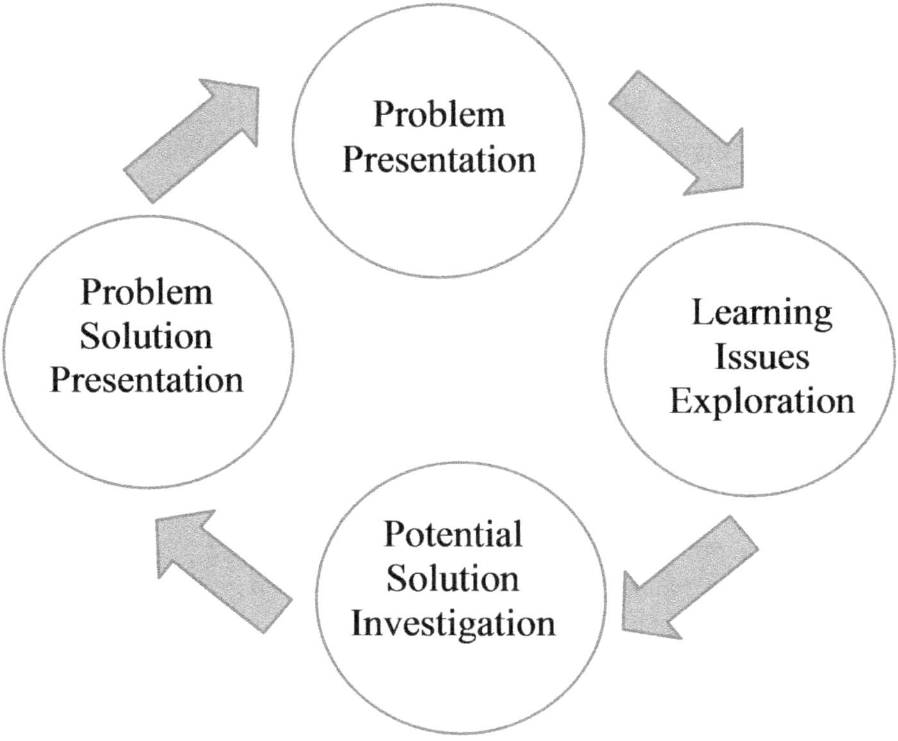


**Stage 4:** *"Connections of New and Prior Knowledge"* students explore the previous concerns/problems and integrate the newly researched knowledge into the context of the problem. Actually here students are encouraged to review their knowledge and connect new concepts to prior ones, this way as they progress new issues are defined and students will realize that learning is an ongoing process, and there will always be something new to learn.

The educator's role in the PBL process is to guide, support and facilitate students' initiatives. It is important for every educator who decides to incorporate PBL in their courses to realize that students are responsible for their learning and he/she must encourage them to have their protagonist role in their learning. More specifically, students will develop with the educator relevant and meaningful assessments for high quality work. Assessments must be meaningful by having connections to the real world. For the purposes of the current case study general, and at minimum, students were assessed in three areas:

1. **Applied Competence:** Demonstrate the ability to use the latest trends in Information Communication Technologies to identify and examine variables that can influence the effective operation of a business.
2. **Problem-Solving, Critical Thinking, and Communication Capabilities:** Identify problems and/or opportunities in a business context and make specific recommendations. Examine similar existing real-life business case studies to improve the situation by interrupting and solving the problem, and effectively communicating the findings with the group always with a commitment to quality.
3. **Collaborative and Leadership Competence:** Group collaboration, initiative for becoming group leader mainly to identify and solve problems or follow opportunities for learning and improvement within the group. (Purser 2010

## Cloud Computing: Applications and Services

Since the phenomenon "Internet Technology" was introduced to the world tremendous evolutionary technologies evolved with cloud computing perhaps or is considered the most biggest. Cloud computing can be defined as the delivery of hosted services over the Internet. According to Rhoton, cloud computing is currently the only solution that integrates three different types of services into a single framework (Rhoton, J, 2009). These services are broadly divided into three categories:

- **Software-as-a-Service (SaaS):** A software distribution model where applications are hosted by a service provider and made available to users customers over the Internet.
- **Infrastructure-as-a-Service (IaaS):** A provision model where organizations can outsources the equipment used to support operations, including storage, hardware, servers and networking components.
- **Platform-as-a-Service (PaaS):** A service delivery model that rents hardware, operating systems, storage and network capacity over the Internet services to organizations.

SaaS Enables software to be deployed from provider, delivered over the Internet, and accessed by subscriber SaaS is becoming an increasingly prevalent delivery model with more and more web-based services being available to the user over the Internet. Services can be anything from Web-based email to inventory control and database processing; most importantly with ubiquitous access. The delivery of such service could obviously benefit education.

Particularly today's cloud computing providers hosting both the application and the data offer higher education institutions the opportunity to substitute a presence in "the cloud" for their existing data centers, servers, and applications,

## Google Applications for Education

With cloud computing in education, students can have ubiquitous access to powerful software and massive computing resources. The pioneers offering cloud services for education were initially Google with Google Docs (Beswick, 2011) followed by Microsoft with its Cloud Application product Microsoft Azure (Dudley, 2010).

In August 2006 Google Inc. announced the launch of Google Apps for individual domains, a set of hosted applications for organizations that need to provide high quality communication tools to their users without having to do anything related to software and hardware installation and maintenance (Google, 2006). Currently, Google provides one of the best cloud computing applications. Google Apps is currently available in the following editions: Google Apps Standard, Google Apps Business, Google Apps Education, Google Apps Government and Google Apps non-profit. Data in Google Apps are stored in the cloud instead on end-users computers, and multiple users can communicate, collaborate and share documents simultaneously without worrying about using the same operating system, software, or browser. Attachment round-trips are eliminated; storing data in the cloud saves time and reduces frustrations for effective team work. Google is able to efficiently manage security issues across the nearly homogeneous global cloud computing infrastructure by implementing a multi-layered security process protocol designed to help keep users data safe. Furthermore, it provides controls so as administrators can manage which applications their users could access and how they could use each service.

The idea of the classroom being the only outlet for students to learn from and interact with faculty is outdated, as the development of out-of-the-classroom tools continues to skyrocket (Motschnig & Holzinger, 2002). Google applications for education are one of those tools offering constant enterprise innovation saving a university time, money and hassles of managing IT solutions. At present, over 10 million students, staff, faculty and students are actively using Google Apps for Education. In addition to that, 60% of U.S. universities with hosted email use Google Apps for Education (Lardinois, 2010). The Google Apps Education edition provides a set of customizable tools to the entire campus community enabling students, faculty, and staff to share information and ideas easily through the school's custom email addresses, shared calendars, collaborative web sites, shared online documents, safe instant messaging, and more. These tools can be categorized in three groups: 1) Communication: hosted email, shared calendars and integrated video chat, 2) Collaboration: students and teachers can share documents online at any time and location via Google Docs and Google sites, and 3) Customization: IT systems can be easily integrated and adjusted with Google (Google Apps, 2010).

Moreover, all educational accounts are free of charge and the school's registered domain can be used. Time and money savings for the institutions are guaranteed: no assignment loses, no computer platforms and software versions updates. The most important, though, is that all these tools are available to lecturers and students anytime and anywhere that there is an Internet accessible computer. In addition, Google applications can be innovatively applied in higher education curricula, by developing an online environment, meeting students' needs and providing access to learning and higher order thinking skills development. The advanced technology needed is provided for faculty and students to collaborate, and communicate in a totally different dimension, providing numerous learning opportunities and challenges (Eteokleous & Ktoridou, 2011).

For the purpose of this case study Google Applications for Education edition were employed providing a set of customizable tools enabling students, from MIS-151 and MGT-372 courses, to have ubiquitous collaboration, communication and sharing while working on interdisciplinary problem-based tasks. More specifically, students through a) communication tools - hosted email, and integrated video chat could communicate and work virtually on assigned projects. b) Collaboration & Sharing tools - Google Docs and Google sites could develop and share their assigned work on almost any mobile device or tablet anytime, anywhere (Wolf 2010).

## MIS-151 Business Software Applications

MIS-151 Business Software Applications is a required MIS course that introduces the fundamental concepts of Business Information technology components: Computer Systems, Telecommunications and Networks and the Internet; identifies business problems calling for the implementation of the latest software applications; introduce the basic web technologies that support business applications and explore the challenges in bringing businesses on the web. (Web 2.0, Cloud Computing, M-computing, Social Networking, User Generated Content) and examines issues of business data resource management.

*MGT-372 Management of Innovation and Technology* is an elective management course that emphasizes how the future manager or entrepreneur can use strategic management of innovation and technology to enhance firm performance. Students will focus on the business skills needed to develop their ideas and innovative approaches and learn to create and implement solutions to social problems related to technology innovations. More specifically, students will: Understand the significant role social entrepreneurs play in society, develop leadership and organizational skills to strategically manage innovation and technology within a growing social enterprise as well as learn to apply business tools and strategies.

## SETTING THE STAGE

The current work reports the findings from a case study exploring undergraduate students' experiences in an Interdisciplinary Problem-Based Learning (IPBL) cloud environment. More specifically, IPBL was implemented within two independent courses: MIS-151 Business Software Applications and MGT-372 Management of Innovation and Technology, through the integration of Google Applications for education. The two courses, MGT-372 and MIS-151, run in parallel for Spring 2012, Fall 2012 and Spring 2013 semesters with an average of 100 students attending.

Students meet twice a week in sessions of 75 minutes each. Groups of 5-6 students were formulated from both courses. During the courses, each student shared specific knowledge from his/her discipline while at the same time learned from others. The course syllabi for both courses included topics in technology evolution, securing innovations, social networking sites, cloud computing, virtual communities, e-business, and user generated content. The lecturer applied the same assessment strategies for both courses: lectures, case study analysis and discussion, academic paper discussions, in-class and virtual exercises, and presentations. Through in-class and virtual observations, the lecturer tried to investigate how interdisciplinary learning differs from more traditional learning focused on single-subject topics.

According to (Baloche, Hynes, Berger, 1996; Humphreys, Post, Ellis 1981; Jakobs 1999) traditional learning may be too structured and limited in range to meet students' educational goals. In addition, (Baloche, Hynes, Berger, 1996) state that approaches supporting specific disciplines often fail to reveal how a particular discipline interfaces with another.

The data collection methods were the following: in-classroom and online activities observations, reflective journals, focus groups and, peer and self-evaluation, and finally through evaluation of students' assignments and homework activities. Regarding the last-mentioned data collection method, the students were requested to individually analyze a topic/ problem by answering questions and in groups to develop presentations of a topic/ problem.

The observation sheets for the in-classroom and online activities observations were developed based on the principles of interdisciplinary learning (as mentioned above). More specifically, during group discussions the lecturer observed students' group work, (i.e. communication and collaboration processes) in recognizing particularities, thinking critically, tolerating uncertainty, recognizing and appreciating ethical concerns. Along the same lines, the reflective journals sheets were developed.

The lecturer was requested to complete a reflective journal by the completion of each lecture. The reflective journal was developed based on the principles of interdisciplinary learning. Both the observation sheet and the reflective journals aimed in examining how the IL method was employed throughout the lesson's processes: i.e. lecturing, readings, assignments, guidelines as well as in-classroom and online activities.

More specifically groups of 5-6 students were formed consisting of students from both courses (2-3 from MGT-372 and 2-3 students from MIS-151 or vice versa). Through this the lecturer aimed to help students identify insights from the two different course curriculums that contribute to an understanding of the given issue/topic/problem and develop the skills to integrate concepts and ideas from these two courses into a broader conceptual framework of analysis.

Finally, students were asked to present their findings in 8-10 slides. They had the freedom to work in groups to develop the structure of the presentation outlining, though, the most important aspects of the issue/topic/problem. It is significant note that students had to investigate and brainstorm the issue/topic/problem individually and then meet with their team members to discuss their findings and come-up with conclusions. Finally they needed to develop the presentation structure and present it in class.

The lecturer was involved as a consultant to the whole teaching/learning process. After individual and/or group investigation and brainstorming, the lecturer observed their online/in class group work for the preparation of the common presentations. The lecturer made sure that the criticisms to be offered constructively. The lecturer observed and evaluated the presentations based on clarity, role assignment (within the group), design of presentations, timing, style of delivery and ability to answer questions.

Six focus groups (one focus group at the end of each semester) were organized. Specifically, student-participants for each focus group were chosen based on a number of criteria, such as age, gender, specialization, and educational background. The focus group members were students from both courses with 10 students participating in each focus group (5 students from the MIS-151 course and 5 from MGT-372 course). On average, the duration of each focus group was between 1 to 1.5 hours. Open-ended questions were used to encourage students to share their opinions and experiences and freely express themselves.

The focus groups were conducted in order to get insights and to in-depth analyze the concepts under investigation from students' points of view, investigating (Kvale 1986) the distinct educational benefits of interdisciplinary learning including gains in the ability to: recognize particularities, think critically, tolerate uncertainty, recognize and appreciate ethical concerns. More specifically, the focus groups targeted to investigate and explore students' views, experiences, and perceptions in learning in a cloud IPBL cloud environment.

## CASE DESCRIPTION

Findings of the current case study provide evidence that Interdisciplinary Problem-Based Learning practices through Google Applications proved to be a motivating and engaging approach that enhanced students' problem-solving abilities to recognize particularities, think critically, tolerate uncertainty, recognize and appreciate ethical concerns for two distinct courses. The use of cloud resources, asynchronous communication, collaboration and sharing opportunities facilitated and enhanced students experiences, knowledge construction and extend learning in a interdisciplinary problem-based learning environment.

It mirrors positive student perceptions on the IPBL practices in the current courses.

More specifically, the focus groups targeted to investigate and explore students' views, experiences, and perceptions on gains in working in a cloud IPBL environment.

For the analysis of the focus groups a manual approach was followed: category and sub-category heading titles were identified followed by short paragraphs summarizing findings for each sub-category and finally quotes to each sub-section were added.

### Students' First Impressions Using IPBL for the First Time

Even though students had prior experience with PBL, IPBL was a novel approach to them. According to students comments IPBL was not only a motivating approach that engaged them actively in the learning process but also they seemed to enjoy the learning of multiple skills. As two second year students from MIS151 class said:

*I could never imagine myself working with students from another distinct course on a specific topic and come-up with a solution to a complex problem.*

*I believe interdisciplinary topics helped me to broaden my views on approaching a problem even when it is not my subject area and develop my critical thinking skills.*

In addition the selection of topics given to students is of major importance to the IPBL approach. A comment from a third year student from MGT372 class was:

*Working in an IPBL environment helped me to access a given problem from many different aspects and I believe that from now on I will try to work with multiple sources of information when I have to work on a complex problem.*

## Students' Perceptions towards Developing Problem-Solving Abilities in Recognizing Particularities for the Two Distinct Subject Areas

It is evident that students seemed satisfied in working with real life problems and recognized the opportunities, for developing deeper understanding and problem-solving skills in recognizing particularities for two distinct subject areas. As one first year student from MIS-151 course said:

*The case study of "Web 2.0 applications in the sports industry" was difficult for me to understand and relate technology evolution and web 2.0 applications but with group work all rising learning issues were solved.*

Having to deal with complex problems combining two distinct areas intrigued students to use and share prior and new knowledge in a collaborative environment and come up with a solution. A strong comment from a third year student from MGT372 class was:

*We were given a problem to mobilize a business. I believe that was a topic for MGT-372 class. Nevertheless, a first year student from MIS151 class with his knowledge on mobile apps development led the group towards finding the most appropriate solution.*

## Students' Perceptions towards a More In-Depth Exploration of Topics, Issues, and Problems within and across Distinct Subject Areas

An issue mainly for the lecturer and author of this case study was how to motivate students towards in-depth exploration of topics, issues and problems two distinct subject areas. IPBL developed students' confidence in facing challenges motivated and urged them to investigate the given problems and come up with a solution. A student from MIS151 class stated:

*When the lecturer gave us the topic for investigation I thought it was not interested in such topic. But after a long discussion with my group i realized that it was innovative, interesting and I was challenged to offer myself to do the major investigation.*

## Students' Perceptions towards Development of Critical Thinking

IPBL approach managed to promote critical thinking by supporting students to identify the appropriate knowledge and skills needed to investigate the given problem, bring together information and ideas from different subjects, consider alternative ways of solving a problem predict outcomes and give explanations. Two comment from second and third year students from MGT-372 were:

*After we were given the problem we had a group discussion to see who knew what so as to see what background knowledge we had and therefore which learning issues we had to face.*

*The fact that in my group we had first year students from MIS-151 course initially made me question the quality of outcomes. But after collaborating in a ILPBL environment I realized that it was a unique opportunity to consider so many alternative ways to solve the given problem.*

## Students' Perceptions towards Web-Based IBPBL

It is evident that even though most of the students had prior experience with Google Applications, IPBL was benefited and facilitated. Google Applications provided a set of customizable tools enabling students, from the two different courses, to have ubiquitous collaboration, communication and sharing while working on interdisciplinary problems. Comments from two students were:

*Google Talk offered to our group ubiquitous communication and collaboration. We could meet often and at our own pace.*

*While having a video conferencing with my peers I was using Google Search to investigate learning issues for the given problem.*

## Interview with the Lecturer on Students Reactions after Using IPBL

Significant statements from the lecturer revealed insights for the implementation of IPBL in the two aforementioned undergraduate courses. Even though working in an IPBL environment was a new experience for the students it can be said that such an approach can provide challenging and pleasant learning experiences. Two of her comments were:

*Revisiting an idea or skill from different viewpoints deepened their understanding. This motivated them to actively participate suing their existing knowledge and find the most suitable solution.*

*Research has demonstrated that interdisciplinary teaching can increase students' motivation for learning as well as their level of active engagement. By monitoring their web-based collaborations recognize the value of what they are learning and become more involved in it.*

IPBL if is planned well it can eliminate the fragmentation and the learning of isolated skills. It allows students to access a particular issue from different views while working with multiple sources of information and perspectives. Her exact comments were:

*IPBL needs to be planned very well not only to eliminate the fragmentation and the learning of isolated skills and motivate students but also for the educator to better differentiate instruction and create more challenging and rich methods of assessment.*

She continued…

*…the topics, the issues and the problems I have chosen for my students mainly aimed for a more in-depth exploration for a better understanding of different perspectives across the two subject areas.*

## Lecturer's Comments on Challenges Faced Using IPBL

As mentioned in previous sections of this work, students were responsible for their learning and the lecturer played the role of the facilitator. According to lecturers' comments, the major challenge was that IPBL was a new learning experience for the students. More specifically students had no prior experiences with curriculums that included space for learning beyond subject boundaries and therefore have to make connections between different areas of learning. An initial comment on the challenges was:

*Most of the students were ex-students of mine. I use problem-based learning, case-based learning and all my students have Google Apps accounts. I have observed that when assigned them the problem it was evident from their arguments that it was hard for them to realize the commotions between the diverse the two courses MIS-151 and MGT-372.*

Another two comments of the lecturer for the challenges were on the planning process:

*The most difficult part of implementing IPBL in my courses was uniting courses' contents on one relevant topic/issue/problem.*

*When a educator decides to implement IPBL, he/she must organize teaching and learning around problems, or issues, students in such a way for students to be motivated to look for knowledge and skills from multiple disciplines to provide a wide understanding of the learning issues.*

Another significant issue deriving from lecturer's observations is the way students communicated, collaborated and shared their findings within groups online. After they were presented to the problem they had to make individual investigations and then meet online to explore the involved learning issues and define possible problems. The major part of the "Play" was when in a group environment they investigated potential solutions using any prior and newly investigated knowledge necessary for finding a solution. The exact comment was:

*Ubiquitous communication and collaboration facilitated my students' group work and motivated them through the use of Google Apps towards learning and increased their level of active engagement eliminating any fragmentation and the learning of isolated skills.*

It is evident that an approach like IPBL can work as an educational benefit for courses and programs eliminate any fragmentation and the learning of isolated skills for the students. This way students become active learners and motivated towards learning.

## SOLUTIONS AND RECOMMENDATIONS

The current case study examined the challenges and opportunities of implementing Interdisciplinary Problem-Based Learning, through Google Applications, in two diverse undergraduate courses at the university of Nicosia.

Finding connections for diverse knowledge domains provides an in-depth understanding of common features, dimensions, and characteristics. IPBL even though a new learning approach for the students, it proved to be challenging since they could experience a wide spectrum of possible relationships between the two diverse subject areas.

From the theoretical background of PBL and the discussion of the practical implementation it can be concluded that students seemed intrigued and satisfied in working with complex real-life problems that combined two distinct areas, share prior and newly researched knowledge, as well as acquire integrated viewpoint and solution-focused strategies deriving from those disciplines.

Students developed critical thinking skills as they needed to bring together information and ideas from different subjects and consider alternative ways of solving a problem, predict outcomes and give explanations for problem investigation.

A well-planned IPBL can be benefited and facilitated if it is implemented through a web-based environment. Google Applications provided a set of customizable tools enabling students, from the two different courses, to have ubiquitous collaboration, communication and sharing while working on interdisciplinary problems.

As educators, we must continuously seek ways to enhance and enrich out reaching and learning processes and offer our students learning experiences that will influence the kinds of skills and knowledge they develop. Those learning experiences must be based on real-life problems, relevant and transferable to their future learning - the goal of IPBL teaching.

Finally, a key educational benefit of courses and programs that aim to use IPBL approach is in eliminating the fragmentation and the learning of isolated skills allowing students to increase their motivation for learning as well as their level of active engagement.

## REFERENCES

Ackerman, D. B. (Ed.). (1989). Intellectual and practical criteria for successful curriculum integration. In H. H. Jacobs (Ed.), Interdisciplinary curriculum: Design and implementation (pp. 25–38). Alexandria, VA: Association for Supervision and Curriculum Development.

Ackerman, D. B., & Perkins, D. N. (Eds.). (1989). Integrating thinking and learning skills across the curriculum. In H. H. Jacobs (Ed.), Interdisciplinary curriculum: Design and implementation (pp. 77–96). Alexandria, VA: Association for Supervision and Curriculum Development.

Baloche, L., Hynes, J. L., & Berger, H. A. (1996). Moving toward the integration of professional and general education. *Action in Teacher Education*, *18*(1), 1–9. do i:10.1080/01626620.1996.10462817

Barr, J. (2010). Host Your Web Site. In *The Cloud: Amazon Web Services Made Easy: Amazon EC2 Made Easy*. Melbourne: SitePoint.

Beswick, J. (2011). *Google Apps Express: The Fast Way To Start Working in the Cloud*. CreateSpace.

(1994). Cognitive apprenticeship and problem based learning. In *Reflections on Problem Based Learning* (pp. 15–33). Sydney: Australian Problem Based Learning Networ.

Creamer, E. G., & Lattuca, L. R. (2005). *Advancing Faculty Learning Through Interdisciplinary Collaboration: New Directions for Teaching and Learning, No. 102*. San Francisco, CA: Jossey-Bass Publisher.

DeZure D. (1999). Interdisciplinary Teaching and Learning. In *Essays on Teaching Excellence: Toward the Best in the Academy*. Academic Press.

Domin, D. (1999). A review of laboratory instruction styles. *Journal of Chemical Education*, *76*(4), 543–547. doi:10.1021/ed076p543

Duch, B. J. (1995). What is problem-based learning? *About Teaching: A newsletter of the Center for Teaching Effectiveness*, 47. Retrieved October 7 2013, from http://www.udel.edu/pbl/cte/jan95-what.html

Dudley, R. (2010). *Microsoft Azure: Enterprise Application Development*. Birmingham, UK: Packt Publishing.

Eteokleous, N., & Ktoridou, D. (2011). Higher education: A web 2.0 world of communication, collaboration, participation and sharing. In *Proceedings of ICICTE-International Conference on ICT in Education*. Rhodes, Greece: ICICTE.

Field, M., Lee, R., & Field, M. L. (1994). Assessing interdisciplinary learning. In J. T. Klein, & W. G. Doty (Eds.), *Interdisciplinary Studies Today* (pp. 69–84). San Francisco: Jossey-Bass.

Gogle Inc. (2006, August). *Google Launches Hosted Communications Services*. Retrieved Aug 25 2001, from http://www.google.com/intl/en/press/pressrel/gafyd.html

Humphreys, A. H., Post, T. R., & Ellis, A. K. (1981). *Interdisciplinary methods: A thematic approach*. Santa Monica, CA: Goodyear.

Ivanitskaya, L., Clark, D., Montgomery, G., & Primeau, R. (2002). Interdisciplinary Learning: Process and Outcomes. *Innovative Higher Education*, *27*(2), 95–111. doi:10.1023/A:1021105309984

Jacobs, H. H. (Ed.). (1999). The growing need for interdisciplinary curriculum content. In H. H. Jacobs (Ed.), Interdisciplinary curriculum: Design and implementation (pp. 1–11). Alexandria, VA: Association for Supervision and Curriculum Development.

Kavaloski, V. (Ed). (1997). Interdisciplinary education and humanistic aspiration: A critical reflection. In J. Kockelmans (Ed.), *Interdisciplinarity and Higher Education*. University Park, PA: The Pennsylvania State University Press.

Ktoridou, D. (2010, April). Applying an Inductive Method to a New, Multidiciplinary, Management of Innovation & Technology Course: Evidence from the University of Nicosia. In *Engineering Education Conference – The Future of Global Learning in Engineering Education* (pp. 452–460). IEEE. doi:10.1109/EDUCON.2010.5492422

Kvale, S. (1996). *Interviews: An introduction to qualitative research interviewing*. Thousand Oaks, CA: Sage.

Michel, M. C., Bischoff, A., & Jacobs, K. H. (2002). Comparison of problem- and lecture-based pharmacology teaching. *Trends in Pharmacological Sciences, 23*, 168 – 170.

Newell, W. H. (1998). *Interdisciplinarity: Essays from the Literature* (8th ed.). New York: The College.

Newell, W.H. (1990). Interdisciplinary curriculum development. *Issues in Integrative Studies*.

Purser, R. (2010). *Problem-Based Learning*. Retrieved October 7, 2013 from: http://online.sfsu.edu/rpurser/revised/pages/problem.htm

Repko, A. F. (2009). *Assessing Interdisciplinary Learning Outcomes* (Working Paper). School of Urban and Public Affairs, University of Texas at Arlington.

Schmidt, H. G. (1993). Foundations of problem-based learning: Some explanatory notes. *Medical Education*, *27*(5), 422–432. doi:10.1111/j.1365-2923.1993.tb00296.x PMID:8208146

Starting Point Teaching and Learning Economics. (2010). *Interdisciplinary Approaches in Learning*. Retrieved Sept 20 2013, from: http://serc.carleton.edu/econ/interdisciplinary/why.html

Wolf, T. (2010). Google Apps for Education Users Grow to 10 Million. *TMCnet Education Technology*. Retrieved October 1 2011, from: http://education.tmcnet.com/topics/education/articles/109042-google-apps-education-users grow-10-million.htm

## ADDITIONAL READING

ChemConnections Project. (2004). *W.W. Norton & Co. (interdisciplinary modules for introductory college chemistry) University of Michigan's Global Change I Course: A Technology-Enhanced Interdisciplinary Learning Environment*. Author.

Cheng, K. W. E., Cheung, S. C., & Wu, G. (2001). Examination of PBL and Web-based exercises in English language improvement for engineering students. In *Proceedings of the 3rd Asia Pacific conference on problem based learning: experience, empowerment and evidence*, (pp. 49-55). Callaghan, Australia: Australian Problem Based Network.

Donald, J. G. (2002). *Learning to Think: Disciplinary Perspectives*. San Francisco, CA: Jossey-Bass.

Google Apps. (2010). More than 10 million students use Google Apps. Retrieved October 1 2010 From http://www.google.com/a/help/intl/en/edu/index.html#utm_medium=et&utm_source=cath_all

Haynes, C. (Ed.). (2002). *Innovations in Interdisciplinary Teaching*. Wesport, CT: Oryx Press.

Hovland, K. (2006). *Science, Diversity, and Global Learning: Untangling Complex Problems. Diversity Digest, 9 (3).*

Ktoridou, D., & Eteokleous, N. (2013). Interdisciplinary Web-Based Learning Practices in Higher Education. In *Engineering Education – Collaborative Learning & New Pedagogic Approaches in Engineering Education* (pp. 536–539). IEEE. doi:10.1109/EduCon.2013.6530157

Lattuca, V., Voigt, L. J., & Fath, K. Q. (2004). Does Interdisciplinarity Promote Learning? Theoretical Support and Researchable Questions. *The Review of Higher Education*, *28*(1), 23–48. doi:10.1353/rhe.2004.0028

Mansilla, B., & Veronica, D. & Middlebrooks, K. (2004). Building bridges across disciplines: Organizational and individual qualities of exemplary interdisciplinary work (Interdisciplinary Studies Project, Project Zero). Harvard Graduate School of Education.

Nikitina, S. (2002). *Three Strategies for Interdisciplinary Teaching: Contextualizing, Conceptualizing, and Problem-Solving. Interdisciplinary Studies Project, Project Zero*. Cambridge, MA: Harvard Graduate School of Education.

## KEY TERMS AND DEFINITIONS

**Cloud Computing:** The internet representation in network diagrams, in the form of a cloud.

**Google Applications:** An open source suite of hosted email and collaboration applications for: individual users, Apps Business, Educational institutions Google Apps Governments and non-profit editions.

**Google Applications for Education:** A free suite of hosted email and collaboration applications exclusively for schools and universities.

**Infrastructure-as-a-Service (IaaS):** A provision model where organizations can outsources the equipment used to support operations, including storage, hardware, servers and networking components.

**Interdisciplinary Learning:** A learning approach that enables educators and students to make connections of experiences and outcomes from within and across curriculum areas.

**Platform-as-a-Service (PaaS):** A service delivery model that rents hardware, operating systems, storage and network capacity over the Internet services to organizations.

**Problem-Based Learning (PBL):** Students tackle with an open-ended, real-world problem and work in groups to identify learning needs and develop a feasible solution, with instructors acting as facilitators.

**Software-as-a-Service (SaaS):** A software distribution model where applications are hosted by a service provider and made available to users customers over the Internet.

# Chapter 11
# Transdisciplinary Research in Sustainable Scientific Education in the Field of Urbanism and Architecture

**Svetlana Perović**
*University of Montenegro, Montenegro*

## EXECUTIVE SUMMARY

*The chapter presents a case study based on transdisciplinary research, which was conducted at the Faculty of Architecture in Podgorica and is an innovation in architectural and urban practice of higher education in Montenegro. The study is based on the view that autonomous action of disciplines in the case of architecture and urbanism as multidisciplinary activities is limited, and an integrated approach to solving complex problems in the urban system is required. A global approach to research and solving urban issues is an important actor of sustainable development, where universities are central in this process. Collaborative educational discourses with a high degree of cooperation can develop an adequate platform for responses to the complex issues of the urban system. Producing experts with a developed awareness of a comprehensive understanding of the problem and transdisciplinary collaborative knowledge can strongly contribute to sustainable improvement, control, and management of urban spaces.*

DOI: 10.4018/978-1-4666-6375-6.ch011

## ORGANIZATION BACKGROUND

The study was conducted in 2013 at the Faculty of Architecture in Podgorica, Montenegro, a primary montenegrian institution of higher education in the field of architecture and urbanism.

Academic architectural and urban orientation in Montenegro is based on the integral program and urban conceptual strategy and integral methodological platform (Perovic, 2013).

Architectural and urbanism education at the University of Montenegro started in 2002 and has been developing on the systematic, synergetic strategy of urban studies and architectural programs, interacting at different levels of studying, communicative relationships, and tendency towards universal knowledge transfer(Perovic, 2013).

Various authors representing the importance of transdisciplinarity at universities (Andalécio, 2009; Nicolescu, 1998; UNESCO, 1998) start from the premise that the study of the complex issues of the modern world is not possible in a disciplinary context. The globalization of knowledge is essential in the third millennium. Transdisciplinarity at the universities is a condition for sustainable development (Nicolescu, 1998).

Authors who are dealing with transdisciplinarity in architecture and urbanism (Després, Vachon, & Fortin, 2011), indicate the importance of complex understanding of the problem, its complexity, and cooperation among different actors of society and forms of knowledge.

In this regard, there is importance in the implementation of transdisciplinarity as a model to better define the complex problems and identify adaptive solutions for sustainable development. Transdisciplinary research is directed toward coherence, holistic thinking, collaborative methodology, systematic approach, action, and network activity.

The mission of contemporary higher education in the 21st century implies orientation toward a development strategy for integrated knowledge, which as such, can meet challenges of global processes: urbanization, cultural, and social transformations.

Sustainability challenges require integrated forms of knowledge with a research platform. Transdisciplinary research approach can contribute to a better identification and treatment of problems.

In a time of global changes and dramatic loss of viability from the local to the global scale, science needs to take more responsibility for the problems. In a world characterized by rapid changes, uncertainty and increasing interconnection, there is a need for science that will contribute to the solution of complex and persistent problems (Hirsch Hadorn at al.,2008).In this context, science can largely contribute to the sustainable development of the physical structure of the city and is therefore necessary to reexamine current methods of scientific research in order to improve them. Although in recent decades the number of scientific research developed on a transdisciplinary platform has increased in the world, the strategy is not sufficiently developed at global and local levels. Globally, a general universal model has not been developed; there is an unequal representation of transdisciplinary activity at a local level, while simultaneously the smaller local communities face a lack of a high degree of cooperation and activities that are, in the highest degree, conducted in disciplinary terms.

The complexity of architecture and urbanism as disciplines indicates and commits to their layered studying, tailored by the dynamic social changes and modern conditions of life and activities. An important direction in the professional, educational environment is that through integrated engagement, which is woven into the strategic, methodological frameworks oriented toward a research line, new forms of development of urban culture and society are promoted, where the educational process is not only a companion to social change, but is a starter, and this can be achieved solely by an integral and sound comprehension and action.

The 21st century involves flexibility and globalization of knowledge. The complexity of the urban system and the complexification of the needs of users of space require a complex analysis of various factors and influences that shape the developed environment, and are helping the design of architectural and urban programs.

Educational transdisciplinary action aims to avoid the fragmentation of knowledge for the benefit of communication. Transdisciplinarity is a strategy for development of educational engagement that is not exclusively based on theoretical principles, but is also as a way of thinking and acting. A variety of research and experience from elsewhere show that in most universities, even those that are profiled as faculties for transdisciplinary studies, where transdisciplinarity, starting from the name of the faculty and the program itself which is formalized as a study model in theory, the studies are implemented in disciplinary terms.

Modern strategy for architecture and urbanism studies in higher education is diverse, with the dominant model of the implementation of the teaching process is in disciplinary terms. Urbanism and architecture as complex disciplines and professions include a number of other areas that involve collaborative relationship that as such can respond to contemporary challenges in solving complex urban issues of the 21st century. A collaborative theoretical and practical knowledge imply communicative strategy on different levels. The levels of integrality such as multidisciplinary and interdisciplinarity are unable to respond adequately to the challenges of the global time. Local areas are increasingly taking on the title of the "global city", which is why the global view of the local context is an important moment in the development of sustainable physical structures of cities. The physical structure of cities are increasingly complex, urbanism is increasingly oriented toward issues of social housing, brownfield investments, as opposed to greenfield areas which are fewer;

the policy of expanding the city's territory is less and less possible, and it indicates the need for transformation of existing urban structures. It includes an integrated research engagement and integrated knowledge.

Urban Education in the disciplinary framework is a classical form of studying, which results in fragmented knowledge, conventional thinking, and lack of awareness on the collaborativeness. Fragmentary knowledge solves the problem in the disciplinary context, giving less opportunity for long-term problem solving. Generation of experts coming from mostly unified methodology of the study programs with disciplinary knowledge will have more difficulty to respond to global challenges and complex social needs, but also will not have enough understanding for the development of new respond mechanisms to complex issues of sustainable development. Therefore, the cooperation through appropriate programing concepts and methodological strategies is a sustainable orientation in the process of contemporary urban and architectural education at universities.

The academic environment is a particularly sensitive issue and implies the inclusion of all forms of institutional infrastructure and research diagnosis of key issues and sections for its improvement. Academic study programs, personnel policies, transparency, universality and innovative visions of development are the basis of productive knowledge. Acquiring skills and competence, creativity, critical thinking, international cooperation, team work in multicultural contexts, contribute to a more complex overview of the problem.

Educational process at the Faculty of Architecture in Podgorica, although methodologically conceptualized on an integrated platform, is predominantly conducted in disciplinary and interdisciplinary framework. In this regard, the curriculum reform is necessary. The study which was conducted in this paper indicates the need for transdisciplinary forms of engagement in scientific and professional education and for the production of professionals who can respond to the complex challenges of the cities in the 21$^{st}$century.

The aim of this chapter is to promote interdisciplinary research and its application in educational process in the field of architecture and urbanism at universities, for the purpose of sustainable development of the cities of the future.

## SETTING THE STAGE

The study at the Faculty of Architecture was conducted as a part of the project "Interdisciplinarity and urban artifact", where students and mentors from four faculties have been involved: the Faculty of Architecture in Belgrade, the Faculty of Architecture in Ljubljana, the Faculty of Architecture in Podgorica and the Faculty of Philosophy in Niksic. There were a total of forty five students and five mentors.

Space for the research was the city of Niksic. The goal of the project "Interdisciplinarity and urban artefact" is a contribution to a more successful approach to the perception of space from different perspectives and a proposal of concrete solutions for the transformation of spaces through innovative research action. Each group of students approached the task according to their own methodology and sensibilities.

The project was funded by the World Bank and the Ministry of Sustainable Development and Tourism of Montenegro, Land Administration and Management Project in Montenegro - LAMP (LAMP, 2013), within the spatial urban plan of Niksic, which is in the process of development.

Strategy of the LAMP Project is aimed at improving the process of spatial planning and with the emphasis on the participation of citizens in the process of making spatial urban plans. In this sense, the methodological approach of students from the Faculty of Architecture on the project was particularly significant.

Students of the Faculty of Architecture in Podgorica, under the supervision of Svetlana Perovic and Sanja Lješković Mitrović, used the transdisciplinary research methodology. Out of the twenty-three students of the Faculty of Architecture ten teams of two or three members were formed, which included one or two users of space and the same number of students from other fields (sociology, psychology, art, design etc.). Thus, each team consisted of four or five members.

The task of the research project was the incorporation of the new "Transurban center" in the existing urban context of the city of Niksic. The study was performed within course subjects of Urban functions and structure and Landscape architecture on the third year of undergraduate studies. The task has been to identify the characteristic places of the urban structure and implement the new Transurban center that should be developed and become the new engine of urban development of the industrial city. Weakened industry requires new forms of engagement through new programs and structures that represent a new dynamic urban life, transformable, authentic, sustainable with a new visual identity, and new energy that will become part of the collective memory. As a basis for achieving this goal, integral dimension of research was implemented through all stages of the design process.

Adopting the methodology of transdisciplinary basis can strongly contribute to the improvement of the educational process at universities. It can also encourage more innovative, more creative, more humane, future architectural and urban interventions in practice, all with the aim of sustainable development and improving the image of the current urban image of cities.

Flexibility of the transdisciplinary methodological strategy, developed in a dialogue between primarily decision-makers, users and professional people, can contribute to the development of the "new science of space", which is abandoning the conventional and dominant routine practice of planning and design of today's cities. An integrated model of engagement articulates new directions of urban development, which are more universal, flexible, transparent, and sustainable.

Niksic is a leading industrial city in Montenegro, with an industry that dominated the second half of the last century. The city with about 70 000 inhabitants was built by the radial matrix that has determined to a large extent the existing physical structure of the city. A fifty-year long domination of the industry in Niksic, has reflected negatively on humanistic values, visual identity, and continuity of the urban tissue (Perović&Popović, 2013).Subordination of other functions to the production has led to discontinuity in the urban image and to the modern need for reidentification of space through new urban artefacts.

With this study, new Transurban centers, incorporated into the existing context, with its characteristics in the physical, functional, aesthetic, sociological, psychological, anthropological, and other terms, should become interpreters of new cultural paradigm transcending the existing models of understanding architectural and urban space. The new multi-functional urban artefacts developed on a realistic basis, should serve as an example for new standards that move the boundaries of the existing ones. Experimental-research and creative-aesthetic dimension of new structures in the space should be at the same time realistic and innovative and stimulating for each additional, more advanced design action.

In the process of the project, we can distinguish three key stages:

1. Forming a team.
2. Forming a Methodology basis.
3. The process of forming a new Transurban center.

Primary directions in designing the new Transurban center have implied:

- Deliberation on the urban artifact in a specific context
- Reconceptualization of conceptual functional-formal analysis
- The development of new methodological paradigm of design, through transdisciplinary approach (sociological, environmental, economic, psychological, and other aspects)
- Experimental-research and creative-aesthetic dimension

## CASE DESCRIPTION

Research phases in the process of development of the new Transurban center implied complexity and stratification through all segments from theoretical and scientific aspects to design and shaping. Each phase has been studied in correlation with the other phases through the review of previously defined patterns. Overall, the research phases included:

- Literature search - the theoretical platform,
- Analytical-critical arguments of the selected literature,
- Research on current examples of successful practice,
- Analytical -critical arguments of the selected projects,
- Research of Niksic space through immediate sojourn in it,
- Choosing a location and a program for incorporating new structures of Transurban center
- The process of developing ideas and forming new structures in the space,
- Revisiting initial views,
- Design and development of the solutions, and
- Presentation the solutions to professional and cultural community and discussion.

Teams had the opportunity to, among other things, get familiar with the strategic goals of spatial urban plan of Niksic, which is in the process of development, and to look at specific problems in a specific area, in order to make it easier to opt for the contents that are missing in the city, and which should enable users of the space to express their philosophy of life, satisfy their needs and aspirations.

An important component in the research process was the direct contact with the real space, the local community, with the aim that through research team work and critical attitude, perceive the values, potentials and shortcomings of the cultural landscape and find the characteristic places for intervention for the purpose of evaluation, development and improvement of the current natural and built environment of the Municipality of Niksic. The teams in the study opted for a variety of spaces to be reshaped in order to develop the Transurban center; such as areas from brownfield, the city's central square, the attractive locations of the city's central core, the fortress, and peripherals such as banks of the lakes "Krupac" and "Slano", and natural vertical elements of the city, such as the hill "Trebjesa". Interruptions in the continuity of the constructed tissue, areas with an insufficiently distinctive identity but an attractive in a sense of disposition and program, as well as the natural setting, have inspired the research teams to opt for new programs and structures that should be incorporated into the existing physical structure of Niksic.

Innovation, creativity, and transdisciplinarity enabled the development of new urban artifacts, of which the following are characteristic:

1. Reshaping the central city square "Sloboda", the primary site of the social process that is dominant with its disposition in the spatial structure and its size, but visually unattractive. Transforming the public space, elements of collective memory into a new contemporary expression is through the creation of a new Transurban center. New identity, new energy, permeating the natural

and the artificial, through communication, universality, authenticity, with a smaller physical structure and humanistic energy are the main features of the new urban artefact.

2. Natural element, the lake “Slano” 9km$^2$ of surface, water power, energy, islands as elements of inspiration, the development of new structural components on the islands and articulated move that transforms the natural environment and initiates new energy in a space by combining various activities from the diving center, observation tower, catering industry, recreation, to visually attractive public spaces.
3. Research Centre complex on the shores of Lake “Krupac”, with a disposition of independent but related units, and with a primary purpose “to provoke emotion through form”. The feeling of peace, serenity, sublimity, and inspiration, exploring emotion through the synthesis of colors and irregular shapes, countering effect of high and low forms and narrow spaces – feeling of anxiety. Playing with the form completed by a zone for recreation, rest, leisure, and meditation. “Transurban center as the initiator of sentiment, a feeling like the creators of forms.” Engaged space, dynamics of the paths where each leads to the lake, relationships with nature, and the structures that shape the experience of the “gap, color, mysticism, dynamics and mystery.”
4. The central theme of “ Dom revolucije,” a megalomaniac structures, brownfield, “unfinished modernization” on the most attractive location in the city, and the development of the new Transurban center through the transformation of the existing and through the replacement with a new multifunctional center, an artefact of sustainable architecture, values, and identity.
5. The fortress “Bedem”, element of identity and memory image of the city, insufficiently attractive for visitors, activated by the introduction of new modern elements into the space and transposed through a new contemporary expression of a cultural character. With particular sensitivity toward the architectural heritage, the area is transformed into a new artefact of rich content that are tailored to fit different types of events and visual effects.

The main criteria for the evaluation of the project were:

- Methodological dimension of the design process-collaboration, transdisciplinarity
- Theoretical- analytical-philosophical platform
- Creative research platform
- Election of new programs and location of the Transurbancentre in an urban context

- Innovation, creativity, transurbanism, modernity, futurism, associativity, contextuality, solution sustainability
- Functional - formal dimensions
- The integrity of the elements of a whole in an integrated organization
- The visual quality and logistics of the organization in the presentation

In the process of working on projects, used CAD technology has enabled the quality development of a methodological platform and the coordinated cooperation among team members.

Continuous cooperation and flexibility in a transdisciplinary action have reflected positively on creative ideas that emerged during the research process. Integrated ideas are more powerful and encourage creative thinking as well as intellectual awareness at a higher level. Promoting awareness on cooperation, dialogue, refers to sustainability, durability and certainty. Transdisciplinary knowledge knows no boundaries between disciplines, between academic and non-academic structures, between theory and practice, between science and profession. Completeness of problem identification and logic arguments are better captured through transdisciplinary methodological choice.

New artefacts that are proposed for implementation in the existing urban tissue contain elements of Futurism and experiment, but at the same time are deeply correlated with the context, which was largely contributed by a new way of design research process. New promising solutions which were offered by research teams include psychological, sociological, economic, ecological and other meaning and indicate the importance of each. Complex processes in a globalized world require research about each specific architectural and urban intervention in the space along with the analysis of all aspects of urban processes.

Compared with disciplinary and interdisciplinary methodology that was used in research projects in previous years at the Faculty of Architecture and the examples of other cities in Montenegro, the conclusion is that the transdisciplinary methodology is necessary in the process of sustainable educational, scientific, and professional practice of architecture and urbanism activities in the 21$^{st}$ century.

## CURRENT CHALLENGES FACING THE ORGANIZATION

Overall, the University of Montenegro and the Faculty of Architecture in Podgorica as a component unit of the University, still lack a sufficiently developed strategy for the implementation of transdisciplinary methodology in the process of higher education. The curricula are predominantly developed on disciplinary and interdisciplinary basis, resulting in a lack of cooperation in scientific research, but generally in the overall system of curricula implementation. Transdisciplinary research projects are less represented, resulting in a limitation of the research results, as well as pedagogy work. In the process of education of architects and urban planners, lacks adoption of new patterns that encourage creative thinking, research component, technological innovation, new forms of professional and scientific activity, as can be developed with the flexibility of transdisciplinary methodology.

With disciplinary action, autonomous disciplines develop own methodology for problem solving. Interdisciplinary action establishes a connection between independent methodologies, and interdisciplinary action involves a unique methodology developed by teams of experts from different disciplines and forms of knowledge. In this regard, the study conducted at the Faculty of Architecture, shows some limitations in the sense that not all members of the team gave the same contribution, and that not all the teams have reached the same good results. Communication was not at the same level through all phases of the project. The process required more time, due to a constant exchange of ideas, attitude harmonization, adoption and offsets of the different approaches, dilemmas etc.

The innovative methodology has been successfully conducted, but it takes time and more intensive engagement to harmonize the different forms of knowledge that are incurred predominantly in disciplinary terms and re-orient a new way of thinking and acting.

The presented study showed a different willingness of team members to embrace innovation, which requires a longer approval process and achievement of significant results in the practice of transdisciplinary education and planning the cities of the future.

## SOLUTIONS AND RECOMMENDATIONS

Whereas on one hand, the choice of transdisciplinary methodology and research paradigms in education at universities has its advantages, the question remains about the willingness of autonomous disciplines working towards a dominant autonomous methodologies to maximally engage in the formation of the general methodology for addressing complex issues of global society. Most commonly problems are solved in specialized teams and with specific methodology of the discipline and dependent of the case. Architectural and urbanism education, and professional activities in practice, involves cooperation where, in particular, in shaping the environment it is necessary to involve all the interested stakeholders of society and users of space in the focus.

Transdisciplinarity that is present predominantly on theoretical grounds can contribute to solving important issues of urban systems. Urban planning and architecture are disciplines that involve constructive dialogue with other disciplines: natural, social, humanistic, in modern education processes, which includes the development of readiness for the complex challenges that are getting more intense. Universities that stimulate transdisciplinary research methodology can significantly contribute to training of personnel which will make a significant contribution to sustainable practice of planning and designing of modern cities.

The study presented in the chapter shows that it is possible to improve the existing system of study at the University of Montenegro, but also that time and dedication are necessary. It is concluded that a more intensive cooperation among the various forms of knowledge is required, but also that it is needed to invest some energy in order for it to become the dominant practice in educational academic and vocational curricula in higher education. Transdisciplinary methodology is a complex process and generally requires a lot of time, even after adoption, for an appropriate implementation in practice.

## REFERENCES

Andalécio, A. M. L. (2009). Transdisciplinarity in the university:discourse and practice. *RECIIS -Electronic Journal of Communication, Information & Innovation in Health*, *3*(3), 83–89.

Després, C., Vachon, G., & Fortin, A. (2011). Implementing transdisciplinarity: Architecture and urban planning at work. In I. Doucet, & J. Geneviève (Eds.), *Transdisciplinary knowledge production in architecture and urbanism, towards hybrid modes of inquiry* (pp. 33-51). New York: Springer.

Hirsch Hadorn, G., Hoffmann-Riem, H., Biber-Klemm, S., Grossenbacher-Mansuy, W., Joye, D., & Pohl, C. et al. (Eds.). (2008). *Handbook of transdisciplinaryresearch*. Berlin, Germany: Springer. doi:10.1007/978-1-4020-6699-3

LAMP. (2013). *Land Administration and Management Project*. Retrieved from http://www.lamp.gov.me/?lang=en

Nicolescu B.(1998). The transdisciplinary evolution of the university, condition for sustainable development. *Rencontres Transdisciplinaires,* 12.

Perović, S. (2013). The Levels of Integrality in Architecture and Urbanism Studies at the University of Montenegro. *Procedia: Social and Behavioral Sciences*, *93*(93), 654–658. doi:10.1016/j.sbspro.2013.09.256

Perović, S., & Popović, S. (2013). Reflections of Utopia and the Ideal City in the Development of Physical Structure of Nikšić Aspect of Visual Perception. *World Academy of Science, Engineering and Technology*, *79*, 1247–1255.

UNESCO. (1998). *Transdisciplinarityvtimulating synergies, integrating knowledge*. UNESCO. Retrieved from http://unesdoc.unesco.org/images/0011/001146/114694eo.pdf

## ADDITIONAL READING

Fry, G. L. A. (2001). Multifunctional landscapes—Towards transdisciplinary research. *Landscape and Urban Planning*, *57*(3–4), 159–168. doi:10.1016/S0169-2046(01)00201-8

Land Administration and Management Project (LAMP). (n.d.). Retrieved from http://www.lamp.gov.me/?lang=en

Lawrence, R. J., & Després, C. (2004). Futures of transdisciplinarity. *Futures*, *36*(4), 397–405. doi:10.1016/j.futures.2003.10.005

Mendez, J. (2013). *Enhancing graduate landscape architecture education through transdisciplinary research approaches: A case study at the University of Connecticut*. (Master's Thesis). University of Connecticut.

*Parsons the new school for design*. (n.d.). Retrieved from http://www.newschool.edu/parsons/masters-architecture/

Stokols, D. (2011). Transdisciplinary action research in landscape architecture and planning: Prospects and challenges. *Landscape Journal: Design, Planning, and Management of the Land*, *30*(1), 1-5.

Thering, S., & Chanse, V. (2011). The scholarship of transdisciplinary action research: Toward a new paradigm for the planning and design professions. *Landscape Journal*, *30*(1), 6–18. doi:10.3368/lj.30.1.6

Wickson, F., Carew, A. L., & Russell, A. W. (2006). Transdisciplinary research: Characteristics, quandaries and quality. *Futures*, *38*(9), 1046–1059. doi:10.1016/j.futures.2006.02.011

## KEY TERMS AND DEFINITIONS

**Sustainable Architecture and Urban Education:** Education that develops on the methodology of integrated knowledge and produces experts who are able to respond to the complex problems of urban systems in order to support sustainable development of cities.

**Sustainable Scientific Education:** Education with a methodology that promotes integrated knowledge and as such can respond to the complex challenges of modern society.

**Transdisciplinary Knowledge:** An integrated knowledge of various disciplines and forms of knowledge.

**Transdisciplinary Research:** A collaborative research process with a methodology developed on the basis of integrated knowledge.

**Transdisciplinary Research in Architecture and Urbanism:** A collaborative research process with a methodology that includes integrated knowledge, includes a variety of disciplines, as well as academic and non-academic knowledge for the purpose of sustainable urban development.

# Chapter 12
# Pre-Service Teachers' Self-Efficacy and Attitudes toward Learning and Teaching Science in a Content Course

**Cindi Smith-Walters**
*Middle Tennessee State University, USA*

**Heather L. Barker**
*Middle Tennessee State University, USA*

## EXECUTIVE SUMMARY

*Science teaching is approached with hesitation by many PreK-8 teachers. This chapter explores the research on attitudes toward science and learning science as well as the perceived science efficacy of elementary pre-service teachers. It also describes a content-based, pedagogically rich life science course for pre-service preK-8 teachers that incorporates active and interactive teaching techniques in lieu of the traditional science methods course. Using evidence from this project and other research studies, the chapter argues for the inclusion and modeling of these approaches when preparing teachers of science and proposes that this non-traditional approach for teaching content-based courses for preparing teachers be considered in place of traditional science methods courses.*

DOI: 10.4018/978-1-4666-6375-6.ch012

## ORGANIZATION BACKGROUND

Over twenty years ago, a large public university in the southeastern United States took a bold step and changed their science methods course from a traditional, pedagogically focused format to a duo of content-based courses: one in biology, and one in chemistry/physics. This unusual move was in response to a growing body of research indicating that increasing teachers' content knowledge of science leads to increased achievement of their students (Druva & Anderson, 1983; Wayne & Youngs, 2003). Teacher preparation programs were being criticized for their superficial curriculum that lacked appropriate emphasis on preparing pre-service teachers to teach rigorous content (National Commission on Teaching and America's Future, 1996). This university felt it was imperative to increase the content knowledge preparation of its PreK-8 teacher graduates and thus changed their requirements.

Teacher preparation program design and requirements vary throughout the United States. Additionally, states have different requirements for obtaining a teaching certification. Typically, secondary level teacher candidates must hold a degree in a specific field of study (English, music, science, mathematics, etc.) and a minor in education. However, students seeking elementary certification are required to take fewer courses in each field of study and more courses in education. These candidates usually take a minimum number of college credit hours in science content courses along with an additional science education methods or integrated methods course to prepare them for the classroom (U.S. Department of Labor, 2014). At this university, preK-8 pre-service teachers complete eight hours of content-based science, but in lieu of the typical science methods course they take an additional eight hours of content-based courses specifically designed for elementary education majors. Four of these additional hours consist of the course, Biology 3000, Life Science for Elementary Teachers. The first author has taught this course for over twenty years with a focus on providing the deep understanding of science content needed by elementary teachers, through reform-oriented, research-based pedagogical techniques. This study examines whether this life science content course for pre-service teachers experiences results in increased science attitudes and increased self-efficacy.

## SETTING THE STAGE

Elementary teachers are expected to be all things for all students: content delivery experts, special education providers, learning disabilities specialists, guidance counselors, health advisors, and pedagogy authorities, as well as content specialists in all traditional subject and skill areas. This is a tall order for anyone, particularly a new teacher who is years away from becoming a master educator (Berliner, 1988).

The National Commission on Teaching and America's Future (1996) reported that one-third of all new teachers leave after three years, and 46 percent are gone within five years. Ingersoll (2003) estimated that 40-50% of teachers leave within the first five years and that rate has remained fairly consistent (Ingersoll & Perda, 2012). Ingersoll and Perda also noted that there is more pre-retirement attrition of mathematics and science teachers than those in any other subject areas.

The quest for excellence in all aspects of education is perhaps most pressing in the area of science. High quality science educators, using reform-based instructional methods, are vital to the development of future generations of scientifically literate citizens. Teacher preparation programs have a duty to effectively prepare candidates to meet this goal and to successfully handle the expectations and demands of this challenging career. With the increased emphasis on in-depth science content knowledge and delivery methods in K-12 classrooms, teachers now more than ever must be fully prepared, self-confident, and accomplished in conveying science content and skills to their students.

The development of national standards detailing content to be taught in all subjects at all grade levels has helped to elevate and standardize the educational process in the United States. The National Science Education Standards (NSES) provided teachers with clear student goals and administrators with professional development requirements (NRC, 1996). The NSES influenced various states' own science learning and associated state-wide standardized testing. National and state entities have continued to expand and mandate rigorous standards which teachers must ensure students attain. The Next Generation Science Standards (NGSS Lead States, 2013) require that teachers be well-versed in science subject matter and able to directly apply that knowledge through hands-on activities and project-based learning (NRC, 2011). A science-specific section within the Common Core Science Standards has set clear expectations for the reading, writing, speaking, listening, and language of science, while enabling students to communicate information and results (Common Core Standards Initiative, 2010). In this demanding climate, it is especially critical for teacher preparation programs to recognize and address the interacting factors that support effective teacher development. Wayne and Youngs' (2003) review reinforced the critical role teacher preparation programs play. They found that the quality rating of the teacher college was positively correlated with their pre-service teachers' future students' scholastic achievement.

## Science Attitudes, Science Self-Efficacy, and Science Knowledge of Teachers

Regrettably, the teaching of science is often approached with hesitation and even fear by pre-service teachers and even experienced educators (Riggs & Enochs, 1990;

Yuruk, 2011). Science anxiety is common in elementary science methods students (Appleton, 2007; Epstein & Miller, 2011; Finson, 2001; Gunning & Mensah, 2011; Westerback, 1982). Practicing elementary teachers with greater apprehension rely upon teacher-directed instructional strategies, whereas those who are less anxious are more likely to use reformed strategies such as open-ended inquiry and student-centered instructional strategies (Czerniak & Shriver, 1994; Lumpe, Haney, & Czerniak, 1998).

Teachers' level of self-efficacy impacts their success and the subsequent academic achievement of their students (Bandura, 1993; Fencl & Scheel, 2005; Hackett, Betz, Casas, & Rocha-Singh, 1992; Haney, Lumpe, Czerniak, & Egan, 2002; Marshall, Horton, Igo, & Switzer, 2009; Palmer, 2006; Ross, 1992). Self-efficacy can be defined as, "People's judgments of their capabilities to organize and execute courses of action required to attain designated types of performances" (Bandura, 1986, p. 391). In the area of teacher preparation, self-efficacy is specifically understood as teachers' belief in how well they are able to influence their students' motivation for learning and subsequent achievement (Tschannen-Moran, Hoy, & Hoy, 1998). Educators who possess high self-efficacy approach their role of teacher in a more confident and assured manner (Bandura, 1977, 1993; Bursal, 2008; Tosun, 2000; Yuruk, 2011). Improving pre-service teachers' attitudes toward science and teaching science, as well as increasing their self-efficacy in key scientific concepts, can create positive outcomes regarding future teaching ability. Alternatively, low self-efficacy results in less confidence and assurance in conveying subject matter. What steps can teacher preparation programs take in order to support the development of positive attitudes and increased self-efficacy? Two important components emerge from a review of the literature: increasing science content knowledge, and modeling innovative, reform-oriented teaching methods.

A solid foundation of content knowledge is an important factor contributing to overall teacher success (Druva & Anderson, 1983; Monk, 1994; Wenglinsky, 2002). Greenwood and Scribner-MacLean (1997) expressed this with the statement that teachers cannot teach what they do not themselves understand. Tosun (2000) found that pre-service teachers with adequate backgrounds in content-specific science courses ultimately held more positive beliefs and attitudes towards teaching that particular subject at the elementary level. Teachers confident in their knowledge of the subject material have less anxiety, exhibit more positive attitudes, and have increased assurance in their ability to teach science (Bursal, 2012; Druva & Anderson, 1983; Wenner, 1993).

Research has identified key beliefs and attitudes that impact a teacher's instructional quality. Self-efficacy towards teaching and learning science has been shown to heavily influence the effectiveness of classroom teachers (Bandura, 1977, 1993). Study results drawing conclusions between student achievements and teaching methodologies (Barufaldi, Bethel & Lamb, 1977; Bianchini & Colburn, 2000; Monk, 1994; Shapiro, 1996) point to the fact that the overall attitude of a teacher towards science content affects instructional quality and effectiveness in the classroom. Unfortunately, although teachers have anxiety related to teaching and learning science in the traditional way, they tend to continue to use the same ineffective methods with which they themselves were taught (Michelsohn & Hawkins, 1994; Prawat, 1992). Promisingly, Westerback and Long (1990) suggested that teachers who were more comfortable with science were likely to devote additional time to teaching it and to use more creativity and diversity in their methods. This study puts forward a novel idea: combining a science content course with gorous pedagogical approaches. This is an idea that, when implemented, may well ease fears and discomfort in the subject area of life science for beginning teachers.

## A New Paradigm

Students of any age must do more than simply read the textbook, cruise the internet, watch an online video, or passively listen to a lecture. In order to learn science, students must be engaged in both a minds-on and a hands-on approach (Bonwell & Eison, 1991; Erlauer, 2003; Shymansky, Kyle, & Alport, 1983; see also Anderson, 2002; Bransford, Brown & Cocking, 2000). Much of science practice and research in the real world is based on the concerns, curiosity, and interests of the scientist (Ziman, 2002). Unfortunately, these qualities are not always fostered in science classrooms (King, Shumow, & Lietz, 2001).

Biology 3000, Life Science for Elementary Teachers, uses the interests and current background knowledge of pre-service teachers to guide coursework and to foster interest. Because this content-based, pedagogically rich approach is relatively unique in teacher preparation programs, it is imperative to determine its effect on pre-service educators' attitudes towards science, their attitudes toward the teaching of science, and their life science knowledge self-efficacy outcomes.

With the awareness of the truism that teachers often teach the way they were taught, Biology 3000 course content is delivered via research-based methods with the explicit intention of providing a model for future classroom instruction. A heavy reliance on holistic, brain-based learning fosters the acquisition and retention of life science knowledge. The brain stores information differently depending on how it is packaged. In other words, emotional, vivid and connected information is set in the brain more deeply than abstract, boring, or arbitrary facts (Erlauer, 2003; McGaugh, 2003; Wolfe, 2010). Thus, hard-to-remember facts and concepts should be bundled into more easily recalled packages. With that in mind, the Biology 3000 instructor uses mnemonics, songs, and rhymes as well as memorable activities like carousel brainstorming, word sorts and foldable graphic organizers of information

(Lipton & Wellman, 1998; Zike, 2004). These strategies and others are extensively incorporated into group work, hands-on inquiry and long-term laboratory activities. Fencl and Scheel (2005) reported that the use of instructional strategies requiring students to work actively, creatively, and in a collaborative manner supported the development of self-efficacy. Burgeoning teachers must know a large amount of science content information in addition to having the confidence and appropriate pedagogical skills to deliver that content to students. Biology 3000 conveys rich content information while the instructor models and incorporates the sophisticated skills and behaviors expected of competent and effective educators. What happens in this course is very similar to what should be happening in the preK-8 classrooms of effective science teachers.

## A Focus on Bloom's Taxonomy

When their teachers are following a traditional "I teach, you learn" model, students—particularly at the K-12 level—are habituated to being plainly told what they must know by both textbook and instructor. Memorization reigns, and regurgitation of content information and proper use of vocabulary rule the day. Generating their own questions, exploring ideas, engaging in reflection and metacognition, and personally applying content to real world situations is rarely done in such classrooms. In contrast, Biology 3000 explicitly incorporates Bloom's Taxonomy (Bloom, Engelhart, Furst, Hill, & Krathwohl, 1956) as a framework for building instruction. Revised slightly by Anderson & Krathwohl (2001), Bloom's Taxonomy is divided into six cognitive domain levels arranged from lower-order to higher-order thinking skills: remembering, understanding, applying, analyzing, evaluating, and creating (see Figure 1). The last three categories are often referred to as the higher level thinking skills. The use of Bloom's Taxonomy helps educators focus on developing those skills in their students, thus ensuring that information is learned more thoroughly and retained longer.

Following are illustrative phrases or examples from Biology 3000 that are grouped in the six categories from lower order thinking skills to higher order thinking skills:

1. **Remembering** (retrieving stored information, reciting back):
   - *List* the kingdoms of life from simplest to most complex.
   - *State* the steps in cell division and what happens at each stage.
2. **Understanding** (revealing depth of understanding of the vocabulary and concepts):
   - *Explain* in your own words the concept of photosynthesis.
   - *Paraphrase* the assigned reading from the class website.

*Figure 1. Diagram of Bloom's Taxonomy adapted from Anderson & Krathwohl (2001).*

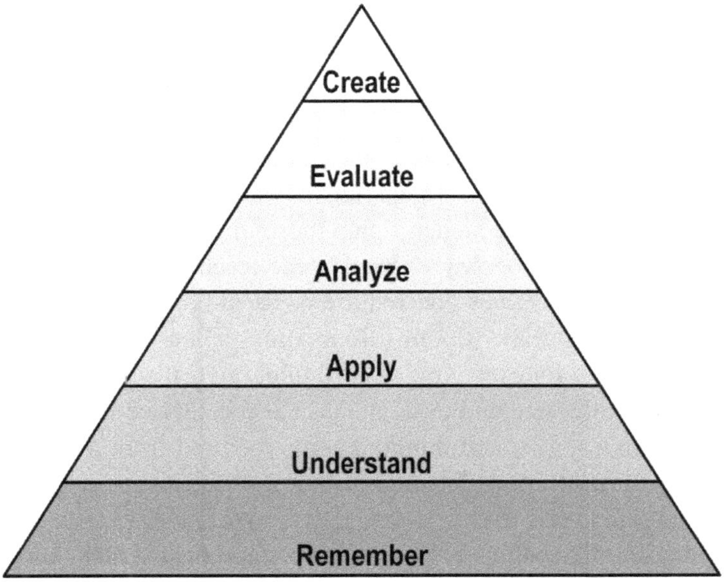


3. **Applying** (solving problems):
    - *Calculate* the population of bean bugs within the study area.
    - *Solve* problem #17 from the reading on the class website.
4. **Analyzing** (unpacking concepts, developing theories or models):
    - *Derive*: State your personal view on hunting. Then using what you know about natural systems, derive an argument opposing your personal viewpoint.
    - *Simulate*: You are a wildlife manager, how would you determine the best number of deer for the habitat given?
5. **Evaluating** (assessing and choosing from various options/possibilities):
    - *Determine* which of the following narratives is the best explanation for the scenario given.
    - *Select* from among available options for genetically modifying a cell, and justify your choice.
6. **Creating** (synthesizing, combining components in new and unique ways):
    - *Design:* After reading about succession, use the information you've gained to design a CD. Be sure to title the CD, name the group, and list the titles of 6-12 tracks, so as to clearly illustrate your understanding of the topic.

- *Make up*: Formulate a multiple choice test question from the content we have covered since the last exam on an index card. On the back of each card, identify the correct answer to your question, and explain why it is the correct answer and why other choices are not.

Careful attention to Bloom's Taxonomy supports the instructor in planning and delivering effective instruction. Organizing objectives within the framework of this taxonomy ensures that the students gain those important higher order thinking skills. Additionally, the taxonomy can guide the teacher in creating assessments that match the stated objectives and require more in-depth thinking than merely the regurgitation of material. Bloom's Taxonomy enhances student learning by providing opportunities for exercising critical thinking skills and enabling students to apply, analyze, synthesize and evaluate. As the pre-service teachers experience the variety of learning tasks and thinking skills required from a curriculum built around all the levels of Bloom's Taxonomy, they develop an understanding of what types of questions and tasks they will someday need to use in their own classrooms. This helps them start to develop that important pedagogical content knowledge that is so critical to good teaching (Shulman, 1986) and prepares them to become more intuitive teachers by increasing their critical thinking skills, building upon current knowledge and enabling them to begin to apply, analyze, synthesize, and evaluate as it also prepares them to become more intuitive teachers.

## Situated Learning

There is also heavy reliance on situated learning in Biology 3000. Lave (1993) suggested that learning is highly situated, meaning it is connected to the context and community within which it is applied. This contrasts with many typical classroom learning activities which involve ingesting knowledge that is often abstract, out of context, and foreign to the learner. Social interaction is a critical component of situated learning – learners become part of a "community of practice" (Wenger, 1998).

As the novice pre-service educators move through Biology 3000, they become more active and engaged within the classroom culture and move from practicing to proficient. Examples of situated learning occur throughout the course. In a unit on ecological succession, students are prompted to look out the window to understand how succession occurs in their neighborhood. They are also given application questions to work on together. A sample problem:

*A forest fire burns through Barfield Park's natural area in the western part of our county. Is this primary or secondary succession? How do you know? Explain through a series of posters (using no words) what happens succession-wise to this area in the next 200 years.*

Another group might be given a different scenario such as a newly dug pond in the field behind their own home. Situated learning is also employed during tree leaf identification and nature walk learning experiences on the campus. In addition, students collaborate to investigate specific questions and learn scientific process skills by growing plants from garlic cloves, or germinating bean and corn seeds. Students compare data collected, ask questions, and solve problems in relation to these investigations. Another favorite situated learning activity occurs when groups of students choose biomes they may have visited and research that biome's climate and other factors such as rainfall, temperature, vegetation types, indigenous animals, soils, and more. Each team then makes a poster depicting the biome using no words, labels, titles, or narrative. After posting their work, groups visit each poster and complete a biome matrix of information gained. When time is called, students ask questions of each expert group to fill in any gaps in their knowledge about each biome. Throughout the semester, student pairs, trios, and quads are formed and re-formed so that each student works with every other student multiple times. Many of the assignments support or require partner or group work. This ensures that students are a true community of learners with each invested in the success of their peers.

## Use of Active Learning Techniques

Students of all ages have difficulty remaining focused for long periods of time. McKeachie (1986) reported that immediately following a lecture students remember 70% of the information presented in the first ten minutes and only 20% of that from the last ten minutes. The use of active learning techniques can help maintain student attention and support retention. Both interactive and collaborative learning experiences are emphasized. Activities such as Think-Pair-Share and Small Group Exercise are commonly used in Biology 3000, along with many others (Angelo & Cross, 1993; Bonwell & Eison, 1991; Brent & Felder, 1992; Johnson, Johnson & Smith, 1998; Lyman, 1987; Meyers & Jones, 1993).

In Biology 3000, a Book of Biology (BoB) incorporating modified Cornell note-taking format (Pauk & Owens, 2005) is used extensively to reinforce content and prepare students for their own science classes. BoB is a composition book organized with a student-generated table of contents and index, along with allocated sections for major content topics. Students are required to take lecture notes (input) which they record on the right side pages. Left side pages are reserved for output. Output can take many forms such as student generated questions, student responses to class questions (either individually or after talking with a partner or group), lists of examples, diagrams which explain and expand knowledge, data tables and conclusions, problem solving, predictions, sketches of schematics or flow charts, arguments supporting or refuting a statement, or paraphrasing lecture notes in their own words. This results in an easily navigable resource to use in preparing for exams in the course as well as a template of content material and lesson ideas for use later in their own science classroom. BoB is a simple way to make sure students stay on topic and on task. An unintended bonus is that former students often cite BoB as their go-to resource for life science information, ideas for lesson introductions, instructional strategies, and more.

An example of another active learning activity used in Biology 3000 is an activity called Write Around. At the beginning of class students are asked to write a fact related to a topic from the previous lecture or from an assigned reading. Papers are then passed around to a neighbor who must either add to the current fact or write something new. After 8-10 rounds, papers are returned to their owners. At that point students compare the Write Around papers to notes previously taken in BoB, check for discrepancies, clarify information, and/or choose to add information to BoB which may be missing. Write Arounds allow students to review, actively analyze, and use knowledge on their own. They arrive at conclusions about what they don't know and are able to own the knowledge they do possess. Carousel Brainstorming is also commonly used. A favorite variation of this technique is when each of six posters is titled with one of the six kingdoms of life. Each small group is given a different colored marker and assigned one of the six kingdoms. Each group then writes bulleted statements or words about their kingdom topic. Groups then rotate to the next poster and must add at least 2 items as well as place a '+' beside statements they strongly agree with and a '?' beside statements about which they remain unsure. This activity allows an instructor to quickly identify areas of misconceptions and areas of agreement among the groups.

Activities such as these foster a community of learners, situate the learning within context, and keep students engaged in the material. One student commented after completing Biology 3000, "I am so thankful for teaching in such a visual and interactive way. It was helpful to a learner like me…I hope you are still teaching the same way you taught me!" Another student said, "Alone we don't know everything, but together we do know everything!" Although an exaggeration, this statement highlights the impact of truly empowering teaching techniques that build, enhance, and foster the cooperative aspects of science learning.

## Non-Traditional Course Materials

Adhering to the relatively new course content source ideology, there is no textbook requirement. Science content information is not static, and the use of technology in schools is increasing exponentially. Thus Biology 3000 replaces the traditional text with a content-heavy class website, a variety of informational handouts, and semester long access to the website BrainPop (www.brainpop.com), where students can access educational videos, activities and sample lessons on science content. In addition, internationally recognized and award-winning resource guides are provided. These guides include Project WILD (2012), Project WET (2011), Project Aquatic WILD (2011) and Project Learning Tree (2013), along with curriculum from Population Connection (www.populationeducation.org). The materials all have been cross-referenced to K-12 science standards for each state in the United States (NRC, 1996), the Next Generation Science Standards (NGSS Lead States, 2013), and the Common Core Science Standards (Common Core Standards Initiative, 2010). The guides are used in several ways within Biology 3000. The instructor uses them to deliver portions of the course. Guide activities supplement and reinforce lecture and content acquisition. An example from Project WILD is the activity entitled "Oh, Deer" (WILD, 2012). In this simulation, students learn how population size changes over time while directly participating in a hands-on learning activity in the out-of-doors. Biological concepts including population dynamics, limiting factors and growth curves are creatively and effectively covered in this activity. Another major use of the guides is in aiding the development of a resource box, a requirement for all the enrolled students. They must choose a minimum of two lessons from each guide and collect all reusable materials required to create ready-to-use activity packs. Therefore, all the pre-service educators will leave the class with a variety of high-quality lessons that are ready to put into practice immediately. Based upon responses from former students, the resource boxes have proved to be invaluable assets in preparing and implementing lessons in their own classrooms.

## CASE DESCRIPTION

This research project aimed to verify that content-based, pedagogically rich science courses improve science and science teaching attitudes while increasing the self-efficacy of students enrolled in the course. Participants in this study were 166 undergraduate elementary education students (139 women, 27 men) from a large (~25,000) public university in the southeastern United States. The sample included all students from three consecutive sections of a biology content course for pre-service teachers, Biology 3000, Life Science for Elementary Teachers. All sections were taught by the same tenured faculty member from the Biology Department. Pre- and post-surveys were completed by students on the first and last days of attendance. Students received no incentives or grade for completing the surveys.

## Research Questions

Specifically, this study sought to address the following four challenges in elementary science teacher preparation:

1. Does Biology 3000's approach emphasizing Bloom's Taxonomy, situated learning, and active learning strategies improve pre-service teachers' *general attitudes toward science*?
2. Does Biology 3000's approach emphasizing Bloom's Taxonomy, situated learning, and active learning strategies improve pre-service teachers' *attitudes toward teaching science*?
3. Does Biology 3000's approach emphasizing Bloom's Taxonomy, situated learning, and active learning strategies increase pre-service teachers' *self-efficacy in specific life science content areas* (methodology and philosophy, the cell, genetics, evolution, diversity of life, plants, animals, and ecology) covered on the Middle School Praxis Examination?
4. How can data obtained in this study *inform the future development of the course or similar courses* which prepare elementary teachers?

## Research Methodology

Students completed surveys that covered three areas of focus: their general attitudes about science, their attitudes toward the teaching of science, and their self-efficacy with respect to understanding and teaching specific life science content. Two attitude surveys were obtained from the book *Elementary Science Methods: A Constructivist Approach* by Martin (2012). The two surveys contained a total of 40 items measuring science-related attitudes, 20 of which pertained to attitudes about science (e.g., dangerous/safe and sad/happy) and 20 pertaining to attitudes about teaching science (e.g., disorderly/orderly and unsuccessful/successful). Students rated each item using a 5-point scale (1 = negative attitude, 5 = positive attitude) which correlated with the negative/positive words (e.g., sad/happy). Higher scores indicated more positive attitudes.

The third survey administered was the Praxis Readiness scale; a 43-item measure of self-efficacy related to specific biological content. The measure addressed biological topics covered on the middle school science Praxis examination. The Praxis is one of a series of American teacher certification exams developed and administered by the Educational Testing Service (ETS, 2013). Various Praxis tests are usually required before, during, and after teacher training in the United States. Usually the participating students are required to pass the Praxis I, which covers reading, writing, and mathematics in order to be admitted into the teacher education program. The Praxis II examination covers specific content areas for certification at the middle and high school levels and is usually taken toward the end of the teacher training coursework.

Items on the Praxis Readiness scale given in Biology 3000 were developed by reviewing the major topics and themes on the middle school science Praxis II examination (ETS, 2013). The instrument included 43 items, which students rated using a 5-point confidence scale. Actual wording for the rating scale were as follows:

1 = EEEKKK!!! I need help with this content info and how to teach about it.
2 = I don't understand this and would need lot of background before I am comfortable teaching it.
3 = I am not sure I understand it and would not know how to teach about it.
4 = I understand this, but need some review. I am not sure how to teach about it.
5 = GOOD TO GO! I understand this and can teach this now.

Sample survey items included, "Distinguish between prokaryotic and eukaryotic cells," and, "Demonstrate understanding of Mendelian inheritance." From this total of 43 individual items, ten content area categories were identified. These Praxis knowledge content categories were assembled by grouping questions on similar topics. For example, a category entitled "Methodology and Philosophy" was created by grouping questions on general concepts of science and common scientific practices. A complete list of categories along with the number of items appears in Table 2. Higher scores indicated more positive self-efficacy and confidence about teaching the content.

## Findings

In the current study, the attitude measures showed acceptable internal consistency for both pre-test and post-test (see Table 1). The subscales used to

*Table 1. Results of paired t-test on students' attitude toward science and teaching science*

| | Pre-test | | Post-test | | Paired t (165) |
|---|---|---|---|---|---|
| | Mean(SD) | α-coefficient | Mean(SD) | α-coefficient | |
| **Attitudes about Science** | 55.43(9.85) | 0.899 | 63.0211.06) | 0.936 | 8.48* |
| **Attitudes about Teaching** | 57.84(12.23) | 0.932 | 66.49(11.38) | 0.957 | 8.91* |
| *N = 166* | | | | | |
| ** Indicates a significant value (p < 0.0001)* | | | | | |

*Table 2. Results of paired t-test on students' self-efficacy in specific science content areas on the Praxis examination. All means show a positive change.*

| | Pre-test | | Post-test | | Paired t (165) |
|---|---|---|---|---|---|
| | Mean(SD) | α-coefficient | Mean(SD) | α-coefficient | |
| Methodology and Philosophy | 8.56(1.89) | 0.803 | 11.2(1.14) | 0.651 | 18.25* |
| Math, Measurements and Data | 8.89(1.95) | 0.781 | 11.27(1.10) | 0.672 | 15.31* |
| Basic Principles | 1.88(0.96) | - | 3.65(0.58) | - | 20.46* |
| The Cell | 13.46(3.25) | 0.867 | 18.04(2.30) | 0.828 | 18.16* |
| Genetics | 20.73(6.91) | 0.933 | 28.29(5.49) | 0.889 | 14.37* |
| Evolution | 9.62(3.27) | 0.902 | 13.64(2.32) | 0.816 | 15.31* |
| Diversity of Life | 8.39(1.90) | 0.831 | 11.33(1.32) | 0.83 | 17.40* |
| Plants | 12.57(3.19) | 0.824 | 17.27(2.65) | 0.798 | 16.86* |
| Animals | 2.3(2.25) | 0.877 | 5.28(1.82) | 0.785 | 15.92* |
| Ecology | 18.27(5.37) | 0.91 | 29.25(2.93) | 0.768 | 25.06* |
| *N = 166* | | | | | |
| ** Indicates a significant value (p < 0.0001)* | | | | | |

measure groupings of content areas by survey items showed acceptable internal consistency for both pre-test and post-test, ranging between 0.651 and 0.933 respectively (see Table 2).

In examining the two attitude surveys and the Praxis Readiness Survey content areas, two-tailed paired *t*-tests were used to compare pre- and post-test attitude and self-efficacy data (see Tables 1 and 2). Analyses were performed using SPSS version 20.0 statistical software (IBM Corp., 2011). It was expected that over the course of the semester students developed more positive attitudes about science and teaching science and higher levels of self-efficacy about teaching all content areas assessed. As the data in Tables 1 and 2 show, these expectations were strongly supported.

## DISCUSSION

We recognize this study is limited by the fact that it was conducted in only one institution of higher learning on a single course and instructor. Nonetheless, results are encouraging considering the apprehension with which many pre-service and in-service teachers approach science in general. While the instructor has multiple years of teaching at the college level in biology and holds a Ph.D. in Environmental Science, her training for a Master's degree in Curriculum and Instruction was over 30 years ago and she is not considered an education specialist by her peers. Also, students enrolled in the class were on a variety of different career paths within education. Students included those seeking grades PreK-3, grades 5-8, or grades K-6 certification as well as individuals seeking accreditation in early childhood, special education, and specialists in other related jobs such as speech pathology or audiology. For significant gains to have occurred across this diverse population is remarkable.

The main findings are discussed in terms of the four research questions:

## 1. Does Biology 3000 Improve Pre-Service Teachers' General Attitudes toward Science?

Students' attitudes toward science became more positive by the end of the semester. Biology 3000 increased developing teachers' interest in science and reduced their anxiety toward learning and teaching science-related content. This increase in general science attitudes and enthusiasm for the subject supports pre-service teachers in becoming more effective science teachers (Patrick, Hisley, & Kempler, 2000; Tobin, Tippins, & Gallard, 1994) although it should be noted that not all studies have demonstrated such a relationship (Munck, 2007). This student's comment regarding Biology 3000 reveals the course's impact on science attitudes, "Science wasn't my thing, but I gave it all [in the class] and now I love it!!! Thank you, thank you, and thank you!"

## 2. Does Biology 3000 Improve Pre-Service Teachers' Attitudes toward Teaching Science?

The findings indicated that Biology 3000 helped pre-service teachers develop improved attitudes toward teaching science. Research has shown that many pre-service teachers feel hesitant and fearful toward teaching science (Appleton, 2007; Epstein & Miller, 2011; Finson, 2001; Gunning & Mensah, 2011; Mallow, 1982). This course helped them gain confidence in their own teaching ability, thus supporting their development into quality science educators. The following student comment encapsulates the growth that many experienced as a result of this class: "This was the subject that made me most nervous about teaching. I feel much more confident now."

## 3. Does Biology 3000 Increase Pre-Service Teachers' Self-Efficacy in Specific Life Science Content Areas (Methodology and Philosophy, the Cell, Genetics, Evolution, Diversity of Life, Plants, Animals, and Ecology) Covered on the Middle School Praxis Examination?

A significant positive change in confidence was revealed from pre-test to post-test for all life science content areas. The pre-service teachers increased their self-efficacy related to critical life science content they are required to know in order to pass the Praxis examination. Students gained comfort with the life science materials and developed a deeper understanding of the content. A number of pre-service teachers who have completed the Biology 3000 course have taken the time to report on their experience with the Praxis exam:

*I just wanted to let you know that your class was challenging and it has truly been one of the best I have taken. I have taken all my Praxis II tests and consistently scored the highest on the science portion of the tests…I had no problems answering the questions. I have used a great deal of the classroom management and strategies you used during our course. During my student teaching I taught science to second grade using a lot of your methods (I also used them for social studies!). They recently took their benchmark and did so well on their tests! I just wanted to let you know…*

*I took my middle school content knowledge Praxis and received my scores yesterday. I knew I killed the science portion of the test…felt more confident of that than anything else. I got my scores and saw I scored higher in science than in math (and I am a math person!).*

*I just want to THANK YOU again for everything. I took the Content Knowledge Praxis last weekend and blew it out of the water. The only questions I answered with complete [certainty] were biology–based.*

The results from the present study as well as the qualitative responses from the students clearly support the findings of Tosun (2000) that pre-service teachers with a background in a content specific science course ultimately held more positive beliefs and attitudes towards teaching that particular subject at the elementary level. Possessing a greater understanding of the content and a positive attitude towards science and science teaching are key ingredients in the making of a successful and effective science teacher.

### 4. How Can Data Obtained in This Study Inform the Future Development of the Course or Similar Courses which Prepare Elementary Teachers?

The findings of this study validate the current approach used in Biology 3000. The content-focused, pedagogically rich instruction has produced improved science attitudes and self-efficacy in pre-service teachers taking the course. Possessing a positive attitude toward the subject matter and confidence in content mastery are essential affective assets of professional teachers. Methods courses with a less positive effect on student attitudes might benefit by considering an approach like that employed in Biology 3000. However, this study did not include a comparison component, so we cannot attribute the growth unequivocally to the teaching approach. Other hidden factors could be playing a role. Also, there is not yet data to show that the growth in this course is measurably better than that which occurs in any other type of science methods course. These preliminary and promising findings do encourage a closer look into the factors that are making the course successful. Future research should seek to tease out the significant factors in order to guide the continued improvement of this specific course and teacher preparation programs in general.

## CURRENT CHALLENGES FACING THIS RESEARCH

Three core challenges have been acknowledged facing this research and its future expansion. First, the specific aspects of the course leading to the improved science attitudes and self-efficacy remain unclear. We have posited the importance of certain activities and instructional methods employed in this course, but more research is needed to verify the causal link between those factors and the positive outcomes in the pre-service teachers. Second, the current research is limited in that it only examined one course taught by one instructor. A comparison study examining the self-efficacy and science attitudes of pre-service teachers who have taken Biology 3000 with those who have had a similar course taught with a traditional approach would be informative. It is hypothesized the results would further support the implementation of content-based, pedagogically rich courses incorporating active and interactive teaching techniques in lieu of the traditional science methods courses. A last challenge lies in tracking former Biology 3000 students to determine if their positive attitude and heightened self-efficacy remains after one, three, and five years in the classroom. A longitudinal study, while logistically difficult, would provide valuable information regarding how persistent the improved attitudes are, and whether they have an impact on the pre-service teachers' future teaching effectiveness.

## SOLUTIONS AND RECOMMENDATIONS

Classroom teachers usually teach as they were taught. Many content courses taken by pre-service teachers in college are not specifically tailored for those going into teaching, nor are they taught by faculty with a background in pedagogical content knowledge. Therefore, appropriate pedagogical methods are not always modeled effectively (if at all) in these content rich courses. That can make it difficult for prospective teachers to know how to apply and transfer the content knowledge to their future students and future classroom situations. Methods courses are expected to provide the connection between content and delivery. Unfortunately, that connection is sometimes lacking (Raizen & Michelsohn, 1994) and pre-service teachers without a strong understanding of the content can find themselves underprepared for the classroom. It has been said that "relevance is the key that unlocks the door to learning" (Ende, 2012, p. 45). Unfortunately, it is this relevance that is often missing from standard content courses taken by future teachers.

We suggest that the explicit use of Bloom's Taxonomy, active learning instructional strategies, and situated learning when planning instruction is critical for developing teachers who are confident in their understanding of scientific principles and unafraid of learning and teaching science. The pre-service teachers in this study commented on their increased assurance in both self-efficacy and positive attitudes regarding science and the teaching of science. In addition, this research supports the incorporation of both active and collaborative learning techniques which include both written and verbal constructs. These practices foster a community of learners whereby pre-service teachers, alone and together, engage with the material to develop greater understanding of concepts and topics while experiencing appropriate pedagogical techniques.

It is hoped that these preliminary findings will encourage teacher development programs to evaluate the techniques and approaches used in traditional science methods courses and in science content courses in which pre-service teachers enroll. Well-prepared pre-service teachers must have strong science content knowledge, positive science attitudes, and high self-efficacy in learning and teaching science. We contend that the integration of content and pedagogy is imperative to developing excellent teachers who not only understand concepts, but are able to translate those concepts into relevant and easily understandable examples for their students

## REFERENCES

Anderson, L. W., & Krathwohl, D. R. (Eds.). (2001). *A taxonomy for learning, teaching, and assessing: A revision of Bloom's taxonomy of educational objectives*. New York, NY: Longman.

Anderson, R. D. (2002). Reforming science teaching: What research says about inquiry. *Journal of Science Teacher Education*, *13*(1), 1–12. doi:10.1023/A:1015171124982

Angelo, T. A., & Cross, K. P. (1993). *Classroom assessment techniques: A handbook for college teachers* (2nd ed.). San Francisco, CA: Jossey-Bass.

Appleton, K. (2007). Elementary science teaching. In S. Abell, & N. Lederman (Eds.), *Handbook of research on science education* (pp. 493–535). Mahwah, NJ: Lawrence Erlbaum Associates, Inc.

Bandura, A. (1977). Self-efficacy: Toward a unifying theory of behavioral change. *Psychological Review*, *84*(2), 191–215. doi:10.1037/0033-295X.84.2.191 PMID:847061

Bandura, A. (1986). *Social foundations of thought and action: A social cognitive theory*. Englewood Cliffs, NJ: Prentice-Hall.

Bandura, A. (1993). Perceived self-efficacy in cognitive development and functioning. *Journal of Educational Psychologist*, *28*(2), 117–148. doi:10.1207/s15326985ep2802_3

Barufaldi, J. P., Bethel, L. J., & Lamb, W. G. (1977). The effect of a science methods course on the philosophical view of science among elementary education majors. *Journal of Research in Science Teaching*, *14*(4), 289–294. doi:10.1002/tea.3660140404

Berliner, D. (1988). *The development of expertise in pedagogy*. Washington, DC: AACTE Publications.

Bianchini, J. A., & Colburn, A. (2000). Teaching the nature of science through inquiry to prospective elementary teachers: A tale of two researchers. *Journal of Research in Science Teaching*, *37*(2), 177–209. doi:10.1002/(SICI)1098-2736(200002)37:2<177::AID-TEA6>3.0.CO;2-Y

Bloom, B. S., Engelhart, M. D., Furst, E. J., Hill, W. H., & Krathwohl, D. R. (1956). *Taxonomy of educational objective: The classification of educational goals, handbook I: cognitive domain*. New York, NY: McKay.

Bonwell, C. C., & Eison, J. A. (1991). ASHE-ERIC Higher Education Report: Vol. 1. *Active learning: Creating excitement in the classroom*. Washington, DC: George Washington University.

Bransford, J. D., Brown, A. L., & Cocking, R. R. (Eds.). (2000). *How people learn: Brain, mind, experience and school*. Washington, DC: National Academy Press.

Brent, R., & Felder, R. M. (1992). Writing assignments—Pathways to connections, clarity, creativity. *College Teaching*, *40*(2), 43–47. doi:10.1080/87567555.1992.10532264

Bursal, M. (2008). Changes in Turkish pre-service elementary teachers' personal science teaching efficacy beliefs and science anxieties during a science method course. *Journal of Turkish Science Education*, *5*(1), 99–112.

Common Core State Standards Initiative. (2010). *Common core state standards for English language arts and literacy in history/social studies, science, and technical subjects*. Washington, DC: Council of Chief State School Officers and National Governors Association. Retrieved from http://www.corestandards.org/

IBM Corp. (2011). IBM SPSS Statistics for Windows, Version 20.0. Armonk, NY: IBM Corp.

Czerniak, C. M., & Schriver, M. L. (1994). An examination of pre-service science teachers' beliefs and behaviors as related to self-efficacy. *Journal of Science Teacher Education*, *5*(3), 77–86. doi:10.1007/BF02614577

Druva, C. A., & Anderson, R. D. (1983). Science teacher characteristics by teacher behavior and by student outcome: A meta-analysis of research. *Journal of Research in Science Teaching*, *20*(5), 467–479. doi:10.1002/tea.3660200509

Educational Testing Service (ETS). (2013). *The Praxis series*. Retrieved from http://www.ets.org/praxis

Ende, F. (2012). NOT another lab report. *Science Scope*, *35*(5), 44–50.

Epstein, D., & Miller, R. T. (2011). *Slow off the mark: Elementary school teachers and the crisis in science, technology, engineering, and math education*. Retrieved from http://www.americanprogress.org/issues/education /report/2011/05/04/9680/slow-off-the-mark/

Erlauer, L. (2003). *The brain-compatible classroom*. Alexandria, VA: ASCD.

Fencl, H., & Scheel, K. (2005). Engaging students: An examination of the effects of teaching strategies on self-efficacy and course climate in a non-majors physics course. *Journal of College Science Teaching*, *35*(1), 20–24.

Finson, K. D. (2001). Investigating pre-service elementary teachers' self-efficacy relative to self-image as a science teacher. *Journal of Elementary Science Education, 13*(1), 31–41. doi:10.1007/BF03176931

Greenwood, A., & Scribner-MacLean, M. (1997). *Examining elementary teachers' explanations of their science content knowledge*. Paper presented at the annual meeting of the National Association for Research in Science Teaching, Oak Brook, IL.

Gunning, A. M., & Mensah, F. M. (2011). Pre-service elementary teachers' development of self-efficacy and confidence to teach science: A case study. *Journal of Science Teacher Education, 22*(2), 171–185. doi:10.1007/s10972-010-9198-8

Hackett, G., Betz, N. E., Casas, J. M., & Rocha-Singh, I. A. (1992). Gender, ethnicity, and social cognitive factors predicting the academic achievement of students in engineering. *Journal of Counseling Psychology, 39*(4), 527–538. doi:10.1037/0022-0167.39.4.527

Haney, J. J., Lumpe, A. T., Czerniak, C. M., & Egan, V. (2002). From beliefs to actions: The beliefs and actions of teachers implementing change. *Journal of Science Teacher Education, 13*(3), 171–187. doi:10.1023/A:1016565016116

Ingersoll, R. (2003). *Is there really a teacher shortage?* (Document R-03-04). Seattle, WA: Center for the Study of Teaching and Policy at the University of Washington. Retrieved from http://depts.washington.edu/ctpmail/PDFs/Shortage-RI-09-2003.pdf

Ingersoll, R., & Perda, D. (2012). *How high is teacher turnover and is it a problem?* Philadelphia, PA: University of Pennsylvania, Consortium for Policy Research in Education.

Johnson, D. W., Johnson, R. T., & Smith, K. A. (1998). *Active learning: Cooperation in the college classroom* (2nd ed.). Edina, MN: Interaction Press.

King, K., Shumow, L., & Lietz, S. (2001). Science education in an urban elementary school: Case studies of teacher beliefs and classroom practices. *Science Education, 85*(2), 89–110. doi:10.1002/1098-237X(200103)85:2<89::AID-SCE10>3.0.CO;2-H

Lave, J. (1993). Situated learning in communities of practice. In L. B. Resnick, J. M. Levine, & S. D. Teasley (Eds.), *Perspectives on Socially Shared Cognition* (pp. 63–82). Washington, DC: American Psychological Association.

Lipton, L., & Wellman, B. (1998). *Patterns and practices in the learning-focused classroom. Guilford.* VT: Pathways Publishing.

Lumpe, A. T., Haney, J. J., & Czerniak, C. M. (1998). Science teacher beliefs and intentions regarding the use of cooperative learning. *School Science and Mathematics, 98*(3), 123–135. doi:10.1111/j.1949-8594.1998.tb17405.x

Lyman, F. (1987). Think-Pair-Share: An expanding teaching technique. *MAACIE Cooperative News, 1*, 1–2.

Mallow, J. (1982). *Science anxiety: Fear of science and how to overcome it*. New York, NY: Van Nostrand Reinhold.

Marshall, J. C., Horton, R., Igo, B. L., & Switzer, D. M. (2009). K-12 science and mathematics teachers' beliefs about and use of inquiry in the classroom. *International Journal of Science and Mathematics Education, 7*(3), 575–596. doi:10.1007/s10763-007-9122-7

Martin, D. J. (2012). *Elementary science methods: A constructivist approach*. Belmont, CA: Wadsworth.

McGaugh, J. L. (2003). *Memory and emotion: The making of lasting memories*. New York, NY: Columbia University Press.

McKeachie, W. J. (1986). *Teaching tips: strategies, research, and theory of college and university teachers*. Boston, MA: Houghton Mifflin.

Meyers, C., & Jones, T. B. (1993). *Promoting active learning: Strategies for the college classroom*. San Francisco, CA: Jossey-Bass.

Michelsohn, A. M., & Hawkins, S. (1994). Current practice in science education of prospective elementary school teachers. In S. Raizen, & A. Michelsohn (Eds.), *The future of science in elementary schools: Educating prospective teachers* (pp. 151–161). San Francisco, CA: Jossey-Bass.

Monk, D. (1994). Subject area preparation of secondary mathematics and science teachers and student achievement. *Economics of Education Review, 13*(2), 125–145. doi:10.1016/0272-7757(94)90003-5

Munck, M. (2007). Science pedagogy, teacher attitudes, and student success. *Journal of Elementary Science Education, 19*(2), 13–24. doi:10.1007/BF03173660

National Commission on Teaching and America's Future. (1996). *What matters most: Teaching for America's future. Report of the National Commission on Teaching & America's Future*. New York, NY: Author.

National Research Council (NRC). (1996). *National science education standards*. Washington, DC: National Academy Press.

National Research Council (NRC). (2011). *A Framework for K-12 science education: Practices, crosscutting concepts, and core ideas*. Washington, DC: National Academies Press.

NGSS Lead States. (2013). *Next generation science standards: For states, by states*. Washington, DC: The National Academies Press.

Palmer, D. H. (2006). Sources of self-efficacy in a science methods course for primary teacher education students. *Research in Science Education, 36*(4), 337–353. doi:10.1007/s11165-005-9007-0

Patrick, B. C., Hisley, J., & Kempler, T. (2000). "What's everybody so excited about?" The effects of teacher enthusiasm on student intrinsic motivation and vitality. *Journal of Experimental Education, 68*(3), 217–236. doi:10.1080/00220970009600093

Pauk, W., & Owens, R. J. Q. (2005). *How to study in college* (11th ed.). Boston, MA: Houghton Mifflin.

Prawat, R. S. (1992). Teachers' beliefs about teaching and learning: A constructivist perspective. *American Journal of Education, 100*(3), 354–395. doi:10.1086/444021

Project, W. E. T. (2011). *K-12 curriculum and activity guide*. Bozeman, MT: The Watercourse/Project WET International Foundation and the Council for Environmental Education.

Project, W. I. L. D. (2012). *K-12 curriculum and activity guide*. Houston, TX: Council for Environmental Education.

Project Learning Tree. (2013). *Pre K-8 environmental education activity guide*. Washington, DC: American Forest Foundation.

Project WILD Aquatic. (2011). *K-12 curriculum and activity guide*. Houston, TX: Council for Environmental Education.

Raizen, S. A., & Michelsohn, A. M. (Eds.). (1994). *The future of science in elementary schools: Educating prospective teachers*. San Francisco, CA: Jossey-Bass.

Riggs, I. M., & Enochs, L. G. (1990). Toward the development of an elementary teacher's science teaching efficacy belief instrument. *Journal of Science Education, 74*(6), 625–637. doi:10.1002/sce.3730740605

Ross, J. A. (1992). Teacher efficacy and the effect of coaching on student achievement. *Canadian Journal of Education, 17*(1), 51–65. doi:10.2307/1495395

Shapiro, B. L. (1996). A case study of change in elementary student teacher thinking during an independent investigation in science: Learning about the "Face of science that does not yet know.". *Science Education, 80*(5), 535–560. doi:10.1002/(SICI)1098-237X(199609)80:5<535::AID-SCE3>3.0.CO;2-C

Shulman, L. S. (1986). Those who understand: Knowledge growth in teaching. *Educational Researcher, 15*(2), 4–14. doi:10.3102/0013189X015002004

Shymansky, J. A., Kyle, W. C., & Alport, J. M. (1983). The effects of new science curricula on student performance. *Journal of Research in Science Teaching, 20*(5), 387–404. doi:10.1002/tea.3660200504

Tobin, K., Tippins, D. J., & Gallard, A. J. (1994). Research on instructional strategies for science teaching. In D. L. Gabel (Ed.), *Handbook of research on science teaching and learning* (pp. 45–93). New York, NY: Macmillan Publishing Company.

Tosun, T. (2000). The beliefs of pre-service elementary teachers toward science and science teaching. *School Science and Mathematics, 100*(7), 374–379. doi:10.1111/j.1949-8594.2000.tb18179.x

Tschannen-Moran, M., Hoy, A. W., & Hoy, W. K. (1998). Teacher efficacy: Its meaning and measure. *Review of Educational Research, 68*(2), 202–248. doi:10.3102/00346543068002202

U.S. Department of Labor, Bureau of Labor Statistics. (2014, December 8). *Occupational outlook handbook, 2014-15 Edition: Kindergarten and Elementary School Teachers.* Retrieved from http://www.bls.gov/ooh/education-training-and-library/kindergarten-and-elementary-school-teachers.htm

Wayne, A. J., & Youngs, P. (2003). Teacher characteristics and student achievement gains: A review. *Review of Educational Research, 73*(1), 89–122. doi:10.3102/00346543073001089

Wenger, E. (1998). Communities of practice: Learning as a social system. *Systems Thinker, 9*(5), 2–3.

Wenglinsky, H. (2002). The link between teacher classroom practices and student academic performance. *Education Policy Analysis Archives, 10*(12).

Wenner, G. (1993). Relationship between science knowledge levels and beliefs toward science instruction held by pre-service elementary teachers. *Journal of Science Education and Technology, 2*(3), 461–468. doi:10.1007/BF00694428

Westerback, M. E. (1982). Studies on attitude toward teaching science and anxiety about teaching science in pre-service elementary teachers. *Journal of Research in Science Teaching*, *19*(7), 603–616. doi:10.1002/tea.3660190710

Westerback, M. E., & Long, M. J. (1990). Science knowledge and the reduction of anxiety about teaching earth science in exemplary teachers as measured by the science teaching state-trait anxiety inventory. *School Science and Mathematics*, *90*(5), 361–374. doi:10.1111/j.1949-8594.1990.tb17225.x

Wolfe, P. (2010). *Brain matters: Translating research into classroom practice.* Alexandria, VA: ASCD.

Yuruk, N. (2011). The predictors of pre-service elementary teachers' anxiety about teaching science. *Journal of Baltic Science Education*, *10*(1), 17–26.

Zike, D. (2004). Dinah Zike's big book of science for elementary K-6. San Antonio, TX: Dinah-Might Adventures, LP.

Ziman, J. (2002). *Real science: What it is and what it means.* Cambridge, UK: Cambridge University Press.

## ADDITIONAL READING

Abell, S., & Lederman, N. G. (Eds.). (2007). *Handbook of research in science education.* Mahwah, NJ: Lawrence Erlbaum.

Barkley, E. F. (2009). *Student engagement techniques: A handbook for college faculty.* San Francisco, CA: Jossey-Bass.

Darling-Hammond, L., & Bransford, J. (Eds.). (2005). *Preparing teachers for a changing world: What teachers should learn and be able to do. National Academy of Education, Committee on Teacher Education.* San Francisco, CA: Jossey-Bass.

Keeley, P. (2008). *Science formative assessment: 75 practical strategies for linking assessment, instruction, and learning.* Thousand Oaks, CA: Corwin Press.

Silberman, M. (1996). *Active learning: 101 strategies to teach any subject.* Des Moines, IA: Prentice-Hall.

Swartz, R. J., & Parks, S. (1994). *Infusing the teaching of critical and creative thinking into elementary instruction: A lesson design handbook.* Pacific Grove, CA: Critical Thinking Press and Software.

## KEY TERMS AND DEFINITIONS

**Active Learning:** A broad term referring to teaching methods that put the responsibility of learning on the learners themselves.

**Bloom's Taxonomy:** A classification of the main learning objectives arranged from lower-order to higher order thinking skills first proposed by Benjamin Bloom.

**Content Course (Content-Based Course):** A college class for pre-service teachers that focuses more on conveying subject matter knowledge rather than teaching strategies or educational theories.

**Pre-Service Elementary Education:** A course of study offered at a university in order to prepare future elementary education teachers for the profession.

**Pre-Service Teacher:** A student teacher who has not yet completed their training or certification in teaching.

**Science Methods Course:** A teacher preparation course focused on conveying strategies and methods for teaching science content.

**Self-Efficacy:** The amount of confidence an individual has in their ability to complete a given task or reach a specific goal.

**Situated Learning:** Learning that takes place in the same situation within which it can be applied.

# Section 4
# Cases on Research-Based Teaching Methods in Science Education

# Chapter 13
# Traditional Teaching and Its Effect on Research-Based Teaching:
## Science via Online Instruction

**LaToya N. Johnson**
*Walden University, USA*

**Dana-Marie Thomas**
*Walden University, USA*

**K. Y. Williams**
*Walden University, USA & Kaplan University, USA & Department of Defense, USA*

## EXECUTIVE SUMMARY

*Land-based institutions that use traditional teaching methods have very well documented methods for providing students with the necessary skills, experience, and knowledge for becoming extremely productive scientists in different research areas that are traditional (chemistry, biology, and microbiology) and interdisciplinary (biochemistry, bioinformatics, and computational chemistry) in nature, and they have very few problems when transitioning into any research environment. However, online institutions do not have a well-documented history of students transitioning into land-based institution or research intensive environments. Within this case study, the authors express ways to help meet the needs of the students and educate students in becoming better scientists who have been educated in online institutions by using methods from land-based institutions and implementing other forms*

DOI: 10.4018/978-1-4666-6375-6.ch013

*of technology into the classroom. The authors explore instruction, knowledge, and experience, and suggest how online science instruction can be supplemented with experience and technology that can increase their experience and knowledge to allow them to become better scientists.*

## ORGANIZATION BACKGROUND

Land-Based Institutions (LBIs) traditional teaching sessions are composed of one of 5 types: three 50 minute sessions that meet 3 times during the week; two 75 minute sessions that meet 2 times during the week; one 3 hour session that meet once a week; one 8 hour session that meet over the weekend, or complete independent study that is conducted between the instructor and the student. Within each of the types of course material instruction, each has the component of more human interaction and sharing of knowledge that allows time for students to: digest material that is presented to them, allow the student to incorporate the knowledge that comes from open discussion and interaction with a knowledgeable, experienced professional, time for students to make corrections within their understanding of the subject material within the courses, and a chance for the instructor to gauge students' abilities, knowledge, and experience from one-on-one interaction. In each of these sessions, the instructor would cover the subject material based on the syllabus, outline of the course objectives, and learning outcomes of the course. In order to achieve the goals of the course, the instructor has at his/her disposal all of the approved forms of information, teaching material, and technology that is available within the university. Additionally Instructors have the discretion to use the material in the course if they feel it will get the information, understanding, and concepts of the course material across to the student. In many instances when the course is not a lecture course, but it requires more hands-on experience, such as a laboratory course or a recitation, the instructor has more interaction with the student in the form of Teaching Assistants (TA) and face-to-face questioning of the student, and this provides a chance for Instructors to develop the students into better researchers.

In LBIs some science courses are normally taught with three components: a lecture component, a laboratory component, and a recitation component. Within these configurations it is possible for the instructor to present material within the lecture component that covers the background, theory, concepts and information of the material that is being presented. While the laboratory component covers the hands-on experience portion of the course, and the recitation component is used to reinforce information that is introduced in the lecture portion of the course and then correlated with information that is pertinent for the experiments within the laboratory. When courses are taught in this configuration, the student is usually required to take the laboratory course with a recitation session, even if the student does not enroll in the lecture course, and laboratory courses would last anywhere from 3 – 5 hours depending on the course. During these sessions this would give students time to evaluate their work, talk with the TA/Instructor to assist in their knowledge/experience development, and time to correct errors within their knowledge/thinking in regards to theory and problems within the field. This time also allows for trouble shooting issues with regard to experiments, and to correct/improve upon the process by discussing their ideas with classmates and the TA/Instructor. During the recitation session many caveats within the course would be explained: the experiment for the laboratory course, issues that may occur during the experiment may be acknowledged, changes in the type of setup within the laboratory procedure, possible experimental outcomes, time commitments to the experiments based on the known issues with the procedure, report writing, good standard laboratory practices, and final submission of experimental results. Even though the presentation of material would change – depending on the course, laboratory type, and recitation session – the instructor provided material to students in more traditional teaching formats than what is used in teaching sessions in Online Institutions.

Online Institutions (OIs) instructional method is very different from traditional instructional and course delivery methods. OIs teaching sessions can range from courses that are taught within: 5 week sessions, 8 week sessions, 10 week sessions, 12 week sessions which correspond to a quarter term, or a semester long course that is similar to LBIs. Within OIs Instructors usually allow for course material to be presented during the week – based on the course and topic information – with the aide of learning resources, and the information presented is based on the syllabus, outline of the course objectives, and learning outcomes. However, the use of technology within the course is paramount for the delivery of the information within the course as all material for the student (even books) must be online for the students to use and to have access to during the course. This is inclusive of, but not limited to: the course syllabus, book chapters and reading material, instructional video(s), journal articles, asynchronous discussion boards, and library resources for the topics being presented. Within OIs the information when presented is developed and delivered within a self-contained package based on the course material, and such material and information is independent of the instructor that will teach the course or the background of the instructor when the course is being developing. Within this process the presentation of the course material and the information for the course is not dependent on the instructor but more on the goals and outcomes of the course. Therefore changes within the course are not dependent on the instructor but upon the developers of the courses.

Course changes within LBIs traditional class setting, reading material within the book, or even in the book used to teach the course is based on the Instructors experience and the Instructors desired format for teaching the material that must be covered within the syllabus. If changes within the material will be made, based on the lecture, then it can be made at the discretion of the instructor, while larger more drastic course changes or corrections must be approved at a higher level especially if it affects: the course syllabus, what material (will not or) will be covered, the learning outcomes, or the direction of the course. Usually this is an easy process that is governed by the Instructor, and approval is only needed and granted by the department chair. OIs usually have a team of individuals who aide in the development of course material. These teams are composed of: course developers, committees, and Instructors. Each member has various responsibilities and take initiatives to complete aspects of the course development process. Unlike Instructors in LBIs the course development team has to have university approval to make the necessary changes within the development of the course (Williams, 2014); therefore, this process can take longer because the changes must be approved by a committee and the process may not occur by the end of a term, semester or within the same year. With this process, it is possible that the lapse in time between conceptualization of the changes and the deployment of the changes can take anywhere from 12 weeks to 1 year, depending on the type and number of changes that will be implemented. With this process, issues can occur based on the time that can lapse, as time is needed to approve changes, or the issues that can occur can be based on the development of technology within the science field. Development of technology does not allow for students to have large lapses of time due to the rapid changes made with technological innovations within the science field, or for institutions not to be able to keep up with the technology within the science field. In order for students to be properly prepared for transitioning into the field of science students should be exposed to technology via different formats and for different purposes, as well as transitioning into research positions.

Whether in LBIs or OIs, students should be exposed to all forms of technology that would be needed to allow them to develop their expertise in areas such as: developing their ability to conduct independent research, experimentation, hands-on experience, developing the abilities needed to be able to troubleshoot issues within their experiments and research, and knowing how to or being exposed to technology that is used in research, teaching, or what is needed for the educational process. LBIs have excellent experience with research and placing students into their science discipline, and they are able to show that students are able to proceed and succeed within their fields. LBIs have research capabilities within the science fields that they have cultivated, enhanced, and developed by conducting experiments within the field with the latest technology and tools that are available at the university. Students receive the tools and techniques based on their involvement with research laboratories and became better, experienced scientist by being exposed to technology that pertain to their research projects. Use of technology tend to increase based on the length of time students have been exposed to the technology, the use of the technology, the types of research projects, and research experience. Most OIs do not show a clearly defined path that students take to receive the research experience within their fields to give them the hands on experience with technology, or to conduct independent research within their fields once they receive their degrees or while receiving their degrees (Williams, 2014). Due to lack of technology that OIs expose their students to in the courses or in the educational process most students who are trained and educated online do not become better, experienced scientist who conduct independent research projects that incorporate technology within their field.

LBIs have access to technology within the different science fields, but have not used the various forms of technology to instruct students from a distance in their usage and the application of the technology. These technologies are usually included within the various laboratories within the institution; however for various reasons students in OIs have not been trained to implement the different forms of technology within their research as they have not been exposed to them as in LBIs. By being educated on the aspects of technology within their scientific discipline, using them in their experiments, and mastering them within their graduate program, it will allow students within OIs to become better scientists upon completion of their educational programs. Currently, the use of technology and the experience that comes with its use give students of LBIs an advantage over others when they apply for and receive jobs because of the experience they have with technology within their field. By having this experience, it allows for better and more scientific accomplishments, publications within the fields, and research projects that are more scientific and technology based. In order for students to develop these skills, different forms of instruction and delivery of the material is necessary for students in addition to what is used in the course as many OIs use asynchronous discussion boards, electronic journals, eBooks, and instructional video(s) to deliver the course information to students. However additional forms of instructional material is available and should be considered to enhance the learning environment within Land-Based and Online Institutions.

## SETTING THE STAGE

Academia, Industry, and research intensive institutions value graduates who possess experience with technology and who possess research experience as it allows for instant growth of the company, the institution, and the agency that is willing to employ the graduate. Incorporating individuals who possess the experience and knowledge into the company, agency, or research institution allows for the instant addition of new projects; it increases the number and possibility of new ideas on existing projects; and begin to increase and broaden the knowledge and experience base of the organization. In addition, experience adds to the number of ideas that can exist with having experience with the various forms of technology, and for technological advancement that can be used to enhance the already existing forms of technology. Not having the proper experience does not add to the program, department, or the organization as the voids that are created by not staying abreast of the advancements with technology, or the rapid growth of a company or organization can lead to a lapse of time and the institution may not be abreast of the latest forms of technology. In order to lessen the void that exists within the company or the organization, additional faculty will be necessary to reduce the possibility that a lapse in experience, knowledge, and/or technology will exist.

With any change or advancement within the education system that relies on technology, the organizations or institutions should be concerned with the advances and forms of technology that are available to the organization, and how these forms of technology can be used to conduct or increase research initiatives. These organizations should also ask more direct and challenging questions as it relates to advancing the technology within the institution. Essentially the institution should question what is needed to make the necessary changes in their organization that will allow them to effectively ask and answer questions such as:

1. Why should we enhance our technology within our organization and courses?
2. What can be done to improve the educational and research experience of the student, and what will produce better results in terms of their educational and professional growth?
3. What forms of technology is available and what are the limitations of the technology when used within our organization?
4. What other forms of technology is available if we do not use the technology that is currently being used in this field?
5. Can the technology be optimized to effectively deliver the course material for the currently available courses?
6. How can we use the available forms of technology to increase our enrollment within the courses, and which forms of technology give students the abilities that are required and needed to be productive members of the science community within their respective fields?

In addition to being instructed in the science field, it is also imperative that students be able to adapt and use technology within their field upon graduation from their programs of study – undergraduate, masters, or professional. Additionally it is important to incorporate technology into their research programs and be able to conduct independent research beyond graduation. With their various degrees students should be able to formulate hypothesis and construct research questions that can be carried out with the available technology in their fields of study, and be exposed to forms of technology that is available in the educational field that can assist in their education. In order for this to occur within the academic institution, the institution must evaluate the type and forms of technology that is currently in place, and consider all forms of education that is available to deliver the information and course material to the student.

In most online courses, regardless of the course, the forms of technology usually include email, Blackboard (Chasen & Pittinsky, 2014) or eCollege (Pearson, 2014) which has chat features, asynchronous discussion boards, and TurnItIn (TurnItIn, 2014) for similarity checking and assignment submission. However other forms of technology exist that can be used to further the growth of the university, and these forms of technology can assist the student in their understanding and exposure to the knowledge within the field, and also to improve upon the educational and research experience that the students can receive within either Land-Based or Online Institutions.

## CASE DESCRIPTION

The need for better science instruction in an online environment is paramount, as research and trouble-shooting experience has more of an impact than just theoretical knowledge, as research and trouble-shooting experience is necessary for anyone who wishes to gain complete understanding within the science field. As academic institutions begin to improve upon the use of technology associated with different educational platforms, LBIs and OIs should use technology to assist the educational and research experience of their students based on the different forms of learning styles of the student, increase the types and forms of technology that are available to the institution, and enhance the educational and research experience for the student by making alternative forms of information presentation available to the student.

### Technology Concerns

Land-Based Institutions have a hands-on approach to instruction that has been shown to be effective in meeting the students understanding of the material when presented with information in a lecture and/or recitation session; moreover, this type of education and instruction makes the student more dependent on working with a senior professional within their scientific and academic professional career. This builds a sense of collaboration and professionalism that leads to mentoring and an exchange of experience and knowledge that encourages the student to become better researchers. However, this can cause the student to take more time to become the expert within their field and not be as independent as expected upon graduation. Online Institutions have a more hands-off approach with their student education and with the institution, which allows for students to be more independent within their fields earlier on in the educational process, as they must comprehend and synthesize the information in their academic fields. However this independence comes at a price to the student and the organization, as the student usually does not have an accurate understanding of the scientific and technological material when placed in front of them, they are only able to work in short bursts before moving to the next phase of a project, and they possess limited or no research or technology experience. Even though students have more freedom with gathering, interpreting, and reporting information within their field, online instruction does not allow for complete correction, knowledge exchange, or accuracy with regard to scientific education, research experience, or technology exposure.

When LBIs begin to consider incorporating more information and courses online, this means that the institution has taken a step in moving from teaching in a traditional format to including information, subject material, and an education format that has been adopted by most online environments. In many ways this move makes it easier for students who are independent to become more independent while still having the presence and constant exchange of knowledge and experience that comes within LBIs. For OIs it may seem difficult for students to gain the necessary experience and skills that come from the interaction and knowledge that exist within LBIs; however, with the use of other forms of technology, it is possible for students to achieve aspects of this experience. Some of the various forms of technology that exists that Land-Based Institutions and Online Institutions can use include: streaming video, video presentations/pod casts, tutorials, discussion boards, instant messaging/chat, virtual laboratories, virtual office hours, webinars/seminars and tutoring/knowledge exchange sessions. Each of these forms of technology can be adapted to either the land-based or online environment, and deployed to meet the different students' needs based on their program of study.

## Technology Components

Online Institutions readily depend on functional technology to achieve its goals of educating students at a distance. In distance learning environments streaming video allows for instructors to be able to teach students in a different location the normal subject material that is presented in a course. As more courses moved away from streaming video to courses that are self-contained and completely online, the use of streaming video for a course began to diminish. However streaming video does have its advantages as in a lecture/seminar format it allows for interaction with the student, and it gives the instructor a chance to go over, correct, and incorporate new material as needed. Within Online Institutions using streaming video for question and answer periods would be ideal for students to have a chance to interact with the instructors who teach the courses in which they are enrolled; however most online courses are not designed for constant streaming of video. Although some institutions use seminar sessions that allows for students to interact with the instructor as they present a course topic, these sessions are usually offered once-a-week within the course. These sessions do not allow for instructors to impart a lot of their knowledge on the subject material as the sessions are at most one-hour per week, and the instructor can only cover so much material within this given period of time. Within Land-Based Institutions this form of technology is not needed as most instructors have at least 4 hours per week of office hours that allows for interaction and correction of student material. However course lectures can be recorded and archived to allow for students to have the experience that is similar to streaming video of course lectures and question/answer periods of the course.

Online Institutions use podcasts/video presentations to distribute material and examples within the courses, and the use of video presentations supposed to be used as a resource in the student's educational program. These videos can range from 5 – 15 minutes in length and they cover material or examples that is usually presented within the week. Using podcasts/videos allow for easy dissemination of the material to the student, and the student has a means of reviewing the material before, during, or after the week of instruction. The video usually provides a chance for the student to apply a theoretical approach that has been garnered from the course material, but it does not allow the student to have an interactive exchange between the student and the instructor, to allow for better analysis of the video or to better understand the theoretical concepts that are covered within the material. However issues with regard to the material or the video cannot be easily corrected without voice-over or re-recording the material. Video from other sources are available but only what is found within the course is allowed for instructors to use, unless supplemental material is added by the instructor. Within OIs this is not possible as the videos must be material that has been approved by the university.

Online tutorials allow for students to get the experience of being walked through experiments and examples based on the topic of discussion for the week. When presented with the subject matter for the week, the information can enhance the learning experience of the student and allow for better understanding of the material being presented. However when the tutorials are used alone then the material can only be taken as a means of instruction on select topics or as supplemental material. Tutorials on the various topics presented within the course can allow students to receive a different perspective on the topic from other professionals within the field and adds an additional resource for students enrolled within the course.

Asynchronous discussion boards allow for students to respond to discussion board questions in the class, and post responses during the week. When used in conjunction with the learning resources, discussion boards can be used as a means for easy interaction with the student; it can be a means for an instructor to gain knowledge on a student's thoughts on the topic unhindered by a lecture within a normal course; and it can be used as supporting material that students have a hard time comprehending from the learning resources. Even though asynchronous discussion boards does not allow for the instructor to give lectures on the material, it can assist the student in their development and understanding of the material, however instructors are only allowed to correct the student knowledge based on what material the student presents in their discussion posts. With the use of asynchronous discussion boards instructors can develop responses to the discussion board questions, ask additional questions of the students and respond to every post that the students leave within the course; however this process is not one that many instructors are able to perform on a regular basis.

Instant messaging/chat within the course can allow for the exchange of knowledge between instructor and students and between peers within a course. The use of this technology within a course can allow students to get instant clarification on a topic, and to receive information in a confidential format without the knowledge of other students. Instant messaging and chat features can be a less time consuming way for individuals to get the knowledge that they seek without having to travel to meet with individuals in person.

Virtual laboratories and the use of video journals can provide students with partial hands-on experience, exposure to new technology, and the ability to be able to see how various experiments are performed in different areas of science. By using this means of instruction, virtual laboratories and the use of video journals can give students a means of seeing the process occur without actually doing the work themselves, and it can enhance the learning experience that students receive within the course. Virtual laboratories and video journals can greatly enhance the exposure and presence of the experiments when coupled with video streaming or Skype to give the student a presence as the experiment is being performed or as the technology is being displayed. Troubleshooting of failed experiments and practical experimentation gained from the laboratory is limited or non-existent with regard to the experiences of the student in an online institution. Hands-on laboratory experience via recitation periods that include question/answer periods allows for the development of laboratory skills; more so class question/answer periods (where everyone is informed) allows all students present to benefit from the discussion and experience of trial-and-error with regard to experimentation.

Within the online community it is not unheard of to have virtual office hours as an online instructor. These office hours are normally used for conference calls, information gathering, and making oneself available to students during the course of the class. Accommodating students during virtual office hours allows the faculty to interact with their students and to accomplish tasks that are associated with being a faculty member at a university. LBIs usually use actual office hours – even appointment hours – within the academic institution, therefore this means of accommodation would be redundant to the institution.

LBIs and OIs feature webinars and seminars to present information to large groups of students and the general public to increase the numbers of people present for events. By recoding webinars and seminars it creates an archive of information for students who are not able to participate in the event. Whether the webinar or seminar is live or pre-recorded the information can be used as a resource in the course for both LBIs and OIs. Webinars can be a way for students to view previous versions of the course from different instructors, guest presenters, or as a means for reviewing topics that they may have difficulty with understanding during the normal course of a course.

LBIs have tutors that specialize in every subject of the academic university. It is not surprising that tutoring has been a means of gaining knowledge and insight within the scientific and academic areas of interest. However not all OIs have tutors for the different sessions and courses that are taught within the university. If OIs possessed a pool of students and faculty who specialize in tutoring and knowledge exchanges for students within the academic institution, this could be a means for better trouble-shooting and exchange of experience between students and from faculty to student.

With the exception of virtual laboratories, video journals, and streaming video none of the methods of instruction mention above will give students the hands-on or research experience within their fields or grant them the experience that comes from conducting research experiments. The methods can provide students with the aspects of instruction as it relates to research.

## Management and Organizational Concerns

Technology students are exposed to technology throughout the education process; however the educational process does not expose all students to technology. Whether within their field, beyond what is presented within the text book, or within the course students should receive the same or similar education whether in a Land-Based Institution or in an Online Institution. If Land-Based Institutions or Online Institutions become concerned about students receiving the same or similar education it would be easy for Management or the Organization of the academic institution to implement updates and resources needed to make the changes to the programs of study for the student. These changes would allow students to become exposed to the areas that are needed for them to compete in the current employment arena. Online Institutions have always represented themselves as places of higher learning and advancement with the aid of technology, however this means that the organization has to stay abreast of the updates and changes of the technology within the field. These types of updates and changes are paramount to being able to attract students to programs, especially in the online environment.

## CURRENT CHALLENGES FACING THE ORGANIZATION

Many aspects of the educational system and education process within Land-Based Institutions are lost when courses are offered online. From the perspective of the instructor, many parts of the process can be taught remotely, but the lack of in-person interaction and the assessment of student knowledge that is gained and assessed in person are not present due to the lack of interaction with the student, the nature of communication with students in an online environment, and the lack of research participation of the school and student. Although online instruction in science courses gives a theoretical and a historical view of the science field that is being studied, not much is done in the way of hands-on or research experience. Substitutions for the hands-on or research experience and building upon the student's knowledge and experience can include the use of online tutorials, video streaming, and virtual lab experiments. This type of technology exists within some institutions, but actual experience from hands-on experimentation, and learning from troubleshooting from trial-and-error, is not gained from the process offered in the online environment.

As an organization, Land-Based Institutions and Online Institutions should refer to some of the questions and answers provided to assist in the use of technology for research purposes and to improve the educational process for their students. By viewing these questions as challenges and ways to improve upon the educational experience, these questions can assist in the development of programs and can provide a means for management and the organization to have an idea of the technology that can assist in the growth of the organization, enhancement of the research potential of the university and encourage the use of technology in the program.

## 1. Why Should We Enhance Our Technology within Our Organization and Courses?

Natarajan (2006) stated, "Technology such as video and the internet can be combined with hands on activities to enhance critical thinking and support learning skills." By enhancing the technology within the institution or organization, the program or institution can allow for students to stay up-to-date with the technology in their field, and have a better understanding of the issues with technology within their field. For example, Sunoikisis is one of the earlier online learning programs that was developed by LBIs with the hopes of "expanding the scope and curriculum of classics education" across the Associated Colleges of the South. Sunoikisis blends traditional and online learning by giving the students access to "more advanced and diverse subjects than their own departments can offer" (Frost & Olsen, 2006, p. 20). When the software is implemented students would have been able to expand upon their educational program, and from the services within the university that are in place to assist the student. Students reported learning the importance of integrating what they have learned rather than regurgitating it, while the faculty members also reported benefits, such as forming collaborative relationships (Frost & Olsen, 2006). Technology has to meet the demands of the people. To do this the forms of technology within place must make it possible for others to be able to study and collaborate without being limited by time, space constraints, or location; they should be able to easily access information from devices in multiple locations through online data storage, enhanced networking capability to accomplish the research and educational goals, and the institutions would benefit by being able to exponentially expand the resources of the university; additionally this would provide a wider range of archived materials that would allow clarification and sharing, and being part of a global learning community (Sandars, 2013). This can also be a means for generating more independent thoughts and research questions. Furthermore, when used for online learning, technology can be cost-effective, learner focused, measurable, and able to produce "better outcomes in terms of learning and knowledge retention when compared to traditional methods of teaching" (Natarajan, 2006). Therefore, this will allow the organization to be able to improve upon the educational experience of the students.

## 2. What Can Be Done to Improve the Educational and Research Experience of the Student, and What Will Produce Better Results in Terms of Their Educational and Professional Growth?

Academic institutions can improve upon the learning environment and the educational experience that is currently in place in Land-Based Institutions and in Online Institutions by developing the courses in unison. Outlined within this study are means of technology that can assist in attracting different students to the organization and increasing the means of education within the institution. This means incorporating more courses which is diverse in the technology used to include more: streaming video, video presentations/pod casts, tutorials, discussion boards, instant messaging/chat, virtual laboratories, virtual office hours, webinars/seminars, and tutoring/knowledge exchange sessions. Additionally, the use of online instructional tools, adds substance to the environment as a teaching aid, but the experience that comes from the tools are problematic and unrealistic in terms of the outcomes that are achieved, especially when students move from the online environment to the practical setting. Video streaming and lectures that are available online have been incorporated into courses for the learning process, but virtual laboratories are needed and to assist in the development and practical nature of courses in the online environment.

Improving upon the educational experiences of the student would mean making courses consistent and having a consistent design across campuses, departments, and schools in the institution. This would ensure quality and timely interactions between students and professors; availability of technical and technological support; and productive interaction among students are all necessary when it comes to improving the educational experiences of the students (Young & Norgard, 2006).This would also enhance the learning and research environment for faculty, staff, and students within the academic institution. Educational communities can be considered any community that encourages collaboration; sharing of thoughts, ideas, and research; building upon relationships and developing new research relationships; or enhance the educational experience of the students with regards to pursuing their mission, and achieving learning objectives (Murdock & Williams, 2011). In fact, Murdock and Williams (2011) determined that "an instructor's intentionality in adapting assignments and interactions to foster community development can facilitate students' experience of a learning community despite course format." Wang, Shannon, and Ross (2013) noted that "instructors can design course activities in a way that can also help students improve their self-regulated learning strategies and their levels of technology," both of which were demonstrated to enhance the educational experiences of the students. Furthermore, it is important that institutions provide user-friendly learning platforms, as well as workshops and training sessions for both staff and students (Wang *et al.*, 2013). By providing these user friendly platforms it would be possible to meet the needs of all the students within the institution. Examples of such customizable learning platforms would pertain to the use of technology that can increase the enrollment and personalization of technology for the student, program, institution and student's disability if necessary.

With regard to science-based courses, previous research studies determined that the inclusion of "kitchen chemistry" experiments [and other hands-on or field-based activities using simulations, software, and laboratory kits] enhanced distance learning students' appreciation of the relevance of chemistry in their own lives since they were using materials and methods done in familiar surroundings (Mawn, Carrico, Charuk, Stote, & Lawrence, 2011). The incorporation of case studies into the curriculum can be used to create a sense of community and facilitate interaction amongst students in OI science course, while facilitating learning and comprehension and making the students better prepared for their professions (Saleh, Asi,& Hamed, 2013; Williams, 2014).

## 3. What Forms of Technology Are Available and What Are the Limitations of the Technology When Used within Our Organization?

Proper evaluation of the current forms of technology within the organization and testing the limits of the technology, including support personnel, would assist in knowing what features are needed within the next generation of the technology that is currently being used within the organization. Using this information would be helpful to know when to replace the current forms of technology within the organization, classroom, and educational programs with technology that can move the organization forward. The most basic forms of technology include both synchronous and asynchronous discuss board inclusive of Lotus Notes/Domino, First Class, TopClass and WebBoard; the introduction of computer mediated conferences, CD-ROMS and computer-based applications or Apps, virtual laboratories (*i.e.* LateNiteLabs), the use of teleconferencing with hands-on activity sessions, and computer-based quizzes/surveys (Natarajan, 2006). Other technology-based materials that can be used for science teaching include hand-held or microcomputer software; scientific visualization programs, simulations, animations, and/or videos; distributed information sources such as real-time data, online databases, peer groups and mentors/experts within the students field of research; Web-based photo journals and virtual field activities; telecommunication networks that use the resources of the institution, Web-based primary sources, and modeling tools as desired by the Association for Science Teacher Education (ASTE) (ASTE, 2014).

In the past, researchers reported that there were both extrinsic and intrinsic factors associated with integrating technology. These "extrinsic factors include lack of access to computers, insufficient time to plan instruction, inadequate technical and administrative support, while intrinsic factors include beliefs about teaching, beliefs about computers, established classroom practices, and unwillingness to change" (Leonard & Guha, 2001). Although this observation was made more than a decade ago, some of these limitations still remain within LBIs and OIs. Today, however, the most common limitations are those of connectivity associated with transmission of signals and information to the forms of technology, costs associated with the acquisition and implementation of the technology, compatibility across multiple media platforms (e.g. Mac vs. PC or Android vs. iOS), and compatibility with the rapidly changing operating systems (e.g. Windows 7 vs. Windows 8, or Tiger vs. SnowLeopard).

## 4. What Other Forms of Technology Are Available if We Do Not Use the Technology That Is Currently Being Used in This Field?

Evaluating the science field and the technology within each of the disciplines would mean making adjustments within the organization to meet the changes that have occurred within technology. By adjusting, upgrading, updating, or migrating to new forms of technology, this would allow an institution to remain at the forefront of the educational system, and become more aware of the timing and uses of the technology. However, in the event that the organization chooses not to use the current forms of technology in this field, it can still make use of other forms that are available. WISE, Web-based Inquiry Science Environment, is an open-ended learning environment that "promotes students' solving of interdisciplinary science problems and the debating of natural phenomena in scientific ways using the Science Controversy On-line (SCOPE)" (Kim & Hannafin, 2004). There are also several different types of learning management systems, such as Angel, Blackboard, and Blackbaud. Finally, when considering technology, one must also consider the wide range of multimedia tools and resources which can incorporate the material into the scientific research environment, and the courses which are used for instruction. Such tools should include multimedia software that contains web-based programs, hypertext enabled software that allows for image processing and word processing capabilities. These tools can be used to customize the learning experience based on the resources available to the organization and the needs of the students.

## 5. Can the Technology Be Optimized to Effectively Deliver the Course Material for the Currently Available Courses?

The Association for Science Teacher Education (2014) noted, "Technology-integrated materials when used appropriately can enhance science teaching and learning. It is therefore the position of the Association for Science Teacher Education that the qualified science teacher educator should possess a strong knowledge base in understanding how implementing technology in science curricular contexts may be used to promote the teaching and learning of science." In order for the technology to be optimized additional cost and evaluation of the current technology should be undertaken by the organization. This optimization would probably mean a change in the current structure and utilization of the technology within place, a removal of various systems or components of the technology that is in place, or complete shutdown of the system to move to another system. Either way this knowledge can be useful for handing different aspects of the use of technology within the organization.

## 6. How Can We Use the Available Forms of Technology to Increase Our Enrollment within the Courses, and Which Forms of Technology Give Students the Abilities That Are Required and Needed to Be Productive Members of the Science Community within Their Respective Fields?

With any organization or educational system, enrollment increases and enrollment decline is a constant issue. Too many people at any given time period utilizing the same technological system can foster issues within the system, or not enough people utilizing the technology that is in place cannot justify the procurement and cost of the technology. When management and organizations begin to increase enrollment, additional components should be evaluated. One question that must be asked and answered prior to the implementation of the software and system is, what effect would an increase in the number of students have on the system in place? This means that the organization will begin to ask: how, when, where and for how long will this increase in students affect the overall infrastructure within our organization that is already in place.

In terms of the forms of technology that give students the abilities required and needed to be productive members of the scientific community within their respective field, the answer to this question is as vast as the fields of science themselves because it can range from basic technologies that encompass all fields, such as Suamans, Inc, which provide animations of scientific and statistical concepts, to very specific programs like Celestia which is a space simulation program. Additionally, many resources are shared in other established institutes like the University of

Pennsylvania Health System which provides video animations related to medical conditions or free to use on the Internet like Google Body Browser. Therefore, determining which technologies will actually achieve this goal will require collaboration between university officials, curriculum developers, and instructors in both Land-Based Institutions and Online Institutions.

## SOLUTIONS AND RECOMMENDATIONS

Land-Based Institutions have used traditional teaching methods to incorporate information in the lecture in a timely fashion. In the traditional sense instructors would utilize information from the course text and where necessary, supplement the material with information found in research journals and articles with scientific experience/knowledge from prior courses, events, and experimentation. Therefore the means of incorporating the information within the daily active learning environment was simple, straightforward, and without much challenge within the organization. What has changed over the years with regard to traditional teaching is that instructional material within the course has become more standard and the options that the instructor had for including material in the lessons have increased with regard to the presentation of the material, the way it is presented, and how often it can be presented. Where the use of overhead projects, chalk boards, and slide presentations were once used, the use of video, instructional resources such as tutorials, animation, video journals, and websites that walk students through the digital components and examples have replaced a large amount of the work that is done by professors.

With each form of technology discussed within this study, we suggest various things that can be done to assist the academic institution – both Land-Based Institutions and Online Institutions:

1. If institutions would like to incorporate video streaming or Skype into a course, this would be a creative way to allow for guest lectures within a course in Land-Based Institutions, or for Online Institutions to have presentations that are readily available for students during the entire course.
2. Video presentations, webinars, and seminars could be used during the week and presented with the course material which contains weekly readings of chapters, videos of the topics, a journal article, and instructional material on what questions the students must answer during the week's assignments. Advances in the scientific field, changes to experiments, updates on software, and upgrades to computing systems as well as changes in technology and new discoveries would not make it difficult to incorporate video or pod cast into a

course outline or as a teaching method. One of the concerns that institutions would have to keep in mind is changing and updating the videos over time. This material could be easily interchangeable, but course developers would have to be aware of the changes in the field to stay abreast of technology.

3. Online tutorials can be easily housed and archived to accommodate courses that are offered in LBIs and in OIs. Even though tutorials can give a sense of the technology and work that can be performed within technology, it does not replace the hands-on experience, but it can be a means of teaching the material presented to the students.
4. Discussion boards are very effective in students researching the topic, expressing their points of views on the topic and also incorporating the scientific literature. It is our view that the utilization of discussion boards should not be removed from the course. However discussion boards and capabilities within them should be enhanced to include additional material that would give the student a better experience with writing, threading, organization of the students thoughts, and to provide assistance with the presentation of their thoughts on topics discussed in the course.
5. Different versions of discussion boards also have instant messaging and chat features built into the course. This means that it is already a possibility for students to interact with their instructors and their peers; however if these services are not being used within the courses may be an issue. If it is a requirement of the course or a utility used in the course to assist the students with their growth and development, then it could be easily implemented to enhance the educational experience for the student. For instance, the use of customizable discussion boards that allows for font and color changes can give visually impaired students, and students who are color blind an opportunity to have better educational experiences within their respective academic institutions.
6. Virtual resources such as virtual laboratories, video journals, and virtual office hours could be easily incorporated into the learning resources. This would mean a change in the teaching protocols as video experiments could be used to assist in the theoretical approach presented in the class and to include the use of material that students will cover in the class. Virtual office hours coupled with tutoring/knowledge exchanges can be a way to include more technology within the course and to exposure students to material in the course. This could assist students by providing more exposure to their intended field of study and to allow them to become better scientist. We are aware that some students may not utilize the technology or see a need for the technology based on their level of instruction, however by being exposed to the types of technology that a student will use in their intended field would be better than not knowing about its existence at all.

7. Finally, our last recommendation to Land-Based Institutions and Online Institutions would be to consider utilizing all platforms and technology at their disposal to increase enrollment by making user friendly environments that can promote the use of technology that is personalized to the student and the student's program. One example of this is by incorporating tablet computers into the technology within the students program. As an Online Institution, it would not be difficult to have a tablet computer developed that has been personalized with all the communication software and resources needed to connect to the university (Blue-tooth, Wi-Fi); with the preloaded books for the students course (electronic books which requires purchase of the license or rental use); communications, journal, and library access (email programs, Blackboard, Web of Science, PubMed); Apps for video conferencing such as Adobe Connect, Skype, and GotoMeeting; Apps to assist students with science educations and research; and ways for students to connect to the cloud computing environments.

## REFERENCES

Chasen, M., & Pittinsky, M. (2014). *Blackboard Learn* (Version 9.1) [Computer software]. Blackboard Inc. Retrieved from http://www.blackboard.com

Frost, S., & Olsen, D. (2006). Technology, Learning, and the Virtual Liberal Arts Classroom. *Peer Review*, *8*(4), 20–22.

Gates, B. (2014). *Skype* [Computer software]. Skype Communications. Retrieved from http://www.skype.com

Jacqueline, L., & Smita, G. (2001). Education at the Crossroads. *Journal of Research on Technology in Education*, *34*(1), 51–57. doi:10.1080/15391523.2001.10782333

Kim, M., & Hannafin, M. (2004). Designing Online Learning Environments to Support Scientific Inquiry. *Quarterly Review of Distance Education*, *5*(1), 1–10.

Mawn, M. V., Carrico, P., Charuk, K., Stote, K. S., & Lawrence, B. (2011). Hands-on and Online: Scientific Explorations through Distance Learning. *Open Learning*, *26*(2), 135–146.

Murdock, J. L., & Williams, A. M. (2011). Creating an Online Learning Community. *Innovations in Higher Education*, *36*(5), 305–315. doi:10.1007/s10755-011-9188-6

Natarajan, M. M. (2006). Use of online technology for multimedia education. *Information Services & Use*, *26*(3), 249–256.

Pearson Education. (2014). *eCollege* [Computer software]. Pearson Education, Inc. Retrieved from http://www.eCollege.com

Saleh, S. M., Asi, Y. M., & Hamed, K. M. (2013). Effectiveness of integrating case studies in online and face-to-face instruction of pathophysiology: A comparative study. *Advances in Physiology Education*, *37*(2), 201–206. doi:10.1152/advan.00169.2012 PMID:23728138

Sandars, J. (2013). Technology-enhanced learning. *Education for Primary Care*, *24*(4), 300–301. PMID:23906176

The Association for Science Teacher Education. (2014). *ASTE Position Statement on Technology in Science Teacher Education*. Retrieved from http://theaste.org/about/aste-position-statement-on-technology-in-science-teacher-education/

*TurnItIn*. (2014). Retrieved from http://www.turnitin.com

Wang, C.-H., Shannon, D. M., & Ross, M. E. (2013). Students' characteristics, self-regulated learning, technology, self-efficacy, and course outcomes in online learning. *Distance Education*, *34*(3), 302–323. doi:10.1080/01587919.2013.835779

Williams, K. (2014). Research Institutions: Research-Based Teaching through Technology. In Cases on Research Based Teaching Methods in Science Education. Academic Press.

Young, A., & Norgard, C. (2006). Assessing the quality of online courses from the students' perspective. *The Internet and Higher Education*, *9*(2), 107–115. doi:10.1016/j.iheduc.2006.03.001

## KEY TERMS AND DEFINITIONS

**Discussion Boards:** An open forum which allows users/groups to communicate and leave messages for others.

**Instant Messaging/Chat:** Real-time message transfer via the internet or wireless connection.

**Streaming Video:** Compressed viedo that is sent over the internet that allows vierwers to see in real time.

**Tutorials:** An interactive method to assist in the transfer of knowledge.

**Tutoring/Knowledge Exchange Sessions:** The process of instructing someone on a topic that allows for the excahnge of knowlegde between individuals.

**Video Presentations/Podcasts:** A digital multimedia file that can be placed on the internet for download to a computer, portable multimedia player, or other technological device.

**Virtual Laboratories:** A simulated environment used for conducting experiments; connecting to real laboratories via a simulated environment to watch/experience ongoing experiments.

**Virtual Office Hours:** Having a presence for a predetermined time within a virtual classroom or office.

**Webinars/Seminars:** Seminars that are given over the internet or within a classroom that is broadcasted over the internet.

# Chapter 14
# Research Institutions:
## Research-Based Teaching through Technology

**K. Y. Williams**
*Walden University, USA & Kaplan University, USA & Department of Defense, USA*

## EXECUTIVE SUMMARY

*Students that have been educated from online institutions do not readily receive the hands-on experience needed to make an easy transition into research-based institutions, research-intensive laboratories, or into the workforce. Online instruction does not cultivate the knowledge that comes from hands-on experience and experimentation. This type of experience is better facilitated via in-person interaction. In online institutions students only receive interaction via email, online discussion boards, or through phone calls. This does not allow for instructors to sufficiently improve upon the student's skills, assist in the development of their knowledge, or evaluate students' hands-on abilities within the science field. Within this case study, the author outlines some of the basic items that students should have been exposed to within their programs of study and state some of the issues that students in online institutions face when they are educated in an online setting and then transition to research-intensive settings. The author also outlines ways to assist students with these transitions and the types of facilities needed to assist students.*

DOI: 10.4018/978-1-4666-6375-6.ch014

## ORGANIZATION BACKGROUND

The mission of this case study is to address the issues and concerns of student hands-on experience when they transition from Online Institutions to practical settings, and to offer solutions to real-life dilemmas that exist within the science field. Our intent is to outline the issues that students face when they receive an online education from the perspective of four separate science fields, and express the needs of the students when they transition into research-intensive environments, the work force, or into institutions that are research-based. The goals of this case study are to:

1. Outline issues and deficiencies that students who are trained and/or educated online may face when transitioning to research-based institutions.
2. Evaluate and outline the experience and exposure of students to technology in Online Institutions that is necessary when transitioning to research-based institutions.
3. Suggest and state technology experience needed to assist students when transitioning to research-based institutions, research-intensive institutions, and into the workforce from Online Institutions.
4. Outline and detail the research experience of the student when transitioning into their related research-based institution.
5. Suggest methods and programs that exist or the need for implementation of programs to assist in student development to allow for easy transition to research-based or research intensive institutions.

### Popularity of Online Institutions

The popularity of Online Institutions on the Undergraduate, Graduate, and Professional level has grown. With the decrease in the number of students that are enrolled in Land-Based Institutions which has a well-established traditional method of instruction, one can see a proportionate increase in the numbers of online programs because of their different approach to teaching and instruction, flexibility, and programs of study.

Online Institutions have become available to students all over the world, and on many levels of the educational system because they offer teaching methods and options that are not available to students within Land-Based Institutions. With each new Online Institution more and more academic programs offer more flexibility within their programs; scheduling of core and elective courses; a reduction in the time it takes to acquire a degree in the student's intended field; different modes of instruction for the courses; easier access to the course instructors; and the anonymity that comes with being taught from a distance. Students find Online Institutions attractive because they are advertised as being flexible to the student's schedule, comprehensive of the standards that will allow them to meet and achieve their educational and academic goals, it allows students to be able to work on their degree within their free time in their daily lives, and to be able to submit their work in an electronic fashion from remote locations.

Online Institutions tend to offer all the conveniences of Land-Based Institutions: financial aid to the student with normal borrowing limits, career counselling, 100% online instruction based on the program of study, access to experienced professionals within their field of study, academic advising, and streamlined paths to their degrees. With the change in approach to the student, the academic programs have tailored the programs based on the requirements needed to complete the degrees and to show that they can produce scholars who will compete within the workforce, and they also have made changes to the time needed within the various courses. Essentially programs have reduced the time to complete courses within their institutions by allowing courses to begin and end within: 5 weeks, 8 weeks, 10 weeks, and 12 weeks for what is considered a quarter term, or they will present the same material within the traditional 4 or 5 month semester term. These institutions also offer numerous sections of courses with a continuous scheduling of core and elective courses that can fit into any student's schedule. Each course has an asynchronous environment that allows for accessing the institution; submitting assignments when desired; a trusted, monitored environment that allows for student-student and student-teacher interaction; and an asynchronous discussion board that allows for convenient conversations within the course.

One can receive a lot of experience from an Online Institution that most students only dream of when they enter a Land-Based Institution. Based on their educational field, students get the chance to be taught from a mired of professionals working in their intended fields, and the faculty make themselves accessible via phone and office hours that students can use for question and answer periods, advising, information gathering on any course/class issue, assignment concerns, or opportunities to seek assistance.

Online Institutions attract students who desire degrees which will enable them to advance within their current field in which they are studying; students who wish to transition into another field; students who cannot – for whatever reason – leave their current position to attend traditional institutions within the normal 8 AM to 6 PM Monday through Friday time frame; students who may not have been admitted to Land-Based Institutions; professionals who are returning to school after long absences, and students entering college for the first time.

## Organizational Structure of Online Institutions

Similar to Land-Based Institutions, Online Institutions have the similar types of managerial, organizational, and technological structure and issues, with one major difference being that most Online Institutions are "for profit" organizations. Online Institutions are similar to Land-Based Institutions in that they have a University President, Provost, Deans, Department Chairs, Program Directors, Faculty (both Full-Time and Adjunct), and Staff (inclusive of Academic Advising and IT staff). The major difference within Online Institutions organizational structure is shown in the use of Program Coordinators, Academic Coordinators, and Course Developers which is complemented by an even larger pool of Adjunct Faculty who make up the majority of the faculty composition of the academic programs.

Online Institutions hire groups of Course Developers to develop and update courses for the institution. The Course Developers have been trained with backgrounds within the various science fields, and they work closely with committees to coordinate the development and the arrangement of the courses while updating the material within the courses. Course Developers within the various programs are tasked with the development of the courses found within each of the programs, and they in turn have to have the necessary background for the suggesting, preparing, and writing of the learning outcomes while gathering the material that will be used within the courses. Essentially they must have a firm grasp and handle on the type of technology currently within the field, the use of technology that will assist students within their courses, and an idea of the information that should be taught within the courses, as their ideas and thoughts will help move the program forward. Once course development is complete and approved, management within the organizations deploy a learning environment that is self-contained, asynchronous, and available to the students based on the guidelines of the institutions and the educational system in which they are affiliated. However visually monitoring the student's progression, and offering challenges that show their complete understanding of the material is not possible within an interactive environment when taught online. Giving the student assistance with their education and being able to educate them in your presence, or educating the student throughout the education process in a face-to-face setting is minimal – at best. Technology within the science field and the students having access to computing and technological environments is done virtually and only through discussions within courses.

Online Institutions use technology within the courses, but it is limited to computing that is available at home, and the type of technology that students use to conduct science experiments is also available at home. This does not allow students to have access to technology that is used in a workforce environment, as exposure to technology should be constant and instructional by experimentation. Various forms of technology exist as technology is needed for programming, code writing, software development and for overall field awareness within the computing industry as technology has improved greatly within the biological, chemistry, and mathematical field. To develop the skills needed for their field, the student has to be exposed to the technology that has advanced within their field and the technology needed to succeed beyond the academic environment. If the course developers do not introduce the use of technology within the course during its development, then it is left up to the Faculty – often referred to as Instructors – to develop the information needed in the course and to introduce the technology to the students.

In traditional Land-Based Institutions, the overall strength and development of the organization and department is based on the backgrounds and expertise of the faculty. Faculty, both full-time and adjunct contributes to the knowledge, skills, and areas of expertise that makes the programs known for its strength and direction. However if the Instructor's knowledge and expertise is not a major part of the course development but is a part of the student resources when it comes to dissertation writing and development, it makes it difficult for the student to comprehend the information when they are in the final stages of their education plan. Even though a student's dissertation committee will lend its experience and expertise to the students research project and area of study, based on the student's field and the direction of the dissertation, not possessing the experience will usually decrease the potential for the student to use their degree in a manner in which they intend.

## SETTING THE STAGE

Technological advancement within the science field has changed greatly over the last 25 years with the introduction of new genetic sequencing technology, methods to assist in chemical and biological techniques, and advances within the mathematical modeling and biological prediction fields. Instruction within the science field in Land-Based Institutions has also changed to meet the changes with technology in the fields of chemist, biochemistry, statistical genetics, and bioinformatics. Additionally, the types of courses that are offered have also changed to meet the changes in the science field, as the courses have now incorporated more technology, hands-on experience, and instructors incorporate information based on the changes within the science field. In many ways one can predict the changes based on the needs of the science fields, the research results that are found within scientific journals, and the types of scientists needed for various projects based on the trends within the respective research fields.

### *Research Institutions*

Traditional instruction of students within Land-Based Institutions have changed to meet the needs of industry, and with that change we see changes within the research fields and in the areas that will make students more competitive when they graduate from Land-Based Institutions. However Online Institutions have not changed in their approach to education. Very few Online Institutions have changed to meet and assist students in meeting the challenges of the world outside of their educational institution. Students enter into the educational system and they are asked to produce results within their field based on the academic program in which that are enrolled to show their competency within their field of study based on their education level. Land-Based Institutions have changed their means of instructing and exposing students to the ever changing and advancing forms of technology within their field, and they have the means to grant them opportunities to have exposure to different forms of technology. Additionally Land-Based Institutions have more of a hands-on approach that leaves the student with the skills and knowledge to easily transition into the work force and to conduct research outside of the educational setting. However, most Online Institutions have not made the transitions to include such changes and this leaves the students without the skills, knowledge, or experience to have the same hands on a experience with students who hold the same type of degree from Land-Based Institutions.

Online Institutions must have up-to-date technology in order to constantly progress and grow to meet the ever changing needs of the student and the fields in which they study. The proper use of available advanced technology allows for growth of the institution, better instruction in the institution, and enhanced training of the student and the faculty within the institution, thus better technology is needed to facility this growth and instruction. Any lag with progression with regard to the changes in technology means that the institution risk a decrease in the numbers of students as a result of the reduced capabilities offered in the course and a lag in the technology that is needed to facility the growth and instruction within the institution, with regard to the hands-on experience with technology and training for Online Institutions. Traditional Land-Based Institutions use technology within the laboratories throughout the educational process on both the undergraduate and graduate level and students receive experience with such technology during research projects within the educational program during the year and through summer internships. This allows students to gain the necessary hands-on experience which becomes invaluable when looking for employment after graduation.

With regard to introducing students to technology within their field, Online Institutions are placing their students at a disadvantage when compared to students in Land-Based Institutions. These deficiencies occur when students transition from the Online Institution into the positions they were trained and/or educated in in their Institution. Students also face challenges within research-based institutions and research-intensive environments when they have to compete for the same funding, research programs, and in writing publications based on constructed hypothesis and research results. The students' deficiencies become more pronounced when their knowledge of the technology within the field is only based on the theory of the techniques and what is expressed within the course text and resource materials, and not from hands-on experience.

When transitioning to research-based institutions, students need to be able to think, process, synthesize and produce research results in a timely, independent fashion when they transition from their graduate educational program into their field of study. However, from their undergraduate program they should be able to assist, learn, and synthesize information from experience when they transition to a graduate educational program. One of the ways that this is accomplished is by students being evaluated based on their experiences, written recommendations from their programs based on their research experience in summer programs during their undergraduate educational program, and research advisors from research conducted during their years in their undergraduate programs. How they handle technology and what they are exposed to in the technology field states what they have been able to do with regard to their experience and their knowledge. Therefore, it becomes essential that Online Institutions expose their students to technology to assist in their transition to research-based institutions, research-intensive institutions, and into the workforce.

Traditional fields of study that are offered by Land-Based Institutions would have opportunities for students to have experience with technology by offering opportunities to work with their professors in the laboratories, in their research labs, and assisting with their applications for summer internships. Based on the level of education, undergraduate and graduate students majoring in chemistry, biology, genetics, or bioinformatics would have experience in different ways based on their degree. Additionally the theoretical approach with regard to their degree is placed into context with their education.

Undergraduate chemistry majors should have hands-on, technical, and theoretical experience with mass spectroscopy. Biology majors should have hands-on, technical, and theoretical experience with gel electrophoresis and the use of restriction enzymes. Students studying genetics that will lead to the path of becoming a statistical geneticist should have hands-on, technical, and theoretical experience with sequencing methods such as the Sanger sequencing method, while students majoring in Bioinformatics or Statistical Genetics should have hands-on, technical, and theoretical experience with programming languages such as Perl, C/C#, UNIX, Linux or some form of hardware.

On the graduate level, chemistry majors should have hands-on, technical, and theoretical experience with High-Performance Liquid Chromatography. Biology majors should have hands-on, technical, and theoretical experience with Polymerase Chain Reactions (PCR), and Real Time-Polymerase Chain Reactions (qPCR). Graduate students studying in Statistical Genetics programs should be familiar with clinical data and the type of data that can be collected from experiments that produce genetic data from resources such as the 1000 Genomes Project, and programs such as Plink, Plink/Seq, and PBAT, while students that are Bioinformaticians should have hands-on, technical, and theoretical experience with technology such as Transcriptomics, DNA Microarray, and programs such as BLAST.

Although this is what is suggested by instructors in this process, there are different administrators and decision-makers that play a very crucial role in the overall planning, implementation and management of the information technology applications. Decisions makers within their different areas have the choice of whether they should or should not create the desired opportunities, to assist their students in their development, to invest in the areas needed to get students the experience to develop their skillset, or to develop courses that would assist students in getting their hands on experience from either internships, externships, or new courses within the organization.

## CASE DESCRIPTION

This case looks at how Online Institutions have exposed students to technology within their field and suggest what is needed to assist students based on first-hand experience, research, and knowledge of four fields in the area of science. Listed are technologies and capabilities that students should have been exposed to in the different levels of their educational growth and educational programs. Once the students have graduated from their respective institutions – Land-Based or Online – employers in research facilities or research organizations in academic settings, government, industry, or private research groups would expect the student to be familiar with technology listed in the various fields of this case description.

### Technology Concerns and Components

Although numerous forms of technology can be listed, the bare minimum of technology listed here can be seen as a step in the right direction for Online Institutions as these technologies can assist students when transitioning to research-based institutions, research-intensive environments, and into the workforce from Online Institutions. Listed is some of the technology within the field that students in four of the science disciplines should be familiar with upon graduating from their institution in the four science fields and any use of the emerging technologies would be considered in moving in the right direction of solving the problem(s) related to the use of technology. As an undergraduate student, technology should be exposed to students in their field as they progress within their educational plan. Any undergraduate chemistry and biology majors should have experience and exposure to laboratory techniques.

Many techniques within chemistry exists, but one most students should be familiarity with is mass spectroscopy. This technique allows for the determination of DNA, proteins, and composition of chemical structures from the sheer masses of atoms and molecules within a given sample (Cooper and Hausman, 2009).

Gel electrophoresis has been used in many laboratories as a means for separating biological macromolecules of DNA, RNA, and protein. This process uses the size and charge of the macromolecule as a means of separating the components for analysis (Cooper and Hausman, 2009; Haglund, 1971).

*Deoxyribonucleic Acid (DNA) Sequencing:* DNA sequencing gives a detailed ordering of the bases and base pairing of stretches of DNA. Many research projects, inclusive of the 1000 Genome Project and the Human Genome Project, use DNA sequencing techniques to sequence many areas of the Human Genome to understand, locate, and identify mutations and genes that relate to diseases within the human genome (Henry, 2011).

*Restriction Enzymes:* One of the most important types of technology that was discovered in the last century was the use of restriction enzymes or restriction endonucleases that have been used to recognize and cleave DNA at specific recognition points within a DNA sequence. This allows for better separation with the use of gel electrophoresis (Cooper and Hausman, 2009; Nathans and Smith, 1975).

Undergraduate bioinformatics and statistical genetics students should have experience with some of the more general purpose, object-oriented programming languages. As the number of programming languages has increased, the uses and versatility of the members have increased to include Perl, C, and C#. Perl, C, and C# (pronounced 'see sharp') are programming languages that has various uses and applications within the fields of Bioinformatics, Computer Science, and Statistical Genetics – to name a few. Additionally, bioinformatics and statistical genetics students should know how to operate and develop programs in a UNIX or Linux environment. UNIX is a computer operating system that allows for multiple processes, tasks, and users to develop and test scripts and software in a closed environment. Linux is the Unix-like environment that also allows for development and distribution of free and open source software development and deployment.

Graduate students in the various areas of Biology and Chemistry should have experience from a theoretical and experimental perspective with the following technology:

- **High Performance Liquid Chromatography:** A computerized chemical technique that allows researchers to identify and separate the different components of a mixture for analysis and quantification.
- **Polymerase Chain Reactions:** Polymerase Chain Reactions (PCR) is a technique used to amplify small quantities of DNA so it can be used in processes such as DNA fingerprinting, bacterial and viral detection, and diagnosis of genetic disorders (Henry, 2013).
- **Real Time-Polymerase Chain Reactions (qPCR):** Real Time-Polymerase Chain Reaction also knowing as quantitative Polymerase Chain Reaction has the distinctive nature of amplifying DNA and then quantifying the amount of DNA to a predetermined quantity as determined for use in protocols, detection limits, or sample size for experiments (Wilkinson, Cheifetz & De Grandis, 1995).

Graduate students in Bioinformatics and Statistical Genetics should have experience from a theoretical and experimental perspective with the following technology and projects:

- **1000 Genomes Project:** The 1000 Genome Project is a government funded project that uses DNA sequencing technology to sequence the genomes of a large number of participants in an effort to provide a comprehensive resource for human genetic variation (Flicek, 2012).
- **Plink:** Many open source genetic analysis tools exist, but Plink is a whole genome association analysis tool that is used to evaluate the relationship between genotypic and phenotypic data (Purcell, et. al, 2007).
- **Plink/SEQ:** Humans are genetically diverse. To understand the genetic variation that allows for that diversity with regard to genotypic and phenotypic differences, data can be analyzed using analytical tools such as Plink and Plink/SEQ (Purcell, et.al, 2007). These tools allow for working with large genome and exome information from various sources.
- **PBAT:** A computational sciences tool used for the statistical analysis of family-based association studies based on the subjects genetic profile (Lange, 2003).
- **Transcriptomics:** The process of measuring the amount of RNA transcripts within a given cell. DNA is transcribed to RNA, however all DNA within the cell is not transcribed at the same time and these transcripts can be measured to determine the purpose, regulation, and influence within the cell (Henry, 2012).
- **DNA Microarray:** All the genes within the cell are not activated within the cell at the same time, and all the same cells do not activate the same genes

within the same time frame. DNA microarray technology allows for the detection of the active nature of genes within the cell over time (Henry, 2011).

- **Basic Local Alignment Search Tool (BLAST):** A tool that "finds regions of local similarity between sequences. The program compares nucleotide or protein sequences to sequence databases and calculates the statistical significance of matches. BLAST can be used to infer functional and evolutionary relationships between sequences as well as help identify members of gene families," (Haus, 2014).

## Management and Organizational Concerns

Within academic institutions, course oversight committees exist that informs management of necessary changes needed in the university and within the department. Normally this type of committee would be constituted by: Deans, Program Directors, Program Coordinators, Instructors, Course Developers, and students whose entire objective would be to approve and pass along information for the updating of courses within the various scientific fields. Then it would become the job of the university decision makers to accept the recommendations of the committee to make the changes needed.

Normally Deans, Program Directors, and Program Coordinators along with Course Developers would suggest updates to programs based on the documented needs of the student. The Course Developers must evaluate the goals of the university, learning-outcomes of the academic program and the courses, and the capabilities of the students upon completion of the course before they make changes to the courses of interest. Academic Program Directors and Instructors assess the students' skills based on what is needed to assist the student beyond the educational environment, and Course Developers must implement those suggested changes within the course to reflect the desired needs of the students in their respective fields of study. This means deployment of a learning environment that will assist the student in developing the necessary skills to be proficient within their field of study, and creating an environment that students can use for hands-on experience within their research courses that could be updated every three years based on the needs and changes within the fields of science. However students' needs and changes within the field are only going to go unnoticed unless students request that their courses are up-to-date to include the use of new and emerging technology and more hands-on experience which can complement their educational experience.

In many ways, students have to be more concerned about their experience beyond the course and what they endeavor within their proposed fields after they graduate from their institution. This means looking into their intended fields earlier and beyond the courses, and question exactly how their education is preparing them for their futures. However, to do this adequately the student must have some idea of their field before entering their field, by working in closely related fields prior to enrolling in their programs, or having a clear idea of what is required beyond the degree. Students who are within their intended fields usually already have the experience and knowledge from hands-on experience within their field, and are only getting degrees to move forward within their work place.

Therefore Management must show more concerned for the use of technology for student development, more concern for the student within the course, and in their growth after graduation. Currently a large amount of concern is for the student to succeed while within the academic setting, however assisting the student on their track of scientific and educational growth and development after the education portion is completed is not of major concern for many online science programs.

## CURRENT CHALLENGES FACING THE ORGANIZATION

Some of the challenges and problems that Online Institutions face with regard to students transitioning into research-based institutions, research-intensive institutions, and the workforce is due to the lack of hands-on experience, lack of exposure to technology, and lack of experience with technology within their field. Essentially many students from Online Institutions face these challenges and it makes it difficult, or impossible, for students to transition into positions because of their lack of experience, no previous experience or exposure to technology to show their proficiency/ skillset within their field, or the feeling that they possess a more theoretical degree because of their lack of experience.

Within Online Institutions, many schools do not have the laboratory, computational, or the technological resources that are designed to assist students in developing the scientific skills and experience within their respective fields. These skills are usually acquired during the school year when students conduct laboratory experiments and research in their course, or by working in their advisors laboratories during the school year. Additionally students garner experience when they are accepted into summer programs after their freshman, sophomore, and junior year. Without these experiences it leaves the student at a competitive disadvantage. These resources are usually available based on the fact that Land-Based Institutions have space and facilitates that can house the necessary science laboratories, computing facilities, and technological innovation laboratories that have always allowed students to develop their skills within their fields. Although Online Institutions have resources in place for students to use to complete course exams, quizzes, environments designed for discussion boards and the use of facilities to assist with written assignments, not all Online Institutions have ways to show proficiency within regard to the hands-on portions of laboratory work and computer programming.

Several clear alternatives exist with regard to addressing the issues within Online Institutions: increase the use of technology to assist the students in developing their skills for computing and laboratory research; increase the opportunities for students to receive hands-on experience based on the academic programs in which they are enrolled; increase the students' hands-on experience within the courses; invest in technology programs that are partnered with schools to give the students opportunities to garner the experience needed in their field; or update the courses to contain more hands-on experience by partnering with a Land-Based Institution. With each of these options, many pros (benefits) and cons (risks) will exist with regard to the challenges.

If Online Institutions increase the use of technology to assist the student in developing their skills for computing and laboratory research, the benefits of this alternative would become a benefit to the institution as it would increase the use of technology within the courses, and allow the institution to increase the student's capabilities. One of the risks what would exist would be associated with the use of technology that is not already available. Additionally not all students will utilize the system to help them achieve the goals of the programs and/or learning objectives. With regard to the second alternative, one benefit would to be to increase the opportunities for students to receive hands-on experience for the academic programs in which they are enrolled as this would allow students to receive experience with research components that is needed for the development of their skills in the laboratory and in the development of their computing ability. The risk associated with this alternative is in the cost associated with the development of the programs and planning for this portion of the process will require a partnership with organizations and schools that are not able to have this component. The third alternative of increasing the students' hands-on experience within the courses would have the benefit of allowing the students to have more visual and virtual experience via experimentation in a virtual laboratory and with research journals that conduct experiments based on the research topic. The risk with this alterative would exist in the fact that the experience is still purely remote and passive and not active in the students educational program. The fourth alternative of investing in technology programs that are partnered with schools to give the students opportunities to garner the experience needed in their field would have more of a hands-on experience component, but the risk would be that not all students would be in close proximity to the partnered institutions to utilize this opportunity. Finally if Online Institutions did update the courses to contain more hands-on experience by partnering with a Land-Based Institution this would change the entire dynamic of the Online Institution. Essentially this would make the Online Institution partnered with a Land-Based Institution and change the educational policies and environment that are already in place, and it could jeopardize the number of students enrolled in Online Institutions. With each of these options, many pros (benefits) and cons (risks) will exist with regard to the challenges.

The technology that exists within many of the Online Institutions allow for accessing the course to perform normal basic/essential activities with regard to the university educational experience: discussion board posting, chatting with other students and instructors, writing papers for the enrolled course, using the university writing center, accessing library resources, submitting financial aid applications, requesting academic advisement, sending emails, requesting academic transcripts, previewing upcoming course enrollment, submitting papers, video conferencing, and instructors have areas for grading student assignments which can occur within a fraction of the time when compared to the normal educational system. When Online Institutions academic terms are shortened, it leaves the students very little time to assess what was done improperly, and it gives them very little time to correct themselves before the next assignment. Within many Online Institutions, instructors must give substantive feedback, only it does not leave much in the way of interaction to assist the student in increasing their knowledge and giving them feedback that would assist in their exact issue. Primarily this does not assist the student in becoming better scientists or researchers or to better approach their research or their writing. Additionally, this may not always occur or occur as adequately unless time permits because students write more often than they demonstrate their skills within online courses. Therefore students become better writers, and they are able to organize the subject material; however most online programs do not focus on the development of the student, assist the student in thinking like an independent researcher, assist in the development of the student's analytical skills, or to assist the student in their decision making with regard to their research topics. When and where would the instructor or student have time to do this if at all?

If Online Institutions wait until students are making preparations for conducting research then they would be at a disadvantage as many students do not get opportunities to garner research experience beyond the undergraduate education. Therefore getting research experience is usually not presented to students in research institutions beyond the undergraduate years. Graduate institutions within the sciences use research experience as a way of evaluating students when they are admitted to Land-Based Institutions, and when admitted, those students are expected to perform regardless of their educational background, experience, or exposure. Therefore a need exists for students to have hands-on experience that is more project and subject matter specific with regard to technology, laboratory based techniques, and troubleshooting via trial-and-error with regard to experimentation.

## SOLUTIONS AND RECOMMENDATIONS

Solutions and recommendations exist for the issues that have been proposed within this case study. Some of the clear alternatives would have to be grouped into investing in technology centers, course restructuring, and facility development via partnerships with Land-Based Institutions. With our understanding of what is needed to give students the hands-on experience that is needed for them to transition into research-intensive institutions, research-based institutions or into the workforce, one of the simplest solutions would be to invest in technology centers or research institutions.

Research institutions or technology centers can be a source of information for the students and they would have the opportunity to have in-person residencies within the center over the time of the students' education. These centers could be a solution to the issue of meeting with faculty, being taught the theory of the experiment, and then getting the complete hands-on experience that is needed to show competency within the research field. If this is accomplished during the students' undergraduate education, it would provide the student with a means of showing and receiving more hands-on experience that is similar to summer internships that can assist with their matriculation into graduate programs that are Land-Based or Online, as well as garnering the experience that is needed for transitioning into the workforce. This option would allow Online Institutions the chance to increase the use of technology to assist the student in developing their skills for computing and laboratory research. By creating research institutions, Online Institutions could increase the opportunities for students to receive hands-on experience based on the academic programs in which they are enrolled and could bring in additional funding from granting institutions such as the National Science Foundation and the National Institute of Health. These sources of funding would allow the institution to increase the amount and type of technology within the educational environment as a result of program and project grants to build the infrastructure needed to develop the areas of interest. Furthermore, research-based institutions can assist in the development and staffing of the technology centers that can be created by Online Institutions. With regard to the hands-on experience needed for student development, and in the education of students, the number of adjunct faculty that is within Online Institutions can be justified and adjusted to include research staff that can coordinate with the technology centers to include more information on what the courses taught the student and include this information when conducting more hands-on experiments that would correlate with the science courses. By establishing these research institutions, it would be easier to assist students with their programming needs; instructing students on how to carryout experiments; conducting residencies within the institutions to show how the technology work; and it would be easier to operate laboratories for the students to develop their skills while receiving advisement from professionals within their field. This means that students would be able to participate in actual development of their hypotheses, get hands-on experience, and trouble shoot experiments via trial-and-error. Overall this would assist in students getting the necessary experience with technology issues that may relate to the experiments that they would later propose in their master projects or dissertations, and also allow the students the experience needed to easily transition into the workforce.

The second suggestion for Online Intuitions is more internal as Online Institutions could make changes within the institution by increasing the student's hands-on experience within the courses. This would mean having access to virtual experiments and research journals that depict the actual use of the technology and virtual experiments that would allow for the development of the student's technological skillset. This is more difficult of a concept and change to make as the types of technology that is discussed and used within the laboratory would have to be available to each student, and approved by Course Developers and Institutional Committees. However, more assignments and writing of scientific papers that state the use, functionality, and potential projects that the technology can be used in would give the Instructors some evidence and idea about the strength and level of the students' knowledge as it relates to the technology within their field. By changing how the students are exposed to the technology within their field, this would allow for student preparation for conducting master level and dissertation level research. Usually when students are educated in Online Institutions it leaves a void with regard to research exposure, research experimentation, and preparation for conducting research. If students trained within an online environment are not exposed more to technology centers or to environments that allow for better understanding of research, it is possible that the student would not develop the skillset for conducting research independently or with ethical training. Therefore it is not surprising that a need exists for better preparation for conduction independent research on the graduate and post-graduate educational level. These issues suggest the need for the development of training facilities and programs that focuses on exposing students to technology and laboratory equipment to assist in their transition to research-based institutions and research-intensive institutions.

The final solution that can be recommended to the management of the Online Institutions is to consider partnerships and investing in technology programs with Land-Based Institutions to give students enrolled in Online Institutions the opportunities to garner the experience needed in their field. By partnering with Land-Based Institutions students would have the opportunity to get experience with technology by working in the laboratories during their research phase of the PhD dissertation, Master's Thesis, or Bachelor Degree. This is one of the easiest and financially sound recommendations that can be made to assist the students in their development of their research skills, technology experience and with progressing in their field. In many ways it would mean updating the online courses to illustrate points within the course that can be carried out within Land-Based Institutions that could assist students in more hands-on experience when partnering with a Land-Based Institution. Partnering with Land-Based Institutions would give students opportunities to have direct access to technology and make them more competitive with regard to acceptance in summer programs, graduate schools, and in workforce enrichment. Investment in more research institutions would help in resolving the issues that result from students not having the experience from courses in Online Institutions. This investment may mean creating partnerships with companies, other universities, and organizations that would allow students to get the necessary hands-on experience/training needed and to allow them to be able to practice and participate in arenas that would garner them experience that is tailored to their educational backgrounds.

## REFERENCES


Cooper, G. M., & Hausman, R. E. (2009). *The Compositions of Cells. The Cell: A Molecular Approach.* Washington, DC: ASM Press.

Flicek, P. (2012, August 1). *A Deep Catalog of Human Genetic Variation.* Retrieved from http://www.1000genomes.org/

Haglund, H. (1971). Isoelectric focusing in pH gradients--a technique for fractionation and characterization of ampholytes. *Methods of Biochemical Analysis, 19,* 1–104. doi:10.1002/9780470110386.ch1 PMID:4935451

Haus, R. (2014, January 3). *BLAST Assembled RefSeq Genomes.* Retrieved from http://blast.ncbi.nlm.nih.gov/Blast.cgi

Henry, B. (2011, November 15). *DNA Microarray Technology.* Retrieved from http://www.genome.gov/10000533

Henry, B. (2011, December 27). *DNA Sequencing.* Retrieved from http://www.genome.gov/10001177

Henry, B. (2012, June 12). *Transcriptome.* Retrieved from http://www.genome.gov/13014330

Henry, B. (2013, July 11). *Polymerase Chain Reaction*. Retrieved from http://www.genome.gov/10000207

Lange, C. (2003, March 16). *PBAT*. Retrieved from http://www.biostat.harvard.edu/clange/default.htm

Nathans, D., & Smith, H. O. (1975). Restriction endonucleases in the analysis and restructuring of DNA molecules. *Annual Review of Biochemistry*, *44*(1), 273–293. doi:10.1146/annurev.bi.44.070175.001421 PMID:166604

Purcell, S., Neale, B., Todd-Brown, K., Thomas, L., Ferreira, M. A. R., & Bender, D. et al. (2007). PLINK: A toolset for whole-genome association and population-based linkage analysis. *American Journal of Human Genetics*, 599. PMID:17701901

Wilkinson, E.T., Cheifetz, S., De Grandis S.A. (1995). Development of competitive PCR and the QPCR system 5000 as a transcription-based screen. *PCR Methods Applications*, (6), 363-7.

## ADDITIONAL READING

Davis, R. (n.d.). *US National Library of Medicine*. Retrieved from http://www.ncbi.nlm.nih.gov/pubmed/

Emerson, B. (n.d.). *EMC Alliance*. Retrieved from http://www.emc.com/support/training/certification-programs.htm

Young, T. (2013, August 2). *American Education Research Association*. Retrieved from http://www.aera.net/

## KEY TERMS AND DEFINITIONS

**1000 Genomes Project:** The 1000 Genomes Project is an international collaboration to produce an extensive public catalog of human genetic variation, including SNPs and structural variants, and their haplotype contexts.

**Basic Local Alignment Search Tool (BLAST):** A method to determine sequence similarity between DNA, RNA, or proteins.

**C:** A systems programming language.

**C#:** An object oriented programming language that has been used for web development and networking.

**Deoxyribonucleic Acid (DNA) Sequencing:** A laboratory technique used to sequence the nucleotide bases of a DNA molecule.

**DNA Microarray:** A microarray of immobilized single-stranded DNA fragments of known nucleotide sequence that is used especially in the identification and sequencing of DNA samples and in the analysis of gene expression (as in a cell or tissue).

**Plink/SEQ:** An open-source C/C++ library for working with human genetic variation data.

**Plink:** A free, open-source whole genome association analysis toolset, designed to perform a range of basic, large-scale analyses in a computationally efficient manner.

**Practical Extraction and Report Language (PERL):** A computer programming language that is used for scripting.

**Real Time-Polymerase Chain Reactions (qPCR):** Quantitative monitoring of the accumulation of DNA product that is produced from a PCR reaction.

**Transcriptomics:** A genome-wide expression profiling which catalogues the complete set of RNA transcripts produced by the genome.

# Chapter 15
# Application of Information and Communication Technology to Create E-Learning Environments for Mathematics Knowledge Learning to Prepare for Engineering Education

**Tianxing Cai**
*Lamar University, USA*

## EXECUTIVE SUMMARY

*The standards for mathematical practice describe varieties of expertise that mathematics educators should develop in their students, including NCTM process standards (problem solving, reasoning and proof, communication, representation, and connections), NRC's report "Adding It Up" (adaptive reasoning, strategic competence, conceptual understanding, procedural fluency, and productive disposition), common core state standards in mathematics (ICT application) to support mathematics teaching and learning. There is a need to provide effective ways that*

DOI: 10.4018/978-1-4666-6375-6.ch015

*technology can be integrated into mathematics classrooms. Mathematical methods and techniques are typically used in engineering and industrial fields. It can also become an interdisciplinary subject motivated by engineers' needs. Mathematical problems in engineering result in rigorous engineering application carried out by mathematical tools. Therefore, a solid understanding and command of mathematical knowledge is very necessary. This chapter presents the introduction of currently available ICTs and their application of to create e-learning environments to prepare for the students' future engineering education.*

## BACKGROUND

Data integration techniques or information and communication technologies have been intensively used in different data mining applications such as data clustering, classification, association rules mining, sequential pattern mining, outlier detection, feature selection, and information extraction in the industrial and environmental research via air quality monitoring network. A huge increase in the number of papers and citations in the area has been observed in the previous decade, which is clear evidence of the popularity of these techniques. These have included the adoption of such kind of methodologies in the research field of polarization-difference imaging for observation through scattering media(Rowe, Pugh, Tyo & Engheta,1995), biologically inspired self-adaptive multi-path routing in overlay networks (Leibnitz, Wakamiya & Murata, 2006), a biologically inspired system for action recognition, programmable self-assembly using biologically-inspired multiagent control(Jhuang, Serre, Wolf & Poggio, 2007; Nagpal, 2002), biologically inspired growth of hydroxyapatite nanocrystals inside self-assembled collagen fibers(Roveri, Falini, Sidoti, Tampieri, Landi, Sandri & Parma, 2003), biologically inspired cognitive radio engine model utilizing distributed genetic algorithms for secure and robust wireless communications and networking(Rieser, 2004), biomimetics of biologically inspired technologies(Bar-Cohen, 2005), biologically inspired computing(De Castro, & von Zuben, 2005), and biologically inspired algorithms for financial modeling(Brabazon & O'Neill, 2006). Before we start to give the introduction of these techniques in the research field of industrial operation and environment sustainability, the brief introduction will be given for these techniques.

### Artificial Neural Networks

In computer science and related fields, artificial neural networks are models are derived from animal central nervous systems (Wang & Fu, 2008). The biologically neural networks are capable of machine learning and pattern recognition. They can be regarded as systems of internally connected neurons. They can compute values from inputs by feeding information through the network (Stevens & Casillas, 2006). For example, in a neural network for image recognition, a set of input neurons may be activated by the pixels of an input image representing a shape or color. The activations of these neurons are then passed on, weighted and transformed by some function determined by the network's designer, to other neurons, etc., until finally an output neuron is activated that determines which image was recognized. Similar with other methods of machine learning, neural networks have been applied to solve a wide range of jobs which are difficult to solve using ordinary rule-based programming(Yang & Zheng, 2009)..Generally, artificial neural network handles a problem with the combination of simple processing elements which have complex global behavior. A class of statistical models will be called "neural" if they have sets of adaptive weights (numerical parameters that are tuned by a learning algorithm, and are capable of approximating non-linear functions of their inputs) (Patterson, 1998). The adaptive weights are conceptually connection strengths between neurons. They will be activated during the period of model training and prediction. Neural networks can also perform functions collectively and in parallel by the units, which are also similar to biological neural networks. The terminology of neural network usually means the model with the integration of statistics, cognitive psychology and artificial intelligence(Sarle, 1994). They are part of theoretical neuroscience and computational neuroscience. In modern software implementations of artificial neural networks, the approach inspired by biology has been largely abandoned according to statistics and signal processing (Holler, Tam, Castro & Benson, 1989). In some of these systems, neural networks or parts of neural networks can form components in larger systems that combine both adaptive and non-adaptive elements(Cochocki & Unbehauen, 1993). The general approach of such systems is feasible for real-world problem solving while it has been different from the traditional artificial intelligence models, which adopt the principles of non-linear, distributed, parallel and local processing and adaptation.

## Fuzzy Logic System

Fuzzy logic is a logic form of multiple values. It helps to handle the problems which are not fixed and exact. Compared to traditional binary variables, fuzzy logic variables may have a truth value that ranges in degree between 0 and 1. It can be regarded as the concept of partial truth, where the truth value may range between completely true and completely false(Perfilieva & Močkoř,999) The term "fuzzy logic" was introduced with the proposal of fuzzy set theory by Lotfi A. Zadeh.*(Zalta, 2003; Zadeh, 1965) It has been applied in many fields of control theory and artificial intelligence. A basic application might characterize subranges of a continuous variable.

For instance, a temperature measurement for anti-lock brakes might have several separate membership functions defining particular temperature ranges needed to control the brakes properly. Each function maps the same temperature value to a truth value in the 0 to 1 range. These truth values can then be used to determine how the brakes should be controlled.

## Genetic Algorithms

In the field of artificial intelligence, a genetic algorithm (GA) is a search heuristic with the process of natural selection. This heuristic is normally used to generate useful solutions to optimization and search problems. Genetic algorithms is one of the algorithm of evolutionary algorithms (EA) (Whitley,1994). The evolutionary algorithm applies the evolution techniques of inheritance, mutation, selection, and crossover to generate solutions to optimization problems. It can be used in bioinformatics, computational science, engineering, economics, manufacturing, mathematics, physics, chemistry, and other fields (Goldberg & Holland, 1988). In a genetic algorithm, a population of candidate solutions to an optimization problem is evolved to better generations of solutions. Each candidate solution has a set of properties of chromosomes which can be mutated and altered. The procedures usually start from a population of randomly generated individuals. Then the iterative process will provide a new generation of population in each iteration(Vose, 1999). The fitness of every individual in the population is evaluated in each generation. It is usually the value of the objective function in the optimization problem being minimized or maximized. The more fit individuals are selected from the current population, and each individual's genome is changed through recombination and possibly randomly mutation to form a new generation. The new generation of candidate solutions is then used in the next iteration of the algorithm. Commonly, the algorithm terminates when either a maximum number of generations has been produced, or a satisfactory fitness level has been reached for the population. A typical genetic algorithm requires: a genetic representation of the solution domain, a fitness function to evaluate the solution domain. A standard representation of each candidate solution is as an array of bits. The main property that makes these genetic representations convenient is that their parts are easily aligned due to their fixed size, which facilitates simple crossover operations. Variable length representations may also be used, but crossover implementation is more complex in this case. Tree-like representations are explored in genetic programming and graph-form representations are explored in evolutionary programming; a mix of both linear chromosomes and trees is explored in gene expression programming. When the genetic representation and the fitness function are defined, a GA proceeds to initialize a population of solutions and then to improve it through repetitive application of the mutation, crossover, inversion and selection operators. Initially many individual solutions are randomly generated to form an initial population. The population size is based on the nature of the problem, but typically contains thousands of possible solutions. Normally, the population is generated randomly in order to cover the entire range of possible solutions in the searching space. During each successive generation, a proportion of the existing population is selected to breed a new generation. Individual solutions are selected through a fitness-based process, where fitter solutions are typically more likely to be selected. The selection methods help to rate the fitness of each solution and select the best solutions. The next step is to generate a second generation population of solutions from those selected through genetic operators: crossover and/or mutation. For each new solution to be produced, a pair of parent solutions is selected for breeding from the previously selected pool. By producing a "child" solution using the above methods of crossover and mutation, a new solution is created which typically shares many of the characteristics of its "parents". New parents are selected for each new child, and the process continues until a new population of solutions of appropriate size is generated. These processes ultimately result in the next generation population of chromosomes that is different from the initial generation. Generally the average fitness will have increased by this procedure for the population, since only the best organisms from the first generation are selected for breeding, along with a small proportion of less fit solutions. These less fit solutions ensure genetic diversity within the genetic pool of the parents and therefore ensure the genetic diversity of the subsequent generation of children. The attention should be paid to the tuning parameters such as the mutation probability, crossover probability and population size to find reasonable settings for the problem solving. There are theoretical but not yet practical upper and lower bounds for these parameters that can help guide selection through experiments. This above introduced processes are repeated until a termination condition has been reached. Common terminating conditions are the achievement of the identification of solution to satisfy criteria, fixed number of generations, allocated budget of computation time and the highest ranking solution's fitness(Deb, 2001).

## Ant Colony Optimization

The ant colony optimization algorithm (ACO) is a probabilistic technique for solving computational problems which can be reduced to finding good paths through graphs(Dorigo & Di Caro, 1999). This algorithm is one of the ant colony algorithms family of swarm intelligence methods and some meta-heuristic optimizations(Dorigo & Birattari, 2010). Ant colony optimization algorithms have been applied to many combinatorial optimization problems, ranging from quadratic assignment to protein folding or routing vehicles and a lot of derived methods have been adapted to dynamic problems in real variables, stochastic problems, multi-targets and parallel implementations. It has also been used to produce near-optimal solutions to the travelling salesman problem(Dorigo, 2007). They have an advantage over simulated annealing and genetic algorithm approaches of similar problems when the graph may change dynamically; the ant colony algorithm can be run continuously and adapt to changes in real time. This is of interest in network routing and urban transportation systems(Dorigo, 2006).

## Particle Swarm Optimization

Particle swarm optimization (PSO) helps to optimize a problem by iteratively trying to improve a candidate solution with regard to a given measure of quality(Kennedy, 2010). PSO optimizes a problem by having a population of candidate solutions of dubbed particles and moving these particles around in the search-space according to simple mathematical formulae over the particle's position and velocity(Poli, Kennedy & Blackwell, 2007; Venter & Sobieszczanski-Sobieski, 2003). Each particle's movement is determined by its local best known position and guided toward the best known positions in the search-space. They are updated as better positions found by other particles. PSO is a meta-heuristic due to its characterization with few or no assumptions about the problem being optimized and can search very large spaces of candidate solutions. However, meta-heuristics such as PSO do not guarantee an optimal solution is ever found. More specifically, PSO does not use the gradient of the problem being optimized, which means PSO does not require that the optimization problem be differentiable as is required by classic optimization methods such as gradient descent and quasi-Newton methods(Clerc, 2006). A basic variant of the PSO algorithm works by having a population (called a swarm) of candidate solutions (called particles). These particles are moved around in the search-space according to a few simple formulae(Clerc, 2006). The movements of the particles are guided by their own best known position in the search-space as well as the entire swarm's best known position. When improved positions are being discovered these will then come to guide the movements of the swarm. The process is repeated and by doing so it is hoped, but not guaranteed, that a satisfactory solution will eventually be discovered. PSO has also been applied to multi-objective problems, in which the objective function comparison takes Pareto dominance into account when moving the PSO particles and non-dominated solutions are stored so as to approximate the Pareto front(Lazinica, 2009).

## Artificial Immune System

Artificial immune systems (AIS) are a class of computationally intelligent systems inspired by the principles and processes of the vertebrate immune system. The algorithms typically exploit the immune system's characteristics of learning and memory to solve a problem. The field of Artificial Immune Systems (AIS) is concerned with abstracting the structure and function of the immune system to computational systems, and investigating the application of these systems towards solving computational problems from mathematics, engineering, and information technology(Hofmeyr & Forrest, 2000). AIS is a sub-field of Biologically-inspired computing, and Natural computation, with interests in Machine Learning and belonging to the broader field of Artificial Intelligence. Artificial Immune Systems (AIS) are adaptive systems, inspired by theoretical immunology and observed immune functions, principles and models, which are applied to problem solving. AIS is distinct from computational immunology and theoretical biology that are concerned with simulating immunology using computational and mathematical models towards better understanding the immune system, although such models initiated the field of AIS and continue to provide a fertile ground for inspiration. Finally, the field of AIS is not concerned with the investigation of the immune system as a substrate computation, such as DNA computing. The common techniques are inspired by specific immunological theories that explain the function and behavior of the mammalian adaptive immune system(Timmis & Neal, 2001;DasGupta, 1999; Coello & Cortés, 2005; Lei, 2002; Coello & Cortés, 2002).

Clonal Selection Algorithm: A class of algorithms inspired by the clonal selection theory of acquired immunity that explains how B and T lymphocytes improve their response to antigens over time called affinity maturation. These algorithms focus on the Darwinian attributes of the theory where selection is inspired by the affinity of antigen-antibody interactions, reproduction is inspired by cell division, and variation is inspired by somatic hypermutation. Clonal selection algorithms are most commonly applied to optimization and pattern recognition domains, some of which resemble parallel hill climbing and the genetic algorithm without the recombination operator.

Negative Selection Algorithm: Inspired by the positive and negative selection processes that occur during the maturation of T cells in the thymus called T cell tolerance. Negative selection refers to the identification and deletion (apoptosis) of self-reacting cells, that is T cells that may select for and attack self tissues. This class of algorithms are typically used for classification and pattern recognition problem domains where the problem space is modeled in the complement of available knowledge. For example in the case of an anomaly detection domain the algorithm prepares a set of exemplar pattern detectors trained on normal (non-anomalous) patterns that model and detect unseen or anomalous patterns.

Immune Network Algorithms: Algorithms inspired by the idiotypic network theory proposed by Niels Kaj Jerne that describes the regulation of the immune system by anti-idiotypic antibodies (antibodies that select for other antibodies). This class of algorithms focus on the network graph structures involved where antibodies (or antibody producing cells) represent the nodes and the training algorithm involves growing or pruning edges between the nodes based on affinity (similarity in the problems representation space). Immune network algorithms have been used in clustering, data visualization, control, and optimization domains, and share properties with artificial neural networks.

Dendritic Cell Algorithms: The Dendritic Cell Algorithm (DCA) is an example of an immune inspired algorithm developed using a multi-scale approach. This algorithm is based on an abstract model of dendritic cells (DCs). The DCA is abstracted and implemented through a process of examining and modeling various aspects of DC function, from the molecular networks present within the cell to the behaviour exhibited by a population of cells as a whole. Within the DCA information is granulated at different layers, achieved through multi-scale processing.

## Culture Algorithm

Cultural algorithms (CA) are a branch of evolutionary computation where there is a knowledge component that is called the belief space in addition to the population component(Reynold & Sverdlik, 1994; Reynold, 1994)). In this sense, cultural algorithms can be seen as an extension to a conventional genetic algorithm(Coello & Becerra, 2004). The belief space of a cultural algorithm is divided into distinct categories. These categories represent different domains of knowledge that the population has of the search space. The belief space is updated after each iteration by the best individuals of the population. The best individuals can be selected using a fitness function that assesses the performance of each individual in population much like in genetic algorithms. The population component of the cultural algorithm is approximately the same as that of the genetic algorithm. Cultural algorithms require an interface between the population and belief space(Jin & Reynolds, 1996). The best individuals of the population can update the belief space via the update function. Also, the knowledge categories of the belief space can affect the population component via the influence function. The influence function can affect population by altering the genome or the actions of the individuals(Chung & Reynolds, 1996).

## MOTIVATION

The Standards for Mathematical Practice describe varieties of expertise that mathematics educators at all levels should seek to develop in their students. These practices rest on important "processes and proficiencies" with longstanding importance in mathematics education. The first of these are the NCTM process standards of problem solving, reasoning and proof, communication, representation, and connections. The second are the strands of mathematical proficiency specified in the National Research Council's report *Adding It Up*: adaptive reasoning, strategic competence, conceptual understanding (comprehension of mathematical concepts, operations and relations), procedural fluency (skill in carrying out procedures flexibly, accurately, efficiently and appropriately), and productive disposition (habitual inclination to see mathematics as sensible, useful, and worthwhile, coupled with a belief in diligence and one's own efficacy). The Common Core State Standards in Mathematics provide ample opportunities for teachers to use multiple available ICTs(Information and Communication Technologies) to support mathematics teaching and learning. From the grade level content standards to the Standards for Mathematical Practice there is a need to provide examples about effective ways that technology can be integrated into mathematics classrooms.

Based on an available air-quality monitoring network, the data integration technologies will be applied to identify the scenarios of the possible emission source and the dynamic pollutant monitor result, so as to timely and effectively support diagnostic and prognostic decisions. Qualitative and mixed methods researchers have employed a variety of information and communication technology (ICT) tools, simulated or virtual environments, information systems, information devices and data analysis tools in this field. With the collection and representation of information in a range of ways, software tools have been created to manage and store these data. This data management enables more efficient searching ability of various types of digitized information. Various technologies have made the work of research more efficient. The results of the qualitative or mixed methods research may be integrated to reach the research target. Right now, a lot of software tools are available for the analysis to identify knowledge patterns and represent new meanings. The programs extend the capabilities of the researcher in terms of information coding and meaning-making. Machine-enhanced analytics has enabled the identification of aspects of interest such as correlations and anomalies from large datasets.

In this chapter, we will will present the introduction of currently available Information and Communication Technologies(ICTs) and their application of to create e-learning environment to prepare for the students' future engineering education. Actually mathematical methods and techniques such as ordinary and partial differential equations, stochastic processes, calculus of variations, and nonlinear analysis are typically used in engineering and industrial fields of, in particular, aerospace engineering, bioengineering, chemical engineering, computer engineering, electrical engineering, industrial engineering and manufacturing systems, and mechanical engineering are of interest. Along with fields like engineering physics and engineering geology, it can also become an interdisciplinary subject motivated by engineers' needs both for practical, theoretical and other considerations with their specialization, and to deal with constraints to be effective in their work. Mathematical problems in engineering result in rigorous engineering application carried out using mathematical tools. Contributions containing formulations or results related to applications have become very common. Therefore the solid understanding and command of mathematical knowledge is very necessary.

## DATA SOURCE

The basic mission of the industrial and environment research with web service is to preserve and improve the air quality of our living environment. To accomplish this, we must be able to evaluate the status of the atmosphere as compared to clean air standards and historical information. The following are some of the topics associated with monitoring air pollution.

In USA, the Clean Air Act requires every state to establish a network of air monitoring stations for criteria pollutants, using criteria set by OAQPS for their location and operation. The monitoring stations in this network are called the State and Local Air Monitoring Stations (SLAMS). The states must provide OAQPS with an annual summary of monitoring results at each SLAMS monitor, and detailed results must be available to OAQPS upon request. To obtain more timely and detailed information about air quality in strategic locations across the nation, OAQPS established an additional network of monitors: the National Air Monitoring Stations (NAMS). NAMS sites, which are part of the SLAMS network, must meet more stringent monitor siting, equipment type, and quality assurance criteria. NAMS monitors also must submit detailed quarterly and annual monitoring results to OAQPS.

Between the years 1900 and 1970, the emission of six principal pollutants increased significantly. These six pollutants, also called criteria pollutants, are: particulate matter, sulfur dioxide, carbon monoxide, nitrogen dioxide, ozone, and lead. In 1970, the Clean Air Act (CAA) was signed into law. The CAA and its amendments provides the framework for all pertinent organizations to protect air quality. EPA's principal responsibilities under the CAA, as amended in 1990 include:

- Setting National Air Quality Standards (NAAQS) for pollutants considered harmful to the public health and environment;

- Ensuring the air quality standards are met or attained (in cooperation with the States) through national standards and strategies to control air emission standards from sources;
- Ensuring the sources of toxic air pollutants are well controlled;
- Monitoring the effectiveness of the program.

One way to protect and assess air quality was through the development of an Ambient Air Monitoring Program. Air quality samples are generally collected for one or more of the following purposes:

- To judge compliance with and/or progress made towards meeting ambient air quality standards.
- To activate emergency control procedures that prevent or alleviate air pollution episodes.
- To observe pollution trends throughout the region, including non-urban areas.
- To provide a data base for research evaluation of effects: urban, land-use, and transportation planning; development and evaluation of abatement strategies; and development and validation of diffusion models.

With the end use of the air quality samples as a prime consideration, the network should be designed to meet one of four basic monitoring objectives listed below:

- To determine highest concentrations expected to occur in the area covered by the network;
- To determine representative concentrations in areas of high population density;
- To determine the impact on ambient pollution levels of significant sources or source categories;
- To determine general background concentration levels.

These four objectives indicate the nature of the samples that the monitoring network will collect which must be representative of the spatial area being studied.

The EPA's ambient air quality monitoring program is carried out by State and local agencies and consists of three major categories of monitoring stations, State and Local Air Monitoring Stations (SLAMS), National Air Monitoring Stations (NAMS), and Special Purpose Monitoring Stations (SPMS), that measure the criteria pollutants. Additionally, a fourth category of a monitoring station, the Photochemical Assessment Monitoring Stations (PAMS), which measures ozone precursors (approximately 60 volatile hydrocarbons and carbonyl) has been required by the 1990 Amendments to the Clean Air Act.

## METHODOLOGY AND APPLICATION

### Temperature Prediction

#### Climate Prediction

The first priority and prerequisite for a region's development is appropriate climate condition because the climate can affect almost all the aspects of our life such as economy, society, culture, education, food supply or even transportation safety. For example, the frequency of traffic accident will increase definitely during the snowing or raining days; the city cannot become the economic center if it always encounter extreme weather. The city's prosperity relies heavily on the climate with high adaptation to the people who live there so that there is no need for them to worry about the unexpected conditions the extreme meteorology condition. Therefore, it is very important for us to make a prediction on the climate change trend if we want to make a decision for municipal development plan in the future. It is very necessary to investigate the climate change trend, which will let us know whether it will have a great change in the weather condition such as temperature increase. Furthermore, such kind of climate prediction result can also become the data input for the other models such as pollutant dispersion model and flight scheduling in their further evaluation and risk assessment. Up to now a lot of professional scientists have dedicated their research to meteorology simulation and modeling and their great work has helped to provide multiple model tools in microscale, mesoscale, synoptic scale and global scale. These meteorological models rely on the application of meteorological principles (boundary layer meteorology and dynamic meteorology) and typically require powerful computers to produce output on winds, temperature, and precipitation at different horizontal locations and vertical levels in the atmosphere. The high requirement of computation hardware and professional understanding of modeling language cannot always be satisfied by the communities even though they can provide relatively more accurate result. To the majority of people, a better way for common application to predict the regional climate change is a statistical tool by data regression and analysis instead of those complex system provided by national and international research laboratories and organizations. There are two major benefits for this choice: the first benefit is data availability which means the historic climate date will always be available from websites and the second benefit is tool availability which means we can even use the common software to proceed the data mining to get the prediction results. Therefore the tool derived from this study can be used by the citizens to get more understanding of their regional climate conditions(NWS and NCEP, 2013).

As Dr. Steve Running, Climate Change Studies Program Director, and a lead author on the Nobel Prize winning Intergovernmental Panel on Climate Change said, "The climate change topic is rapidly evolving from only an earth science issue to a technological, economic and sociological issue." (Climate Change Studies Program,2013).

Climate change describes the change trend of average pattern of weather over the long term. In the previous millions of years, the earth's climate has been warmed and cooled, even long before our appearance on the earth. Even though climate change isn't new, the study of how the climate change is. The exploration of climate change encompasses many fields of study, including physics, chemistry, biology, geology, meteorology, oceanography, and even sociology.

The scientists study natural phenomena of climate change through evidence gathering, theory test, and conclusion generating. There are several general methodologies to study the climate change: simulation experiment, historical data analog, field observation, numerical modeling and computer simulation.

- **Simulation Experiment:** The procedures in order are carried out to achieve the goal of verifying, refuting, or establishing the validity of a hypothesis of climate change. Experiments provide insight into cause-and-effect between multiple natural and anthropogenic factors and the result of climate change. For example this method is used to study the responses of grassland plants to experimentally simulated climate change depend on land use and region (Bütof, et.al., 2012).
- **Historical Data Analog:** Historical recorded data is processed to be in a form to identify its similarity of its original structure of climate change trend in the spatial and temporal distribution. It has been used to conduct multi-field analog prediction of short-term climate fluctuations by the adoption of a climate state vector (Barnett & Preisendorfer, 1978).
- **Field Observation:** Climate field observation provides long-term or short-term, high quality, timely, observational data, information and products in support of climate and weather research communities, forecasters, and other service providers and users. It has included climate observing system according to climate monitoring principles, time-series indicators of climate variability and change. The observation result will be used to develop and maintain standard data sets for initialization and evaluation of climate forecast models, assessments of climate change, and informed risk management. It will be further used to develop informational products, diagnostics, and assessments of observed climate variability and change on different scales. This method has been used in many studies such as indirect radiative forcing

of climate change through ozone effects on the land-carbon sink (Sitch, Cox, Collins, & Huntingford, 2007).

- **Numerical Modeling:** The results of a numerical model consist of numbers, which represent particular "cases" or "realizations." There are particular examples of what the (model) atmosphere can do: for example, we can "run" a numerical model to create a weather forecast, which consists of a large set of numbers. The application of numerical modeling for climate change has been introduced in a lot of books, such as atmosphere circulation, meteorology dynamics and three dimensional climate modeling (Chang, 1977; Durran, 1999; Haltiner & Williams, 1980; Kalnay, 2003; Manabe, 1985; Mesinger & Arakawa, 1976; Randall, 2000; Washington & Parkinson, 1986)
- **Computer Simulation:** It is a computer program which is run on a single computer, or a network of computers in order to simulate a particular system of climate. For example, the Weather Research and Forecasting (WRF) model(UCAR, 2010) is a specific computer program with a dual use for forecasting and research. It was created through a partnership that includes the National Oceanic and Atmospheric Administration(NOAA), the National Center for Atmospheric Research(NCAR), and more than 150 other organizations and universities in the United States and abroad(NCAR, 2006; Wikipedia, 2013; NCDC & NOAA, 2013).

The section will present the data analysis results based on the statistical processing of the historical temperature monitor results. During data analysis, several years of records have been summarized to characterize the climate so that a large data record of long time period will help to achieve a relatively stable change trend and statistical results. During the seasonal analysis, four seasons with respective to months are defined in Table 1: spring (March ~ May), summer (June ~ August), autumn (September ~ November) and winter (December ~ February).

The section will present the linear regression for the above data analysis results of seasonal and annual average temperatures of Cincinnati. In the linear regression model, the independent variable will be the year, $t$ and the dependent variable will be the temperature, $Y_t$. The linear regression will generate the linear function expression with respective to the year, $t$. The linear equation which is used in our regression is $Y_t = m + k * t$. The methodology of least square will be used to get the linear regression results, which is the value of coefficient of the above equations: $k$ and $m$. Here, $k$ is the slope which represents the trend rate of temperature change by time interval (year or decade) and $m$ is the intercept which is a constant term. The change tendency can be identified by the sign of slope $k$ :if the slope $k$ is larger than zero, there will be a rising upward trend of average temperature; if the

*Table 1. Summary of monthly mean temperature(1973-2012)*

| Year/Month | Jan | Feb | Mar | Apr | May | Jun | Jul | Aug | Sep | Oct | Nov | Dec |
|---|---|---|---|---|---|---|---|---|---|---|---|---|
| **1973** | 0.8 | 1.0 | 10.5 | 11.2 | 15.7 | 23.3 | 24.4 | 23.6 | 21.0 | 15.2 | 8.8 | 2.1 |
| **1974** | 2.9 | 1.8 | 8.5 | 13.4 | 17.5 | 20.4 | 24.2 | 23.4 | 17.0 | 11.8 | 7.5 | 2.6 |
| **1975** | 2.6 | 3.0 | 4.9 | 10.3 | 19.8 | 22.8 | 24.1 | 25.0 | 17.8 | 13.5 | 9.5 | 3.0 |
| **1976** | -1.4 | 6.4 | 10.4 | 13.0 | 16.6 | 22.3 | 24.0 | 21.9 | 18.0 | 9.9 | 3.4 | -1.3 |
| **1977** | -9.0 | -0.2 | 9.3 | 14.2 | 20.8 | 21.8 | 26.0 | 23.6 | 21.3 | 11.9 | 8.5 | -0.2 |
| **1978** | -5.8 | -5.9 | 3.8 | 12.8 | 16.6 | 22.8 | 23.9 | 23.1 | 21.5 | 11.3 | 8.0 | 2.4 |
| **1979** | -4.4 | -4.4 | 8.2 | 11.1 | 16.9 | 21.8 | 22.9 | 23.0 | 19.1 | 12.7 | 6.8 | 2.8 |
| **1980** | 0.3 | -2.4 | 4.0 | 10.8 | 18.2 | 21.4 | 25.7 | 25.5 | 21.0 | 11.6 | 6.1 | 1.8 |
| **1981** | -2.6 | 1.7 | 5.7 | 15.3 | 15.7 | 23.3 | 24.4 | 22.8 | 18.9 | 12.8 | 6.9 | 0.0 |
| **1982** | -3.4 | -0.1 | 6.8 | 10.2 | 20.2 | 20.1 | 24.7 | 21.6 | 18.7 | 13.9 | 8.6 | 6.2 |
| **1983** | 0.5 | 2.3 | 7.4 | 10.7 | 15.9 | 22.2 | 26.3 | 25.9 | 20.3 | 14.0 | 7.9 | -2.6 |
| **1984** | -3.0 | 4.3 | 2.8 | 11.3 | 16.2 | 24.1 | 23.0 | 23.6 | 18.5 | 16.3 | 5.7 | 5.9 |
| **1985** | -3.8 | -1.1 | 8.6 | 15.1 | 18.6 | 21.3 | 24.4 | 23.1 | 20.3 | 15.5 | 10.6 | -1.4 |
| **1986** | 0.1 | 2.7 | 7.8 | 13.6 | 19.1 | 23.3 | 25.8 | 22.6 | 21.5 | 14.1 | 6.9 | 1.8 |
| **1987** | 0.1 | 3.2 | 7.4 | 12.1 | 20.6 | 23.9 | 24.9 | 24.3 | 20.3 | 10.3 | 9.3 | 3.6 |
| **1988** | -1.3 | 0.6 | 6.3 | 12.2 | 18.5 | 23.0 | 25.9 | 25.5 | 19.8 | 9.7 | 7.6 | 1.8 |
| **1989** | 3.6 | 0.0 | 7.2 | 11.7 | 15.9 | 21.8 | 24.6 | 22.9 | 18.7 | 12.4 | 6.6 | -5.1 |
| **1990** | 4.4 | 5.1 | 8.9 | 11.9 | 16.4 | 22.5 | 24.2 | 23.0 | 20.3 | 13.1 | 9.0 | 4.1 |
| **1991** | 0.3 | 3.2 | 8.1 | 14.1 | 21.3 | 23.7 | 25.0 | 23.3 | 20.0 | 14.2 | 6.2 | 3.6 |
| **1992** | 1.2 | 4.5 | 6.8 | 12.6 | 16.4 | 20.6 | 23.9 | 21.3 | 18.7 | 12.0 | 7.7 | 2.7 |
| **1993** | 2.2 | -0.6 | 5.1 | 11.7 | 18.0 | 21.7 | 26.0 | 24.4 | 18.9 | 12.4 | 7.1 | 1.7 |
| **1994** | -3.6 | 1.0 | 6.5 | 13.9 | 15.9 | 23.8 | 24.4 | 22.7 | 18.6 | 13.4 | 10.2 | 4.1 |
| **1995** | 0.8 | 0.0 | 7.8 | 12.3 | 17.3 | 22.7 | 25.1 | 26.0 | 18.2 | 13.7 | 4.3 | -0.3 |
| **1996** | -1.2 | 1.0 | 3.6 | 10.9 | 17.6 | 22.2 | 22.9 | 23.2 | 19.1 | 13.6 | 4.2 | 4.2 |
| **1997** | -1.0 | 4.3 | 8.1 | 10.4 | 15.3 | 21.2 | 24.7 | 22.5 | 18.9 | 12.9 | 5.9 | 2.3 |
| **1998** | 4.0 | 4.8 | 7.1 | 11.8 | 19.2 | 22.4 | 23.9 | 23.8 | 21.5 | 13.1 | 7.5 | 4.1 |
| **1999** | 1.1 | 3.7 | 4.1 | 13.7 | 18.2 | 23.1 | 27.0 | 23.2 | 19.4 | 12.6 | 9.0 | 2.3 |
| **2000** | -0.6 | 4.8 | 8.9 | 11.8 | 19.2 | 22.5 | 23.0 | 22.2 | 18.3 | 13.9 | 6.2 | -3.4 |
| **2001** | -0.3 | 3.6 | 5.0 | 14.9 | 17.9 | 20.9 | 23.4 | 23.5 | 17.4 | 12.3 | 8.9 | 4.2 |
| **2002** | 2.4 | 2.6 | 5.9 | 13.1 | 15.6 | 22.6 | 25.4 | 24.3 | 20.9 | 12.1 | 5.7 | 1.5 |
| **2003** | -3.3 | -1.4 | 7.1 | 12.7 | 16.8 | 19.9 | 23.1 | 23.1 | 18.0 | 12.3 | 9.2 | 2.0 |
| **2004** | -1.9 | 1.0 | 7.8 | 12.4 | 19.7 | 21.8 | 23.0 | 21.4 | 19.6 | 13.2 | 8.9 | 1.4 |
| **2005** | 1.7 | 3.2 | 4.3 | 13.0 | 15.5 | 23.3 | 24.9 | 24.7 | 20.7 | 13.0 | 7.9 | -0.1 |
| **2006** | 5.6 | 1.8 | 5.9 | 13.9 | 16.4 | 20.8 | 24.6 | 24.8 | 17.6 | 11.7 | 7.5 | 4.8 |

*continued on following page*

*Table 1. Continued*

| Year/Month | Jan | Feb | Mar | Apr | May | Jun | Jul | Aug | Sep | Oct | Nov | Dec |
|---|---|---|---|---|---|---|---|---|---|---|---|---|
| **2007** | 3.1 | -3.9 | 9.9 | 11.4 | 19.4 | 23.0 | 23.1 | 26.7 | 21.7 | 16.0 | 7.5 | 3.3 |
| **2008** | 0.3 | 1.3 | 5.7 | 12.3 | 15.8 | 22.8 | 23.9 | 23.3 | 20.5 | 12.6 | 5.6 | 2.1 |
| **2009** | -2.3 | 2.7 | 8.9 | 13.2 | 18.1 | 22.5 | 21.3 | 22.1 | 19.1 | 11.0 | 7.8 | 1.4 |
| **2010** | -1.8 | -1.9 | 7.4 | 14.4 | 19.0 | 23.8 | 25.0 | 24.6 | 20.4 | 13.7 | 6.6 | -2.0 |
| **2011** | -2.1 | 2.7 | 7.1 | 14.0 | 17.1 | 22.3 | 26.5 | 23.8 | 18.2 | 12.0 | 9.4 | 4.5 |
| **2012** | 2.3 | 3.8 | 12.8 | 12.5 | 19.9 | 22.5 | 26.4 | 23.0 | 18.6 | 11.9 | 4.7 | 4.6 |

slope $k$ is less than zero, there will be a descending downward trend of average temperature; if the slope $k$ is equal to zero, there will be no change trend of average temperature. The absolute value of slope $k$ will reflect the increased or decreased degree. The linear regression model for annual average temperature is constructed based on available dataset. The trend analysis plot for annual temperature has also been provided. The linear regression model for average temperature of spring, summer, autumn and winter can be constructed based on available data set. The trend analysis plot for annual temperature has also been provided.

## Inverse Emission Identification

The second application is to develop a systematic methodology to reversely detect the emission conditions from a list of candidates (local chemical plants) according to the abnormal air-quality measurements from an available monitoring network, so as to support diagnostic and prognostic decisions timely and effectively. The outcome of the developed methodology should provide information of emission source location, starting time, time duration, total emission amount, and dynamic emission rate and pattern from the abnormal emission sources. The methodology will firstly determine the background normal emission rates for a given list of candidate emission sources in the region. Next, an optimization model will be employed for reverse emission source detection based on abnormal air-quality measurements. For clarity, the problem statements are summarized below.

### *Assumptions*

1. An air-quality event in a region is caused by abnormal emissions from one and only one emission source based on a given list of candidate emission sources, whose abnormal emission pattern belongs to one of those shown in Table 1;
2. Each candidate emission source has a constant emission rate during its normal operational conditions;

3. Emission transportation follows Gaussian dispersion model and there is no secondary consumption or generation of the pollutant during its air transportation;
4. When the emitted pollutant reached ground through dispersion, it will be absorbed, i.e., there is no pollutant reflection from the ground during its air transportation;
5. Meteorological conditions (e.g., local wind direction and wind speed) during the considered scheduling time horizon are constant (or near constant) in the region.

### *Given Information*

1. Spatial locations of each emission sources and monitoring stations;
2. Emission source stack parameters, such as stack height and outlet temperature;
3. Dynamic monitoring results at each monitoring station;
4. Meteorological conditions in the studied region during the event time period.

### *Information to be Determined*

1. Which emission source caused the investigated pollutant concentration pattern;
2. What are the emission pattern and dynamic emission rate for the identified emission source.

The input parameters at the modeling stage include geographical information (locations of every possible emission source and monitoring station), meteorological condition (e.g., wind direction, speed, and atmospheric stability), measurements at each monitoring station, and emission source data (e.g., stack height, exit diameter and outlet temperature). The next step is to map locations of candidate emission sources and monitoring stations into a rectangular coordinator system (see Appendix). Then, the first stage modeling aims at the determination of normal emission rates from every emission source. The task is accomplished through a regression model based on Gaussian-dispersion model to minimize the sum of squared error (SSE) between the model calculated results and monitoring results from multiple monitoring stations. Since the normal emission rate of each emission source is the background emission during plant steady-state operational conditions, it is also called steady-state emission rate.

This regression model is to identify the normal emission rate for each candidate emission source by minimizing SSE between the model predicting results and the monitoring results from multiple monitoring stations at normal emission status (without emission events).

### *Objective Function*

$$\varphi_1 = \min_{m_i^S} \sum_{t \in T^D} \sum_{j \in J} \left( \bar{C}_j - C_{j,t}^S \right)^2 \tag{2}$$

where $j$ represents the index of monitoring stations grouped by set $J$; $T^S$ represents a selected steady-state time set when each emission source has a normal emission rate. $T^S$ contains multiple time instants indexed by $t$. $\bar{C}_j$ and $C_{j,t}^S$ respectively represents model calculated and measured pollutant concentrations at the $j$-th monitoring station at time $t$. Equation (2) suggests the objective function is to minimize SSE between $\bar{C}_j$ and $C_{j,t}^S$.

The model calculated pollutant concentration at the $j$-th monitoring station ($\bar{C}_j$) should be the cumulative of $\bar{C}_{i,j}$ from all the emission sources, which can be formulated by Equation (3).

$$\bar{C}_j = \sum_{i \in I} \bar{C}_{i,j}, \ \forall\, j \in J \tag{3}$$

### *Dispersion Transportation Principle*

Note that $\bar{C}_{i,j}$ should be calculated by the following Equation (4). It represents pollutant dispersion from emission sources to monitoring stations under the impact of meteorological conditions. The associated details of Equation (4) can be referenced in the Appendix.

$$\bar{C}_{i,j} = \frac{m_i^S f\left(Z_{i,j}, H_i\right)}{2U_i \pi \sigma_{Y_{i,j}} \sigma_{Z_{i,j}}} \exp\left[ -\frac{1}{2} \left( \frac{Y_{i,j}}{\sigma_{Y_{i,j}}} \right)^2 \right], \ \forall\, i \in I,\ j \in J \tag{4}$$

where $\bar{C}_{i,j}$ is the pollutant concentration at the $j$-th monitoring station caused by the emission from the $i$-th emission source; $Y_{i,j}$ is the projection of $d_{i,j}$ along $Y$ direction (Here, $X$ direction is the same as the wind direction; $Y$ direction is horizontally perpendicular to $X$ direction); $Z_{i,j}$ is the ground height difference between the $i$-th emission source to the $j$-th monitoring station; $H_i$ is the plume height above the ground for source $i$; $\sigma_{Y_{i,j}}$ and $\sigma_{Z_{i,j}}$ are the standard deviations of the emission plume's probability distribution function along $Y$ and $Z$ directions, respectively; $f\left(Z_{i,j}, H_i\right)$ is a function with respect to $Z_{i,j}$ and $H_i$; $m_i^S$ represents the constant emission rate at the $i$-th emission source in the normal condition.

The related equations and procedures to calculate the parameters shown in Equation (4) has been included in the Gaussian dispersion model area (Pasquill, 1961 and 1974; Turner, 1979 and 1994; Hanna et al., 1982; Seinfeld, 1986; Slade, 1986; Halitsky, 1989; Griffiths, 1994; Turner et al., 1989; Bowman, 1996)[47-59].

### *Artificial Neural Network Model*

Among all the above mentioned techniques, one of the commonly used methodologies is artificial neural network. In computer science and related fields, artificial neural networks are models inspired by animal central nervous systems that are capable of machine learning and pattern recognition. They are usually presented as systems of interconnected "neurons" that can compute values from inputs by feeding information through the network. Like other machine learning methods, neural networks have been used to solve a wide variety of tasks that are hard to solve using ordinary rule-based programming, including computer vision and speech recognition.

Generally, it involves a network of simple processing elements exhibiting complex global behavior determined by the connections between the processing elements and element parameters. Commonly, though, a class of statistical models will be called "neural" if:

1. They consist of sets of adaptive weights, i.e. numerical parameters that are tuned by a learning algorithm, and
2. They are capable of approximating non-linear functions of their inputs.

The adaptive weights are conceptually connection strengths between neurons, which are activated during training and prediction.

Neural network models in artificial intelligence are usually referred to as artificial neural networks (ANNs); these are essentially simple mathematical models defining a function of $f : X \rightarrow Y$ or a distribution over X or both X and Y, but sometimes models are also intimately associated with a particular learning algorithm or learning rule. A common use of the phrase ANN model really means the definition of a class of such functions where members of the class are obtained by varying parameters, connection weights, or specifics of the architecture such as the number of neurons or their connectivity.

The word network in the term 'artificial neural network' refers to the inter–connections between the neurons in the different layers of each system. An example system has three layers. The first layer has input neurons, which send data via synapses to the second layer of neurons, and then via more synapses to the third layer of output neurons. More complex systems will have more layers of neurons with some having increased layers of input neurons and output neurons. The synapses store parameters called "weights" that manipulate the data in the calculations.

An ANN is typically defined by three types of parameters:

1. The interconnection pattern between different layers of neurons.
2. The learning process for updating the weights of the interconnections.
3. The activation function that converts a neuron's weighted input to its output activation.

To train the artificial neural network we obtain a generalized transfer function of the values of emission rate and the pollutant concentrations which are normalized in the procedure of model preparation. The normalized features were used to give the training on the neural network with the Lavenberg - Marquardt back propagation algorithm. Different numbers of hidden neurons were used and we found a good approximation and generalization with 1000 neurons. Riva-Rocci based measurement is used as the target values. The measurement of the dynamic emission rate changes is not easy because the systolic and diastolic values are measured at different time instances. There are total 7000 samples have participated the modeling of the artificial neural network. 70% of the samples are training samples, which are presented to the network during training, and the network is adjusted according to its error.15% of the samples are validation samples, which are used to measure network generalization, and to halt training when generalization stops improving. The rest 15% of the samples are testing samples. These have no effect on training and so provide an independent measure of network performance during and after training. (see Figures 1, 2, & 3)

15% of the samples are validation samples, which are used to measure network generalization, and to halt training when generalization stops improving. The rest 15% of the samples are testing samples. These have no effect on training and so provide an independent measure of network performance during and after training.

## CASE STUDY

To demonstrate the efficacy of the developed systematic methodology, two case studies including the detection of a real $SO_2$ emission event are conducted.

As shown in Figure 4, the studied involves five chemical emission sources (E1, E2, E3, E4, and E5 represented by red dots) and four monitoring stations (S1, S2, S3, and S4 represented by green dots) distributed in a squared region (30 km×30

*Figure 1. Percentage selection of the samples for validation and test data*

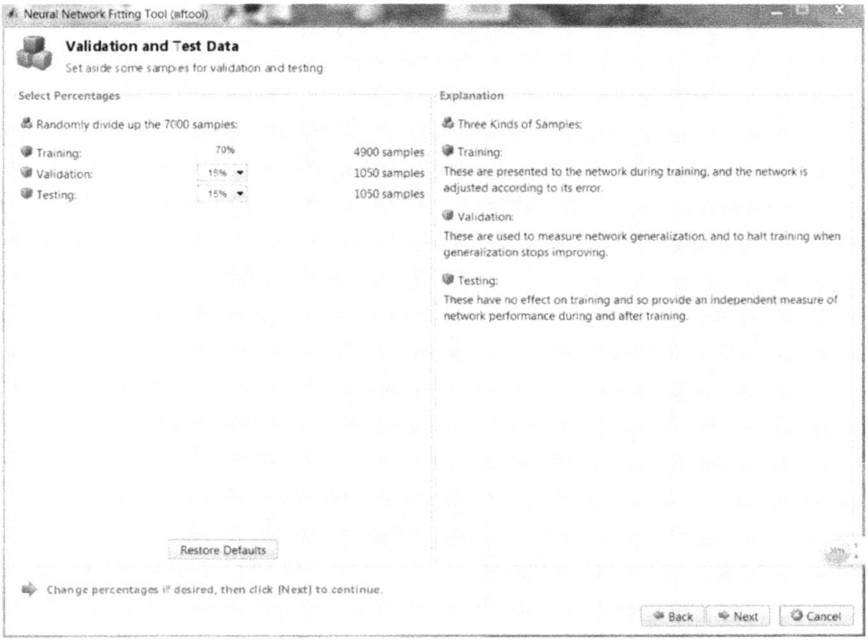


*Figure 2. Setting for the hidden neurons*

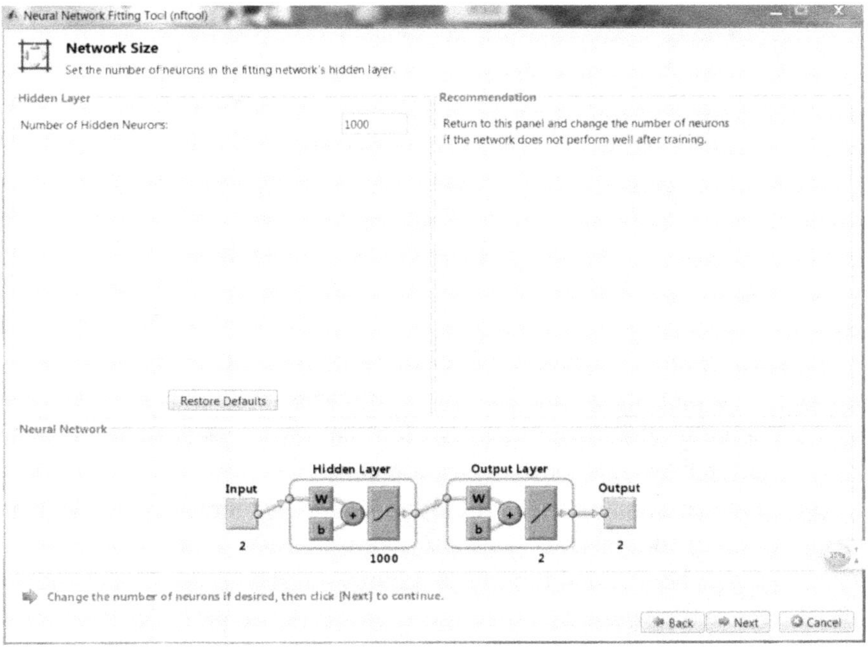

*Figure 3. Training neural network*

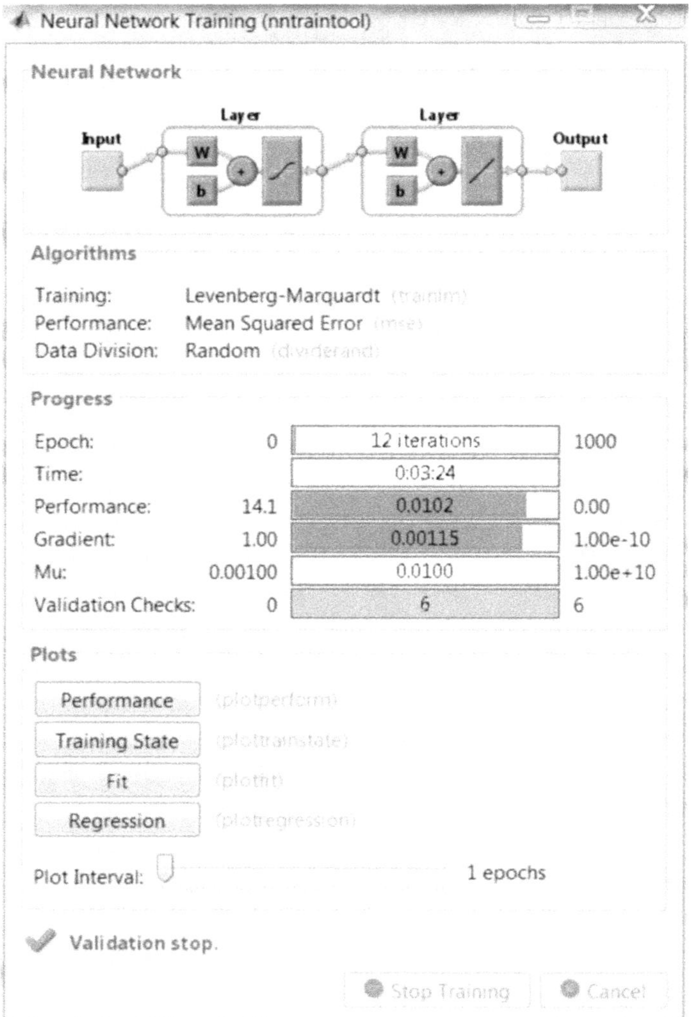


km). The entire region is gridded and the edge length of each gridded cell is 1 km. The surface wind blows from the southwest to northeast as shown in Figure 9. A common pollutant emitted from these five emission sources is monitored by the monitoring stations through hourly measurements.

The plume and stack parameters of the five emission sources are given in Table 2. During the event time period, the lapse rate was 4 K/km, the wind speed was 1.6m/s at 10-meter height, and the ambient temperature was 20 °C. For the investigation, the entire time period has been separated into two parts: steady-state time

*Figure 4. Spatial scope of case study*

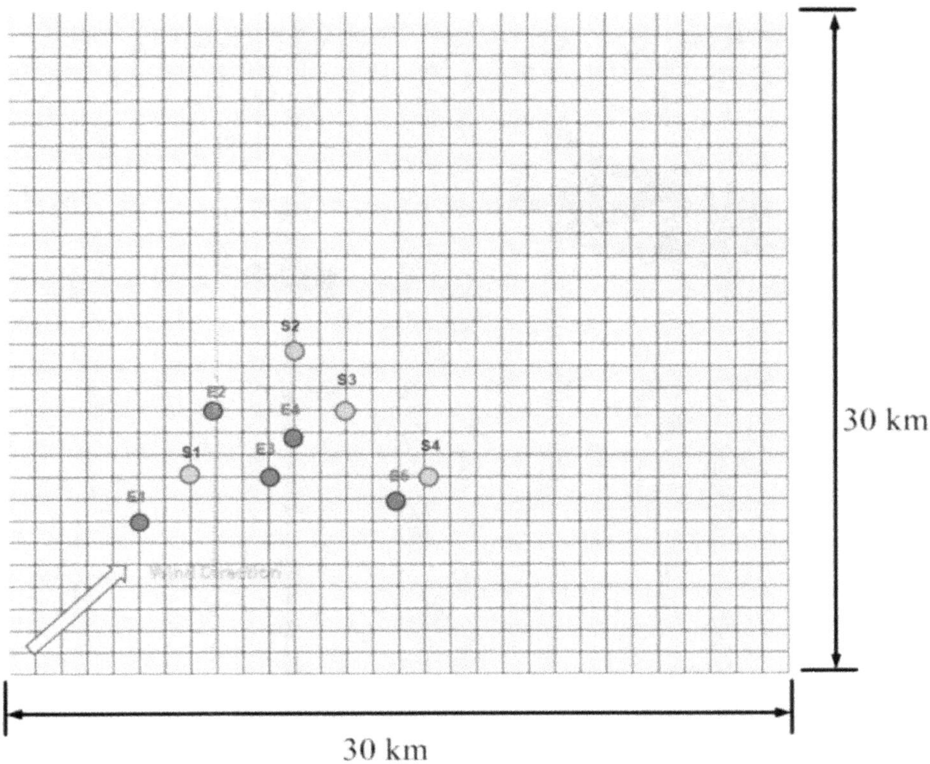


*Table 2. Plume and stack parameters for each emission source*

| Chemical Emission Sources | E1 | E2 | E3 | E4 | E5 |
|---|---|---|---|---|---|
| Stack Height $H_i$ *(m)* | 80 | 110 | 95 | 100 | 105 |
| Stack Exit Temperature $T_{s,i}$ *(K)* | 480 | 400 | 460 | 440 | 430 |
| Stack Exit Velocity $V_{s,i}$ *(m/s)* | 17.5 | 13.0 | 15.6 | 14.2 | 16.1 |
| Stack Exit Diameter $D_{s,i}$ *(m)* | 1.6 | 1.9 | 1.5 | 1.7 | 1.8 |

period and dynamic time period. The data in the steady-state has been applied to determine the normal emission rates for each emission source.

*Figure 5. Regression Result for Training, Validation and Test*

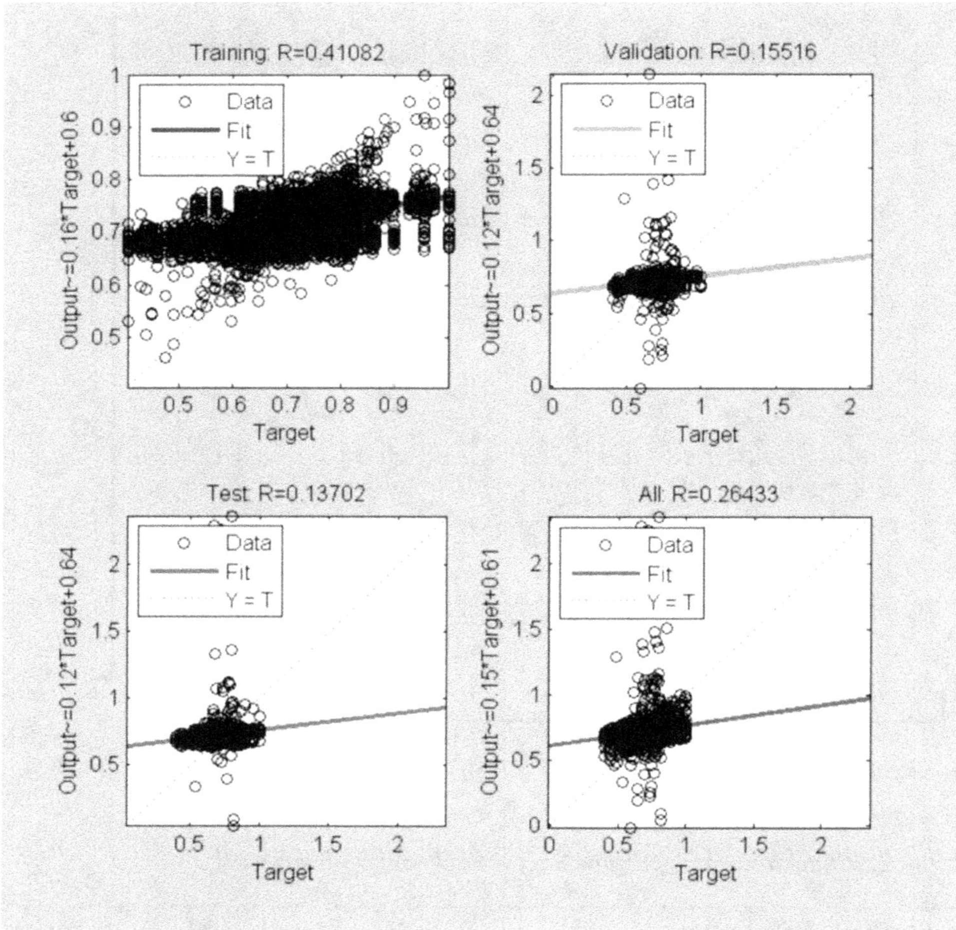


The modeling result for the developed artificial neural network has been plot in Figure 5. It can be seen that the training has the R square value of 0.41802 while the validation and test has the R-square of 0.15516 and 0.13702 respectively. It can be seen that the R square value among the periods of training, validation and test are quite similar and the overall R square value has been 0.16433.This has shown that the artificial neural network model can give the quantitative identification of the normal emission rate. The results of $m_1^S$ through $m_5^S$ are identified as 10.8, 10.5, 10.1, 9.2, and 9.8 kg/h, respectively.

## CONCLUSION

With the introduction of ICTs and their application in the math education, we hope this chapter will help the educators to create e-learning environment to prepare for the students' future engineering education. It is the most important to provide effective ways to ask the students to combine technology with mathematics through an interdisciplinary project activity.

## REFERENCES

*African Virtual University*. (2013). Retrieved from http://www.avu.org/Certificate/Diploma/ ict-integration-in-education-option-mathematics.html

Bar-Cohen, Y. (2005). *Biomimetics: Biologically inspired technologies*. CRC Press.

Brabazon, A., & O'Neill, M. (2006). *Biologically inspired algorithms for financial modelling*. Berlin: Springer.

Chung, C. J., & Reynolds, R. G. (1996, February). A testbed for solving optimization problems using cultural algorithms. In Evolutionary Programming (pp. 225-236). Academic Press.

Clerc, M. (2006). *Particle swarm optimization* (Vol. 243). London: ISTE. doi:10.1002/9780470612163

Cochocki, A., & Unbehauen, R. (1993). *Neural networks for optimization and signal processing*. John Wiley & Sons, Inc.

Coello, C. A. C., & Cortés, N. C. (2002, September). An approach to solve multiobjective optimization problems based on an artificial immune system. In *Proceedings of 1st International Conference on Artificial Immune Systems (ICARIS)*. University of Kent at Canterbury.

Coello, C. A. C., & Cortés, N. C. (2005). Solving multiobjective optimization problems using an artificial immune system. *Genetic Programming and Evolvable Machines*, *6*(2), 163–190. doi:10.1007/s10710-005-6164-x

Coello Coello, C. A., & Becerra, R. L. (2004). Efficient evolutionary optimization through the use of a cultural algorithm. *Engineering Optimization*, *36*(2), 219–236. doi:10.1080/03052150410001647966

Dasgupta, D. (Ed.). (1999). *An overview of artificial immune systems and their applications*. Springer. doi:10.1007/978-3-642-59901-9

De Castro, L. N., & von Zuben, F. J. (Eds.). (2005). *Recent developments in biologically inspired computing*. IGI Global.

Deb, K. (2001). Multi-objective optimization. In *Multi-objective optimization using evolutionary algorithms*, (pp. 13-46). Academic Press.

Dorigo, M. (Ed.). (2006). *Ant colony optimization and swarm intelligence: 5th International workshop, ANTS 2006,* (Vol. 4150). Springer-Verlag.

Dorigo, M. (2007). Ant colony optimization. *Scholarpedia*, *2*(3), 1461. doi:10.4249/scholarpedia.1461

Dorigo, M., & Birattari, M. (2010). Ant colony optimization. In Encyclopedia of machine learning (pp. 36-39). Springer US.

Dorigo, M., & Di Caro, G. (1999). Ant colony optimization: a new meta-heuristic. In *Proceedings of the 1999 Congress on Evolutionary Computation* (*Vol. 2*). IEEE.

Goldberg, D. E., & Holland, J. H. (1988). Genetic algorithms and machine learning. *Machine Learning*, *3*(2/3), 95–99. doi:10.1023/A:1022602019183

Hofmeyr, S. A., & Forrest, S. (2000). Architecture for an artificial immune system. *Evolutionary Computation*, *8*(4), 443–473. doi:10.1162/106365600568257 PMID:11130924

Holler, M., Tam, S., Castro, H., & Benson, R. (1989, June). An electrically trainable artificial neural network (etann) with 10240'floating gate'synapses. In *Proceedings of Neural Networks,* (pp. 191-196). IEEE.

Jhuang, H., Serre, T., Wolf, L., & Poggio, T. (2007, October). A biologically inspired system for action recognition. In *Proceedings of Computer Vision,* (pp. 1-8). IEEE. doi:doi:10.1109/ICCV.2007.4408988 doi:10.1109/ICCV.2007.4408988

Jin, X., & Reynolds, R. G. (1999). Using knowledge-based evolutionary computation to solve nonlinear constraint optimization problems: a cultural algorithm approach. In *Proceedings of the 1999 Congress on Evolutionary Computation* (*Vol. 3*). IEEE.

Kennedy, J. (2010). Particle swarm optimization. In Encyclopedia of machine learning (pp. 760-766). Springer US.

Lazinica, A. (Ed.). (2009). *Particle swarm optimization*. InTech. doi:10.5772/109

Lei, X. R. B. W. (2002). Artificial immune system: Principle, models, analysis and perspectives. *Chinese Journal of Computers, 12*, 000.

Leibnitz, K., Wakamiya, N., & Murata, M. (2006). Biologically inspired self-adaptive multi-path routing in overlay networks. *Communications of the ACM*, *49*(3), 62–67. doi:10.1145/1118178.1118203

Nagpal, R. (2002, July). Programmable self-assembly using biologically-inspired multiagent control. In *Proceedings of the First International Joint Conference on Autonomous Agents and Multiagent Systems: Part 1* (pp. 418-425). ACM. doi:doi:10.1145/544838.544839 doi:10.1145/544838.544839

Patterson, D. W. (1998). *Artificial neural networks: theory and applications*. Prentice Hall PTR.

Perfilieva, I., & Močkoř, J. (1999). *Mathematical principles of fuzzy logic*. Springer.

Poli, R., Kennedy, J., & Blackwell, T. (2007). Particle swarm optimization. *Swarm Intelligence*, *1*(1), 33–57. doi:10.1007/s11721-007-0002-0

Reynolds, R. G. (1994, February). An introduction to cultural algorithms. In *Proceedings of the third annual conference on evolutionary programming* (pp. 131-139). World Scientific.

Reynolds, R. G., & Sverdlik, W. (1994, June). Problem solving using cultural algorithms. In *Proceedings of the First IEEE Conference on Evolutionary Computation* (pp. 645-650). IEEE.

Rieser, C. J. (2004). *Biologically inspired cognitive radio engine model utilizing distributed genetic algorithms for secure and robust wireless communications and networking*. (Doctoral dissertation). Virginia Polytechnic Institute and State University, Blacksburg, VA.

Roveri, N., Falini, G., Sidoti, M. C., Tampieri, A., Landi, E., Sandri, M., & Parma, B. (2003). Biologically inspired growth of hydroxyapatite nanocrystals inside self-assembled collagen fibers. *Materials Science and Engineering C*, *23*(3), 441–446. doi:10.1016/S0928-4931(02)00318-1

Rowe, M. P., Pugh, E. N. Jr, Tyo, J. S., & Engheta, N. (1995). Polarization-difference imaging: A biologically inspired technique for observation through scattering media. *Optics Letters*, *20*(6), 608–610. doi:10.1364/OL.20.000608 PMID:19859271

Sarle, W. S. (1994). *Neural networks and statistical models*. Academic Press.

Stevens, R., & Casillas, A. (2006). *Artificial neural networks. Automated Scoring of Complex Tasks in Computer Based Testing: An Introduction* (pp. 259–312). Mahwah, NJ: Lawrence Erlbaum.

Timmis, J., & Neal, M. (2001). A resource limited artificial immune system for data analysis. *Knowledge-Based Systems*, *14*(3-4), 121–130. doi:10.1016/S0950-7051(01)00088-0

Timmis, J., Neal, M., & Hunt, J. (2000). An artificial immune system for data analysis. *Bio Systems*, *55*(1-3), 143–150. doi:10.1016/S0303-2647(99)00092-1 PMID:10745118

Venter, G., & Sobieszczanski-Sobieski, J. (2003). Particle swarm optimization. *AIAA Journal*, *41*(8), 1583–1589. doi:10.2514/2.2111

Vose, M. D. (1999). *The simple genetic algorithm: foundations and theory* (Vol. 12). The MIT Press.

Wang, L., & Fu, K. (2008). *Artificial neural networks*. John Wiley & Sons, Inc.

Whitley, D. (1994). A genetic algorithm tutorial. *Statistics and Computing*, *4*(2), 65–85. doi:10.1007/BF00175354

Yang, X., & Zheng, J. (2009). Artificial neural networks. In Handbook of Research on Geoinformatics. Academic Press.

Zadeh, L. A. (1965). Fuzzy sets. *Information and Control*, *8*(3), 338–353. doi:10.1016/S0019-9958(65)90241-X

Zalta, E. N. (Ed.). (2003). Stanford encyclopedia of philosophy. Academic Press.

## KEY TERMS AND DEFINITIONS

**Ant Colony Optimization (ACO):** ACO is a probabilistic technique for solving computational problems which can be reduced to finding good paths through graphs.

**Artificial Immune System (AIS):** AISs are a class of computationally intelligent systems inspired by the principles and processes of the vertebrate immune system.

**Artificial Neural Networks:** In computer science and related fields, artificial neural networks are models are derived from animal central nervous systems. The biologically neural networks are capable of machine learning and pattern recognition. They can be regarded as systems of internally connected neurons. They can compute values from inputs by feeding information through the network.

**Clonal Selection Algorithm:** A class of algorithms inspired by the clonal selection theory of acquired immunity that explains how B and T lymphocytes improve their response to antigens over time called affinity maturation.

**Culture Algorithm (CA):** CAs are a branch of evolutionary computation where there is a knowledge component that is called the belief space in addition to the population component.

**Dendritic Cell Algorithms (DCA):** DCA is an example of an immune inspired algorithm developed using a multi-scale approach, which is based on an abstract model of dendritic cells (DCs).

**Fuzzy Logic System:** Fuzzy logic is a logic form of multiple values. It helps to handle the problems which are not fixed and exact. Compared to traditional binary variables, fuzzy logic variables may have a truth value that ranges in degree between 0 and 1. It can be regarded as the concept of partial truth, where the truth value may range between completely true and completely false.

**Genetic Algorithms (GA):** In the field of artificial intelligence, a GA is a search heuristic with the process of natural selection. This heuristic is normally used to generate useful solutions to optimization and search problems. Genetic algorithms is one of the algorithm of evolutionary algorithms (EA).

**Immune Network Algorithms:** Algorithms inspired by the idiotypic network theory proposed by Niels Kaj Jerne that describes the regulation of the immune system by anti-idiotypic antibodies (antibodies that select for other antibodies).

**Negative Selection Algorithm:** Inspired by the positive and negative selection processes that occur during the maturation of T cells in the thymus called T cell tolerance.

**Particle Swarm Optimization (PSO):** PSO helps to optimize a problem by iteratively trying to improve a candidate solution with regard to a given measure of quality. PSO optimizes a problem by having a population of candidate solutions of dubbed particles and moving these particles around in the search-space according to simple mathematical formula over the particle's position and velocity.

# Compilation of References

Abdeldayem, H. (2012). *Brain plasticity: Slideshare presentation*. Retrieved on 14th June 2013 from http://www.slideshare.net/husseindayem/brain-plasticity-12531271

Abrahams, I. (2011). *Practical work in school science: A minds-on approach*. London: Continuum.

Abrahams, I., & Millar, R. (2008). Does Practical Work Really Work? A study of the effectiveness of practical work as a teaching and learning method in school science. *International Journal of Science Education*, *30*(14), 1945–1969. doi:10.1080/09500690701749305

Acher, A., Arcà, M., & Sanmartí, N. (2007). Modeling as a teaching learning process for understanding materials: A case study in primary education. *Science Education*, *91*(3), 398–418. doi:10.1002/sce.20196

Ackerman, D. B. (Ed.). (1989). Intellectual and practical criteria for successful curriculum integration. In H. H. Jacobs (Ed.), Interdisciplinary curriculum: Design and implementation (pp. 25–38). Alexandria, VA: Association for Supervision and Curriculum Development.

Ackerman, D. B., & Perkins, D. N. (Eds.). (1989). Integrating thinking and learning skills across the curriculum. In H. H. Jacobs (Ed.), Interdisciplinary curriculum: Design and implementation (pp. 77–96). Alexandria, VA: Association for Supervision and Curriculum Development.

ACT. (2013). *The condition of college & career readiness 2013 national*. Retrieved from ACT Improve Yourself website: http://www.act.org/research/policymakers/cccr13/pdf/CCCR13-NationalReadinessRpt.pdf

Adolph, K. E., Bertenthal, B. I., Boker, S. M., Goldfield, E. C., & Gibson, E. J. (1997). Learning in the development of infant locomotion. *Monographs of the Society for Research in Child Development*, 162. PMID:9394468

*African Virtual University*. (2013). Retrieved from http://www.avu.org/Certificate/Diploma/ ict-integration-in-education-option-mathematics.html

Amabile, T. M. (1996). Creativity and innovation in organizations. Harvard Business School.

Amabile, T. M. (1982). Social psychology of creativity: A consensual assessment technique. *Journal of Personality and Social Psychology*, *43*(5), 997–1013. doi:10.1037/0022-3514.43.5.997

Amabile, T. M. (1988). A model of creativity and innovation in organizations. *Research in Organizational Behavior*, *10*, 123–167.

American Association for the Advancement of Science (AAAS). (2011). *Vision and Change in Undergraduate Biology Education*. Washington, DC: Author.

Amir, N., & Subramaniam, R. (2012). Fostering inquiry in science among kinaesthetic learners through design & technology. In L. C. Lennex & K. F. Nettleton (Eds.), Cases on Inquiry Through Instructional Technology in Math and Science: Systemic Approaches (pp 221 - 257). Hershey, PA: IGI Global Publishing.

Amir, N., & Subramaniam, R. (2007). Making a fun Cartesian diver: A simple project to engage kinaesthetic learners. *Physics Education*, *42*(5), 478–480. doi:10.1088/0031-9120/42/5/004

Amir, N., & Subramaniam, R. (2009). Making a low cost candy floss kit gets students excited about learning physics. *Physics Education*, *44*(4), 420–428. doi:10.1088/0031-9120/44/4/013

Andalécio, A. M. L. (2009). Transdisciplinarity in the university:discourse and practice. *RECIIS -Electronic Journal of Communication, Information & Innovation in Health*, *3*(3), 83–89.

Andersen, H. M., & Nielsen, B. L. (2013). Video-based analyses of motivation and interaction in science classrooms. *International Journal of Science Education*, *35*(6), 906–928. doi:10.1080/09500693.2011.627954

Anderson, N. (2004). Work with passion: How to do what you love for a living (Revised and Expanded ed.). Novato, CA: New World Library.

Anderson, L. W., & Krathwohl, D. R. (Eds.). (2001). *A taxonomy for learning, teaching, and assessing: A revision of Bloom's taxonomy of educational objectives*. New York, NY: Longman.

Anderson, R. D. (2002). Reforming science teaching: What research says about inquiry. *Journal of Science Teacher Education*, *13*(1), 1–12. doi:10.1023/A:1015171124982

Angelo, T. A., & Cross, K. P. (1993). *Classroom assessment techniques: A handbook for college teachers* (2nd ed.). San Francisco, CA: Jossey-Bass.

Ang, K. C. S., & van Reyk, D. (2013). 'Teach Me Chemistry Like a Ladder and Make it Real' - Barriers and Motivations Students Face in Learning Chemistry for Bioscience. *International Journal of Innovation in Science and Mathematics Education*, *21*(2), 1–12.

Appleton, K. (2007). Elementary science teaching. In S. Abell, & N. Lederman (Eds.), *Handbook of research on science education* (pp. 493–535). Mahwah, NJ: Lawrence Erlbaum Associates, Inc.

Armstrong, N., Chang, S. M., & Brickman, M. (2007). Cooperative learning in industrial-sized biology classes. *CBE Life Sciences Education*, *6*(2), 163–171. doi:10.1187/cbe.06-11-0200 PMID:17548878

Arnold, M., & Millar, R. (1987). Being constructive: An alternative approach to the teaching of introductory ideas in electricity. *International Journal of Science Education*, *9*(5), 553–563. doi:10.1080/0950069870090505

Association of Graduate Recruiters Report. (1995). *Skills for Graduates in the 21st Century*. Cambridge, UK: Association of Graduate Recruiters.

Atkins, P. (2003). *Atkins' Molecules* (2nd ed.). Cambridge, UK: Cambridge University Press.

Austin, L. B., & Shore, B. M. (1995). Using concept mapping for assessment in physics. *Physics Education*, *30*(1), 41–45. doi:10.1088/0031-9120/30/1/009

Australian Curriculum, Assessment and Reporting Authority (ACARA). (2013). *Australian curriculum: Science*. Retrieved on 14th June 2013 from http://www.australiancurriculum.edu.au/Science/Curriculum/F-10

Ausubel, D. P. (2000). *The Acquisition and Retention of Knowledge: a cognitive view*. Dordrecht, The Netherlands: Kluwer Academic Publishers. doi:10.1007/978-94-015-9454-7

Baddeley, A. D. (2003). Working memory: looking back and looking forward. *Nature Reviews Neuroscience, 4*(10), 829-839.

Balchin, T. (2005). A creativity feedback package for teachers and students of design and technology (in the UK). *Design and Technology Education: An International Journal.*, *10*(2), 31–43.

Baloche, L., Hynes, J. L., & Berger, H. A. (1996). Moving toward the integration of professional and general education. *Action in Teacher Education*, *18*(1), 1–9. doi:10.1080/01626620.1996.10462817

Bandura, A. (1977). Self-efficacy: Toward a unifying theory of behavioral change. *Psychological Review*, *84*(2), 191–215. doi:10.1037/0033-295X.84.2.191 PMID:847061

Bandura, A. (1986). *Social foundations of thought and action: A social cognitive theory*. Englewood Cliffs, NJ: Prentice-Hall.

Bandura, A. (1993). Perceived self-efficacy in cognitive development and functioning. *Journal of Educational Psychologist*, *28*(2), 117–148. doi:10.1207/s15326985ep2802_3

Barak, M., & Doppelt, Y. (2000). Using portfolios to enhance creative thinking. *Journal-of-Technology-Studies*, *26*(2), 16–25.

Bar-Cohen, Y. (2005). *Biomimetics: Biologically inspired technologies*. CRC Press.

Barlex, D. (2007). Creativity in school design and technology in England: A discussion of Influences. *International Journal of Technology and Design Education*, *17*(2), 149–162. doi:10.1007/s10798-006-0006-x

Barmby, P., Kind, P. M., & Jones, K. (2008). Examining Changing Attitudes in Secondary School Science. *International Journal of Science Education*, *30*(8), 1075–1093. doi:10.1080/09500690701344966

Barr, J. (2010). Host Your Web Site. In *The Cloud: Amazon Web Services Made Easy: Amazon EC2 Made Easy*. Melbourne: SitePoint.

Barrows, H. S. (1996). Problem-based learning in medicine and beyond: A brief overview. *New Directions for Teaching and Learning*, (68), 3-12.

Barrows, H., & Kelson, A. (1980). *Problem-based learning: A total approach to education*. New York: Springer.

Barufaldi, J. P., Bethel, L. J., & Lamb, W. G. (1977). The effect of a science methods course on the philosophical view of science among elementary education majors. *Journal of Research in Science Teaching*, *14*(4), 289–294. doi:10.1002/tea.3660140404

BBC News. (2006). *Decline in student science uptake*. Retrieved 20 December 2013, from http://news.bbc.co.uk/1/hi/scotland/5124108.stm

Bell, B., & Cowie, B. (2001). *Formative Assessment and Science Education*. Dordrecht, The Netherlands: Kluwer Academic Publishers.

Belt, S. T., Clarke, M. J., & Phipps, L. E. (1999). Exercises for chemists involving time management, judgment and initiative. *University Chemistry Education*, *3*(2), 52–58.

Bennett, J., & Hogarth, S. (2009). Would You Want to Talk to a Scientist at a Party? High school students' attitudes to school science and to science. *International Journal of Science Education*, *31*(14), 1975–1998. doi:10.1080/09500690802425581

Bennett, J., Lubben, F., & Hampden-Thompson, G. (2013). Schools That Make a Difference to Post-Compulsory Uptake of Physical Science Subjects: Some comparative case studies in England. *International Journal of Science Education*, *35*(4), 663–689. doi:10.1080/09500693.2011.641131

Berliner, D. (1988). *The development of expertise in pedagogy*. Washington, DC: AACTE Publications.

Besemer, S. P. (2010). Review of fostering creativity by Arthur and David Cropley. *Creativity Research Journal*, *22*(2), 236–237. doi:10.1080/10400419.2010.481543

Beswick, J. (2011). *Google Apps Express: The Fast Way To Start Working in the Cloud*. CreateSpace.

Bianchini, J. A., & Colburn, A. (2000). Teaching the nature of science through inquiry to prospective elementary teachers: A tale of two researchers. *Journal of Research in Science Teaching*, *37*(2), 177–209. doi:10.1002/(SICI)1098-2736(200002)37:2<177::AID-TEA6>3.0.CO;2-Y

Bidell, T. R., & Fischer, K. W. (1992). Beyond the stage debate: Action, structure, and variability in Piagetian theory and research. In R. Sternberg, & C. Berg (Eds.), *Intellectual development* (pp. 100–140). New York: Cambridge University Press.

Black, P., & Wiliam, D. (1998). Inside the Black Box: Raising Standards Through Classroom Assessment. *Phi Delta Kappan*, *80*(2), 139–148.

Black, P., & Wiliam, D. (2009). Developing the theory of formative assessment. *Educational Assessment, Evaluation and Accountability*, *21*(1), 5–31. doi:10.1007/s11092-008-9068-5

Bloom, B. S., Engelhart, M. D., Furst, E. J., Hill, W. H., & Krathwohl, D. R. (1956). *Taxonomy of educational objective: The classification of educational goals, handbook I: cognitive domain*. New York, NY: McKay.

Bondelli, K. (2007). *An Evaluation of the Traditional Education System*. Scribd. Retrieved from http://www.scribd.com/doc/38418/An-Evaluation-of-the-Traditional-Education-System-by-Kevin-Bondelli

Bonner, D. (2010). Increasing student engagement and enthusiasm: A projectile motion crime scene. *The Physics Teacher*, *48*(5), 324–325. doi:10.1119/1.3393066

Bonwell, C. C., & Eison, J. A. (1991). ASHE-ERIC Higher Education Report: Vol. 1. *Active learning: Creating excitement in the classroom*. Washington, DC: George Washington University.

Bowen, C. W. (2000). A quantitative literature review of cooperative learning effects on high school and college chemistry achievement. *Journal of Chemical Education*, *77*(1), 116–119. doi:10.1021/ed077p116

Brabazon, A., & O'Neill, M. (2006). *Biologically inspired algorithms for financial modelling*. Berlin: Springer.

Bransford, J. D., Brown, A. L., & Cocking, R. R. (2000). *How People Learn: Brain, Mind, Experience, and School Committee on Developments in the Science of Learning*. Washington, DC: National Academy Press.

Brattan, D., Mason, D., & Rest, A. J. (1999). Changing the nature of physical chemistry practical work. *University Chemistry Education*, *3*(2), 59–63.

Brent, R., & Felder, R. M. (1992). Writing assignments—Pathways to connections, clarity, creativity. *College Teaching*, *40*(2), 43–47. doi:10.1080/87567555.1992.10532264

Brook, A., & Driver, R. (1984). *Aspects of Secondary Students' Understanding of Energy: Full Report*. Leeds, UK: Centre for Studies in Science and Mathematics Education, University of Leeds.

Browne, K., & Jackson, D. P. (2007). Simple experiments to help students understand magnetic phenomena. *The Physics Teacher*, *45*(7), 425. doi:10.1119/1.2783151

Brown, T., Rushton, G., & Bencomo, M. (2008). Mighty Molecule Models. *Science and Children*, *45*(5), 33–37.

Bruner, J. (1960). *The process of education*. Cambridge, MA: Harvard University Press.

Budde, M., Niedderer, H., Scott, P., & Leach, J. (2002). 'Electronium': A quantum atomic teaching model. *Physics Education*, *37*(3), 197–203. doi:10.1088/0031-9120/37/3/303

Bursal, M. (2008). Changes in Turkish preservice elementary teachers' personal science teaching efficacy beliefs and science anxieties during a science method course. *Journal of Turkish Science Education*, *5*(1), 99–112.

Cabrera, A. F., Crissman, J. L., Bernal, E. M., Nora, A., Terenzini, P. T., & Pascarella, E. T. (2002). Collaborative learning: Its impact on college students' development and diversity. *Journal of College Student Development*, *43*, 20–34.

Campbell, T., & Campbell, D. (1997). Faculty/Student mentor program: Effects on academic performance and retention. *Research in Higher Education*, *38*(6), 727–742. doi:10.1023/A:1024911904627

Canning, R. (1999). Post-16 Education in Scotland: Credentialism and inequality. *Journal of Vocational Education and Training*, *51*(2), 185–198. doi:10.1080/13636829900200083

Carey, S. (2010). Beyond fast mapping. *Language Learning and Development*, *6*(3), 184–205. doi:10.1080/15475441.2010.484379 PMID:21625404

Carnevale, A. P., Smith, N., & Strohl, J. (2010). *Help wanted: Projections of jobs and education requirements through 2018*. Washington, DC: Georgetown University Center on Education and the Workforce.

Cassels, J. R. T., & Johnstone, A. H. (1984). The Effect of Language on Student Performance on Multiple Choice Tests in Chemistry. *Journal of Chemical Education, 61*(7), 613–615. doi:10.1021/ed061p613

Chapman, S., & Lewis, M. (2001). The McOhm: Using fast food to explain resistance. *Physics Education, 36*(3), 255–260.

Chasen, M., & Pittinsky, M. (2014). *Blackboard Learn* (Version 9.1) [Computer software]. Blackboard Inc. Retrieved from http://www.blackboard.com

Chen, Z., & Siegler, R. (2000). Across the great divide: Bridging the gap between understanding the thought of toddlers and older children. *Monographs of the Society for Research in Child Development, 65*(2).

Childs, P. E., & Sheehan, M. (2009). What's difficult about chemistry? An Irish perspective. *Chemistry Education Research and Practice, 10*(3), 204–218. doi:10.1039/b914499b

Chi, M. T. H. (2008). Three types of conceptual change: belief revision, mental model transformation, and categorical shift. In S. Vosniadou (Ed.), *International Handbook of Research on Conceptual Change* (pp. 61–82). New York: Routledge.

Chi, M. T. H., Slotta, J. D., & de Leeuw, N. (1994). From things to processes; a theory of conceptual change for learning science concepts. *Learning and Instruction, 4*(1), 27–43. doi:10.1016/0959-4752(94)90017-5

Chung, C. J., & Reynolds, R. G. (1996, February). A testbed for solving optimization problems using cultural algorithms. In Evolutionary Programming (pp. 225-236). Academic Press.

Clerc, M. (2006). *Particle swarm optimization* (Vol. 243). London: ISTE. doi:10.1002/9780470612163

Cochocki, A., & Unbehauen, R. (1993). *Neural networks for optimization and signal processing*. John Wiley & Sons, Inc.

Cockman, J. (2002). A bright color mixer. *The Physics Teacher, 40*(9), 553. doi:10.1119/1.1534824

Coello, C. A. C., & Cortés, N. C. (2002, September). An approach to solve multiobjective optimization problems based on an artificial immune system. In *Proceedings of 1st International Conference on Artificial Immune Systems (ICARIS).* University of Kent at Canterbury.

Coello, C. A. C., & Cortés, N. C. (2005). Solving multiobjective optimization problems using an artificial immune system. *Genetic Programming and Evolvable Machines, 6*(2), 163–190. doi:10.1007/s10710-005-6164-x

Coffey, T. (2008). Soda pop fizz-ics. *The Physics Teacher, 46*(8), 473–476. doi:10.1119/1.2999062

College Development Network. (2013). College Development Network. Retrieved 21 December 2013, from http://www.collegedevelopmentnetwork.ac.uk/cdn-homepage.html

Colleges Scotland. (2013). *Colleges Scotland.* Retrieved 21 December 2013, from http://www.collegesscotland.ac.uk/colleges-scotland-homepage.html

Collins, H. (2010). *Tacit and Explicit Knowledge*. Chicago: The University of Chicago Press. doi:10.7208/chicago/9780226113821.001.0001

Common Core State Standards Initiative. (2010). *Common core state standards for English language arts and literacy in history/social studies, science, and technical subjects*. Washington, DC: Council of Chief State School Officers and National Governors Association. Retrieved from http://www.ccrestandards.org/

Connelly, G., & Halliday, J. (2001). Reasons for Choosing Further Education: The views of 700 new entrants. *Journal of Vocational Education and Training*, *53*(2), 181–192. doi:10.1080/13636820100200155

Cooper, G. M., & Hausman, R. E. (2009). *The Compositions of Cells. The Cell: A Molecular Approach*. Washington, DC: ASM Press.

Cooper, J. (2007). *Cognitive Dissonance: Fifty years of a classic theory*. London: Sage.

Corrao, C. T. (2010). The physicist fish. *The Physics Teacher*, *48*(7), 491. doi:10.1119/1.3488204

Costa, A., & Kallick, B. (Eds.). (2000). *Discovering and Exploring Habits of Mind*. Alexandria, VA: ASCD.

Cox, B. E., McIntosh, K. L., Reason, R. D., & Terenzini, P. T. (2011). A culture of teaching: Policy, perception, and practice in higher education. *Research in Higher Education*, *52*(8), 808–829. doi:10.1007/s11162-011-9223-6

Craft, A. (2001). *An analysis of research and literature on creativity in education*. Qualifications and Curriculum Authority. Retrieved September 1, 2013, from http://www.euvonal.hu/images/creativity_report.pdf

Craven, J., & Penck, J. (2001). Preparing new teachers to teach science: The role of the science teacher educator. *Electronic Journal of Science Education*, *6*(1). Retrieved from http://ejse.southwestern.edu/article/view/7670

Creamer, E. G., & Lattuca, L. R. (2005). *Advancing Faculty Learning Through Interdisciplinary Collaboration: New Directions for Teaching and Learning, No. 102*. San Francisco, CA: Jossey-Bass Publisher.

Creswell, J. W. (2005). *Educational Research: Planning, conducting and evaluating quantitative and qualitative research* (2nd ed.). Upper Saddle River, NJ: Pearson Education.

Cronbach, L. J. (1970). *Essentials of psychological testing* (3rd ed.). New York, NY: Harper & Row.

Cropley, D. H., & Cropley, A. J. (2000). Fostering creativity in engineering undergraduates. *High Ability Studies*, *11*(2), 207–219. doi:10.1080/13598130020001223

Cross, K. P. (1981). *Adults as Learners: Increasing Participation and Facilitating Learning*. London: Jossey-Bass.

Cross, N. (2007). *Designerly ways of knowing*. London, U.K.: Springer- Verlag.

Cross, S. E., Jin, Y., Tondre, J., Wong, R., Rao, J., & Gimzewski, J. K. (2008). AFM-based analysis of human metastatic cancer cells. *Nanotechnology*, *19*(38), 384003. doi:10.1088/0957-4484/19/38/384003 PMID:21832563

Crowe, A., Dirks, C., & Wenderoth, M. P. (2008). Biology in bloom: Implementing Bloom's Taxonomy to enhance student learning in biology. *CBE Life Sciences Education*, *7*(4), 368–381. doi:10.1187/cbe.08-05-0024 PMID:19047424

Crown Copyright. (1992). *Further and Higher Education (Scotland) Act 1992*. Retrieved 21 December 2013, from http://www.legislation.gov.uk/ukpga/1992/37/contents

Crown Copyright. (2005). *Further and Higher Education (Scotland) Act 2005*. Retrieved 21 December 2013, from http://www.legislation.gov.uk/asp/2005/6/contents

Csikszentmihalyi, M. (1998). Society, culture and person: a systems view of creativity. In R. J. Sternberg (Ed.), *The nature of creativity* (pp. 325–339). Cambridge: Cambridge University Press.

Czerniak, C. M., & Schriver, M. L. (1994). An examination of pre-service science teachers' beliefs and behaviors as related to self-efficacy. *Journal of Science Teacher Education*, *5*(3), 77–86. doi:10.1007/BF02614577

Dacey, J., & Lennon, K. (2000). *Understanding creativity: the interplay of biological, psychological and social factors*. Buffalo, NY: Creative Education Foundation.

Daley, B. (2004). A project-based approach: Students describe the physics in movies. *The Physics Teacher*, *42*(1), 41–44. doi:10.1119/1.1639969

Daniel, D. B. (2012). Promising principle: Translating the science of learning to educational practice. *Journal of Applied Research in Memory and Cognition*, *1*(4), 251–253. doi:10.1016/j.jarmac.2012.10.004

Danili, E., & Reid, N. (2004, November). Some strategies to improve performance in school chemistry, based on two cognitive factors. *Research in Science & Technological Education*, *22*(2), 203–226. doi:10.1080/0263514042000290903

Dasgupta, D. (Ed.). (1999). *An overview of artificial immune systems and their applications*. Springer. doi:10.1007/978-3-642-59901-9

De Castro, L. N., & von Zuben, F. J. (Eds.). (2005). *Recent developments in biologically inspired computing*. IGI Global.

De Grave, W. S., Dolmans, D. H. J. M., & van der Vleuten, C. P. M. (2002). Student perspectives on critical incidents in the tutorial group. *Advances in Health Sciences Education: Theory and Practice*, *7*(3), 201–209. doi:10.1023/A:1021104201303 PMID:12510142

De Rennes, J. (1999). *Maintaining Space for Adult Learners in Science*. Paper presented at SCUTREA 29th Annual Conference. Warwick, UK.

de Silva, E. (1999). *Understanding the perception of chemistry undergraduates to increase the number of learners*. (Post-graduate Certificate in Education Thesis). University of Manchester, Manchester, UK.

de Silva, E. (2006). *Proceedings from Annual AP Conference: Applied Research in Science and Future Teaching*. College Board.

de Silva, E. (2007, March). *Teaching Physics through Research*. Paper presented at the Annual Physics Day at Commonwealth Governor's School. Fredericksburg, VA.

de Silva, E., & Christian, S. (2009, October). *Application of physics for a homerun*. Paper presented at the Tennessee Academy of Science Annual Conference. Knoxville, TN.

de Silva, E., & de Silva, E. (2009a). *Proceedings fromTennesseeAssociation of Science's Annual Conference: COMSOL as a tool for teaching and research in physics and chemistry*. Knoxville, TN: Tennessee Academy of Sciences.

de Silva, E., & de Silva, E. (2009b). *Proceeding from Tennessee Academy of Sciences East Tennessee Collegiate Division Conference: Teaching Chemistry and Physics Through Software Programs*. Tennessee Academy of Sciences.

de Silva, E., & DeBusk, H. (2008, November). *A Comparative Study of Obesity and Animal Movements*. Paper presented at the Tennessee Academy of Science Annual Conference. Nashville, TN.

de Silva, E., & Garrett, T. (2008, November). *Understanding the Physics Behind the Performance of Roping Horses. Tennessee*. Paper presented at the Academy of Science Annual Conference. Nashville, TN.

de Silva, E., de Silva, E., Horner, J., Campbell, L., & McCamey, W. (2012, June). Multidisciplinary Research and Science Teaching. In *Proceedings presented at the Annual Conference on Chemical Science*. Colombo, Sri Lanka: Academic Press.

de Silva, E., Long, C., & Mennen, M. (2008). *Proceedings from Appalachian College Association's Annual Conference Summit XI: Teaching Science Through Multidisciplinary Research*. Appalachian College Association.

de Silva, E., Long, C., & Mennen, M. (2008, October). *Teaching Science Through Multidisciplinary Research*. Paper presented at the Appalachian College Association's Annual Conference, Summit XI. Abingdon, VA.

de Silva, E., Teitelbaum, H., & Long, C. (2008, November). *A Review of Osteopathic Manipulation among western Physicians and Eastern indigenous practitioners*. Paper presented at the Tennessee Academy of Science Annual Conference. Nashville, TN.

Deb, K. (2001). Multi-objective optimization. In *Multi-objective optimization using evolutionary algorithms*, (pp. 13-46). Academic Press.

Delker, P. V. (1974). Governmental roles in lifelong learning. *Journal of Research and Development in Education*, *7*(4), 24–33.

Denholm, A. (2011). Higher levels of uptake in science subjects. *The Herald*. Retrieved 21 December 2013, from http://www.heraldscotland.com/news/education/higher-levels-of-uptake-in-science-subjects.14629814

Department of Education. (2013). *Science programmes of study: key stage 3*. National curriculum in England: Crown. Retrieved on 15th September 2013 from https://www.gov.uk/government/uploads/system/uploads/attachment_data/file/239134/SECONDARY_national_curriculum_-_Science.pdf

DePino, A. Jr. (2001). "Peeps," cream, heads, and food coloring in a vacuum jar. *The Physics Teacher*, *39*(1), 56–57. doi:10.1119/1.1343436

Designing Curriculum for In-depth Understanding in Science. (n.d.). *PowerPoint presentation*. Retrieved from www.fass.cecs.ucf.edu/.../Designing%20Curriculum%20-%20M.%20Hay

Després, C., Vachon, G., & Fortin, A. (2011). Implementing transdisciplinarity: Architecture and urban planning at work. In I. Doucet, & J. Geneviève (Eds.), *Transdisciplinary knowledge production in architecture and urbanism, towards hybrid modes of inquiry* (pp. 33-51). New York: Springer.

Dewey, J. (1998). *Experience and Education Kappa Delta Pi*. Kappa Delta Pi.

DeZure D. (1999). Interdisciplinary Teaching and Learning. In *Essays on Teaching Excellence: Toward the Best in the Academy.* Academic Press.

DfEE/QCA. (1999). Science: The National Curriculum for England, key stages 1-4. London: Department for Education and Employment/Qualifications and Curriculum Authority.

Dindorf, W. (1999). Strings of pearls resonance. *The Physics Teacher, 39*(4), 251. doi:10.1119/1.1367800

Dishaw, J. P. (2010). Battleship buoyancy. *The Physics Teacher, 48*(4), 242. doi:10.1119/1.3361992

Dolmans, D. H. J. M., & Schmidt, H. G. (2006). What do we know about cognitive and motivational effects of small group tutorials in problem-based learning? *Advances in Health Sciences Education: Theory and Practice, 11*(4), 321–336. doi:10.1007/s10459-006-9012-8 PMID:16953462

Domin, D. (1999). A review of laboratory instruction styles. *Journal of Chemical Education, 76*(4), 543–547. doi:10.1021/ed076p543

Donaldson, G. (2011). *Teaching Scotland's Future: Report of a review of teacher education in Scotland.* Edinburgh, UK: Scottish Government.

Donnelly, J. (2001). Contested terrain or unified project? 'The nature of science' in the National Curriculum for England and Wales. *International Journal of Science Education, 23*(2), 181–195. doi:10.1080/09500690120412

Donovan, J., & Venville, G. (2004). Genes and DNA: What kids and experts think. *SCIOS (Journal of the Science Teachers' Association of Western Australia), 40*(2), 26-32.

Donovan, J., & Venville, G. (2012a). Blood and bones: The influence of the mass media on Australian primary children's understandings of genes and DNA. *Science and Education.* DOI: 10.1007/s11191-012-9491-3

Donovan, J., & Venville, G. (2012b). Exploring the influence of the mass media on primary students' conceptual understanding of genetics. *Education, 40*(1), 75-95.

Donovan, J., & Venville, G. (2005). A concrete model for teaching about genes and DNA to young students. *Teaching Science, 51*(4), 29–31.

Doppelt, Y., Mehalik, M., Schunn, C. D., Silk, E., & Krysinski, D. (2008). Engagement and achievements: A case study of design-based learning in a science context. *Journal of Technology Education, 19*(2), 22–39.

Dorigo, M. (Ed.). (2006). *Ant colony optimization and swarm intelligence: 5th International workshop, ANTS 2006,* (Vol. 4150). Springer-Verlag.

Dorigo, M., & Birattari, M. (2010). Ant colony optimization. In Encyclopedia of machine learning (pp. 36-39). Springer US.

Dorigo, M., & Di Caro, G. (1999). Ant colony optimization: a new meta-heuristic. In *Proceedings of the 1999 Congress onEvolutionary Computation* (*Vol. 2*). IEEE.

Dorigo, M. (2007). Ant colony optimization. *Scholarpedia, 2*(3), 1461. doi:10.4249/scholarpedia.1461

Dorion, K. R. (2009). Science through Drama: A multiple case exploration of the characteristics of drama activities used in secondary science lessons. *International Journal of Science Education, 31*(16), 2247–2270. doi:10.1080/09500690802712699

Driver, R. (1983). *The Pupil as Scientist?* Milton Keynes, UK: Open University Press.

Driver, R., & Oldham, V. (1986). A constructivist approach to curriculum development in science. *Studies in Science Education, 13*(1), 105–122. doi:10.1080/03057268608559933

Driver, R., Squires, A., Rushworth, P., & Wood-Robinson, V. (1994). *Making Sense of Secondary Science: research into children's ideas*. London: Routledge.

Drubach, D. (2000). *The Brain Explained.* Upper Saddle River, NJ: Prentice Hall.

Druva, C. A., & Anderson, R. D. (1983). Science teacher characteristics by teacher behavior and by student outcome: A meta-analysis of research. *Journal of Research in Science Teaching, 20*(5), 467–479. doi:10.1002/tea.3660200509

Duch, B. J. (1995). What is problem-based learning? *About Teaching: A newsletter of the Center for Teaching Effectiveness*, 47. Retrieved October 7 2013, from http://www.udel.edu/pbl/cte/jan95-what.html

Duckett, S., Garratt, B., Lowe, J., & Nigel, D. (1999). Key Skills: What do Chemistry Graduates think? *University Chemistry Education, 3*(1), 1–7.

Dudley, R. (2010). *Microsoft Azure: Enterprise Application Development*. Birmingham, UK: Packt Publishing.

Duit, R. (2009). *Bibliography - Students' and Teachers' Conceptions and Science Education*. Retrieved from http://www.ipn.uni-kiel.de/aktuell/stcse/stcse.html

Duschl, R. A. (2000). Making the nature of science explicit. In R. Millar, J. Leach, & J. Osborne (Eds.), *Improving Science Education: the contribution of research* (pp. 187–206). Buckingham: Open University Press.

Duschl, R. A., Schweingruber, H. A., & Shouse, A. W. (Eds.). (2007). *Taking science to school: Learning and teaching science in grades K-8*. Washington, DC: National Academies Press.

Eagly, A. H., & Chaiken, S. (1993). *The Psychology of Attitudes*. London: Harcourt, Brace Jovanovich College Publishers.

Education Scotland. (2013). *What is Curriculum for Excellence?* Retrieved 21 December 2013, from http://www.educationscotland.gov.uk/thecurriculum/whatiscurriculumforexcellence/

Educational Testing Service (ETS). (2013). *The Praxis series*. Retrieved from http://www.ets.org/praxis

Ellenstein, M. (1982). Magic and physics. *The Physics Teacher, 20*(2), 104–106. doi:10.1119/1.2340960

Ende, F. (2012). NOT another lab report. *Science Scope, 35*(5), 44–50.

Epstein, D., & Miller, R. T. (2011). *Slow off the mark: Elementary school teachers and the crisis in science, technology, engineering, and math education*. Retrieved from http://www.americanprogress.org/issues/education/report/2011/05/04/9680/slow-off-the-mark/

Erlauer, L. (2003). *The brain-compatible classroom*. Alexandria, VA: ASCD.

Eteokleous, N., & Ktoridou, D. (2011). Higher education: A web 2.0 world of communication, collaboration, participation and sharing. In *Proceedings of ICICTE-International Conference on ICT in Education*. Rhodes, Greece: ICICTE.

Featonby, D. (2005). Toys and physics. *Physics Education*, *40*(6), 537–543. doi:10.1088/0031-9120/40/6/005

Featonby, D. (2010). Magic physics? *Physics Education*, *45*(1), 24–31. doi:10.1088/0031-9120/45/1/001

Feldman, D., Csikszentmihalyi, M., & Gardner, H. (1994). *Changing the World: A Framework for the Study of Creativity. West Point, CN*. Praeger.

Fencl, H., & Scheel, K. (2005). Engaging students: An examination of the effects of teaching strategies on self-efficacy and course climate in a non-majors physics course. *Journal of College Science Teaching*, *35*(1), 20–24.

Feynman, R. (1965). *The Character of Physical Law*. Cambridge, MA: MIT Press.

Field, M., Lee, R., & Field, M. L. (1994). Assessing interdisciplinary learning. In J. T. Klein, & W. G. Doty (Eds.), *Interdisciplinary Studies Today* (pp. 69–84). San Francisco: Jossey-Bass.

Fink, L. D. (2002). Beyond small groups: Harnessing the extraordinary power of learning teams. In L. Michaelsen, A. Knight, & D. Pink (Eds.), *Team based learning: A transformative use of small groups*. Sterling, VA: Stylus Publishing.

Finson, K. D. (2001). Investigating pre-service elementary teachers' self-efficacy relative to self-image as a science teacher. *Journal of Elementary Science Education*, *13*(1), 31–41. doi:10.1007/BF03176931

Fischer, K. W., & Bidell, T. R. (2006). Dynamic development of action and thought. In R. M. Lerner (Ed.), Handbook of child psychology: Vol. 1. *Theoretical models of human development* (6th ed., pp. 313–399). New York: Wiley.

Fishbein, M., & Ajzen, I. (1975). *Belief, Attitude, Intention and Behavior: An Introduction to Theory and Research*. London: Addison-Wesley Publishing Company.

Fisher, N. (2004). Creativity breeds best physicists. *Physics Education*, *39*(6), 472.

Fleer, M. (2009). Understanding the Dialectical Relations Between Everyday Concepts and Scientific Concepts Within Play-Based Programs. *Research in Science Education*, *39*(2), 281–306. doi:10.1007/s11165-008-9085-x

Flicek, P. (2012, August 1). *A Deep Catalog of Human Genetic Variation*. Retrieved from http://www.1000genomes.org/

Flynn, E., O'Malley, C., & Wood, D. (2004). A longitudinal, microgenetic study of the emergence of false belief understanding and inhibition skills. *Developmental Science*, *7*(1), 103–115. doi:10.1111/j.1467-7687.2004.00326.x PMID:15323122

Fortus, D., Dershimer, R. C., Krajcik, J., Marx, R. W., & Mamlok-Naaman, R. (2004). Design-based science and student learning. *Journal of Research in Science Teaching*, *41*(10), 1081–1110. doi:10.1002/tea.20040

Fortus, D., Krajcik, J., Dershimer, R. C., Marx, R. W., & Mamlok-Naaman, R. (2005). Design-based science and real-world problem-solving. *International Journal of Science Education*, *27*(7), 855–879. doi:10.1080/09500690500038165

Froehle, P. (2008). Quick way to float a paper clip on water. *The Physics Teacher*, *46*(2), 70. doi:10.1119/1.2834520

Frost, S., & Olsen, D. (2006). Technology, Learning, and the Virtual Liberal Arts Classroom. *Peer Review*, *8*(4), 20–22.

Gallacher, J., Crossan, B., Leahy, J., Merrill, B., & Field, J. (2000). Education for All? Further Education, Social Inclusion and Widening Access. Glasgow, UK: Centre for Research in Lifelong Learning (CRLL), Glasgow Caledonian University.

Gardner, G. E. (Manuscript submitted for publication). The effects of constructing collaborative groups using motivation on undergraduate student attitudes and perceptions of biology and biologists. *CBE Life Sciences Education*.

Gardner, M. (1999). The jumping pencil. *The Physics Teacher*, *37*(3), 178. doi:10.1119/1.1527660

Gates, B. (2014). *Skype* [Computer software]. Skype Communications. Retrieved from http://www.skype.com

Gauld, C. (1986). Models, meters and memory. *Research in Science Education*, *16*(1), 49–54. doi:10.1007/BF02356817

Gauld, C. (1989). A study of pupils' responses to empirical evidence. In R. Millar (Ed.), *Doing Science: images of science in science education* (pp. 62–82). London: The Falmer Press.

Gentner, D. (1983). Structure-Mapping: A Theoretical Framework for Analogy. *Cognitive Science*, *7*(2), 155–170. doi:10.1207/s15516709cog0702_3

Gilbert, J., & Priest, M. (1997). Models and Discourse: A Primary School Science Class Visit to a Museum. *Science Education*, *81*(6), 749–762. doi:10.1002/(SICI)1098-237X(199711)81:6<749::AID-SCE10>3.0.CO;2-I

Gluck, P. (2008, March06). Experiments for a special day. *Physics Education*, *43*(2), 189–197. doi:10.1088/0031-9120/43/2/009

Glynn, S. M., Brickman, P., Armstrong, N., & Taasoobshirazi, G. (2011). Science motivation questionnaire II: Validation with science majors and nonscience majors. *Journal of Research in Science Teaching*, *48*(10), 1159–1176. doi:10.1002/tea.20442

Gogle Inc. (2006, August). *Google Launches Hosted Communications Services*. Retrieved Aug 25 2001, from http://www.google.com/intl/en/press/pressrel/gafyd.html

Goldacre, B. (2013). *Building Evidence into Education*. Retrieved 22 December 2013, from https://www.gov.uk/government/news/building-evidence-into-education

Goldberg, D. E., & Holland, J. H. (1988). Genetic algorithms and machine learning. *Machine Learning*, *3*(2/3), 95–99. doi:10.1023/A:1022602019183

Goldstein, K. M., & Blackman, J. (1978). *Cognitive Style: Five approaches and Relevant Research*. New York: Wiley.

Goodrum, D., Druhan, A., & Abbs, J. (2011). *The status and quality of Year 11 and 12 science in Australian schools*. Canberra, ACT: Australian Academy of Science.

Goodrum, D., Hackling, M., & Rennie, L. (2001). *The status and quality of teaching and learning of science in Australian schools*. Canberra, ACT: Department of Education, Training and Youth Affairs.

Gopnik, A., Meltzoff, A., & Kuhl, P. (1999). *The Scientist in the Crib: What Early Learning Tells Us About the Mind*. New York, NY: Harper Collins.

Gore, G. R. (2009). Physics fun with jelly marbles. *The Physics Teacher*, *47*(9), 606–607. doi:10.1119/1.3264598

Graf, E. H. (2008). Projectile motion demonstration. *The Physics Teacher*, *46*(9), 553. doi:10.1119/1.3023659

Greenslade, T. B. (2010). Physics is all around us. *The Physics Teacher*, *48*(5), 338–340. doi:10.1119/1.3393071

Greenwood, A., & Scribner-MacLean, M. (1997). *Examining elementary teachers' explanations of their science content knowledge*. Paper presented at the annual meeting of the National Association for Research in Science Teaching, Oak Brook, IL.

Griggs, R. (2012). *Report of the Review of Further Education Governance in Scotland*. Edinburgh, UK: Scottish Government.

Groh, S. E., Williams, B. A., Allen, D. E., Duch, B. J., Mierson, S., & White, H. B. (1997). Institutional change in science education: A case study. In A. P. McNeal & C. D'Avanzo (Eds.), Student-active science: Models of innovation in college science teaching (pp. 83-94). Philadelphia: Saunders College Publishing.

Güémez, J., Fiolhais, C., & Fiolhais, M. (2009). Toys in physics lectures and demonstrations - a brief review. *Physics Education*, *44*(1), 53–64. doi:10.1088/0031-9120/44/1/008

Gunning, A. M., & Mensah, F. M. (2011). Pre-service elementary teachers' development of self-efficacy and confidence to teach science: A case study. *Journal of Science Teacher Education*, *22*(2), 171–185. doi:10.1007/s10972-010-9198-8

Hacker, R. G., & Rowe, M. J. (1997). The impact of a National Curriculum development on teaching and learning behaviours. *International Journal of Science Education*, *19*(9), 997–1004. doi:10.1080/0950069970190901

Hackett, G., Betz, N. E., Casas, J. M., & Rocha-Singh, I. A. (1992). Gender, ethnicity, and social cognitive factors predicting the academic achievement of students in engineering. *Journal of Counseling Psychology*, *39*(4), 527–538. doi:10.1037/0022-0167.39.4.527

Haglund, H. (1971). Isoelectric focusing in pH gradients--a technique for fractionation and characterization of ampholytes. *Methods of Biochemical Analysis*, *19*, 1–104. doi:10.1002/9780470110386.ch1 PMID:4935451

Halpine, S. M. (2004). Introducing molecular visualization to primary schools in California: The STArt! *teaching* science through art program. *Journal of Chemical Education*, *81*(10), 1431–1436. doi:10.1021/ed081p1431

Hambrick, D., & Cannella, A. Jr. (1989). Strategy Implementation as substance and selling. *The Academy of Management Executive*, *3*(4), 278–285. doi:10.5465/AME.1989.4277401

Handelsman, J., Ebert-May, D., Beichner, R., Bruns, P., Chang, A., & DeHaan, R. et al. (2004). EDUCATION: Scientific Teaching. *Science*, *304*(5670), 521–522. doi:10.1126/science.1096022 PMID:15105480

Haney, J. J., Lumpe, A. T., Czerniak, C. M., & Egan, V. (2002). From beliefs to actions: The beliefs and actions of teachers implementing change. *Journal of Science Teacher Education*, *13*(3), 171–187. doi:10.1023/A:1016565016116

Harden, J. (1996). Enlightenment, empowerment and emancipation: The case for critical pedagogy in nurse education. *Nurse Education Today*, *16*(1), 32–37. doi:10.1016/S0260-6917(96)80090-6 PMID:8700068

Harrison, A. G., & Treagust, D. F. (2006). Teaching and learning with analogies: friend or foe? In P. J. Aubusson, A. G. Harrison & S. M. Ritchie (Eds.), Metaphor and Analogy in Science Education (pp. 11-24). Dordecht, The Netherlands: Springer.

Harrison, A. G., & Coll, R. K. (Eds.). (2008). *Using analogies in middle and secondary science classrooms*. Thousand Oaks, CA: Corwin Press.

Haus, R. (2014, January 3). *BLAST Assembled RefSeq Genomes*. Retrieved from http://blast.ncbi.nlm.nih.gov/Blast.cgi

Heacox, D. (2002). *Differentiating instruction in the regular classroom: How to reach and teach all learners, grades 3-12*. Free Spirit Pub.

Health Education England. (2013). *Investing in people, for health and healthcare. Workforce Plan for England: Proposed Education and Training Commissions for 2014/15*. London: Health Education England.

Henry, B. (2011, December 27). *DNA Sequencing*. Retrieved from http://www.genome.gov/10001177

Henry, B. (2011, November 15). *DNA Microarray Technology*. Retrieved from http://www.genome.gov/10000533

Henry, B. (2012, June 12). *Transcriptome*. Retrieved from http://www.genome.gov/13014330

Henry, B. (2013, July 11). *Polymerase Chain Reaction*. Retrieved from http://www.genome.gov/10000207

Hérold, J. F., & Ginestié, J. (2011). Help with solving technological problems in project activities. *International Journal of Technology and Design Education*, *21*(1), 55–70. doi:10.1007/s10798-009-9106-8

Hind, A., Leach, J., Lewis, J., & Scott, P. (n.d.). *Teaching for understanding: Electic circuits: A teahcing scheme developed from research evdience on students' learning about electric circuits*. Academic Press.

Hirsch Hadorn, G., Hoffmann-Riem, H., Biber-Klemm, S., Grossenbacher-Mansuy, W., Joye, D., & Pohl, C. et al. (Eds.). (2008). *Handbook of transdisciplinaryresearch*. Berlin, Germany: Springer. doi:10.1007/978-1-4020-6699-3

Hirsch, E. D., Jr. (2006). Building knowledge: The case for bringing content into the language arts block and for a knowledge-rich curriculum core for all children. *American Educator, Spring* [Electronic version]. Retrieved on October 16, 2010 from http://www.aft.org/newspubs/periodicals/ae/spring2006/editors.cfm

Hodson, D. (2009). *Teaching and learning about science: Language, theories, methods, history, traditions and values*. Rotterdam, The Netherlands: Sense Publishers.

Hofmeyr, S. A., & Forrest, S. (2000). Architecture for an artificial immune system. *Evolutionary Computation*, *8*(4), 443–473. doi:10.1162/106365600568257 PMID:11130924

Holler, M., Tam, S., Castro, H., & Benson, R. (1989, June). An electrically trainable artificial neural network (etann) with 10240'floating gate'synapses. In *Proceedings of Neural Networks,* (pp. 191-196). IEEE.

Howard, J. R., Short, L. B., & Clark, S. M. (1996). Students' Participation in the Mixed-Age College Classroom. *Teaching Sociology, 24*(1), 8–24. doi:10.2307/1318894

Howe, C., Tolmie, A., Thurston, A., Topping, K., Christie, D., & Livingston, K. et al. (2007). Group work in elementary science: Towards organisational principles for supporting pupil learning. *Learning and Instruction, 17*(5), 549–563. doi:10.1016/j.learninstruc.2007.09.004

Humes, W., & Bryce, T. (2003). The Distinctiveness of Scottish Education. In T. G. K. Bryce, & W. M. Humes (Eds.), *Scottish Education post devolution* (2nd ed., pp. 108–118). Edinburgh, UK: Edinburgh University Press.

Humphreys, A. H., Post, T. R., & Ellis, A. K. (1981). *Interdisciplinary methods: A thematic approach.* Santa Monica, CA: Goodyear.

Hussein, F., & Reid, N. (2009). Working memory and difficulties in school chemistry. *Research in Science & Technological Education, 27*(2), 161–185. doi:10.1080/02635140902853632

IBM Corp. (2011). IBM SPSS Statistics for Windows, Version 20.0. Armonk, NY: IBM Corp.

IFF. (2003). *Study of learners in Further Education (No. 1 84478 083 X).* IFF Research Ltd.

Ingersoll, R. (2003). *Is there really a teacher shortage?* (Document R-03-04). Seattle, WA: Center for the Study of Teaching and Policy at the University of Washington. Retrieved from http://depts.washington.edu/ctpmail/PDFs/Shortage-RI-09-2003.pdf

Ingersoll, R., & Perda, D. (2012). *How high is teacher turnover and is it a problem?* Philadelphia, PA: University of Pennsylvania, Consortium for Policy Research in Education.

Ing, M. (2013). Can Parents Influence Children's Mathematics Achievement and Persistence in STEM Careers? *Journal of Career Development.* doi: doi:10.1177/0894845313481672

Inhelder, B., & Piaget, J. (1958). *The growth of logical thinking from childhood to adolescence.* New York: Basic Books. doi:10.1037/10034-000

Instrument, S. (1989). The Education (National Curriculum) (Attainment Targets and Programmes of Study in Science). *Order,* 1989.

Interview, R. J. (n.d.). *Interdisciplinary and Multidisciplinary Research.* Retrieved from http://www.4researchers.org/articles/5213

Ivanitskaya, L., Clark, D., Montgomery, G., & Primeau, R. (2002). Interdisciplinary Learning: Process and Outcomes. *Innovative Higher Education, 27*(2), 95–111. doi:10.1023/A:1021105309984

Jacobi, M. (1991). Mentoring and undergraduate academic success: A literature review. *Review of Educational Research, 61*(4), 505–532. doi:10.3102/00346543061004505

Jacobs, H. H. (Ed.). (1999). The growing need for interdisciplinary curriculum content. In H. H. Jacobs (Ed.), Interdisciplinary curriculum: Design and implementation (pp. 1–11). Alexandria, VA: Association for Supervision and Curriculum Development.

Jacqueline, L., & Smita, G. (2001). Education at the Crossroads. *Journal of Research on Technology in Education*, *34*(1), 51–57. doi: 10.1080/15391523.2001.10782333

Jakab, C. (2013). Small talk: Children's everyday 'molecule' ideas. *Research in Science Education*, *43*(4), 1307–1325. doi:10.1007/s11165-012-9305-2

Jean, A. (2007, July26). Science in comedy: Mmm ... pi. *Nature*, *448*(7152), 404–405. doi:10.1038/448404a PMID:17653163

Jenkins, E. W. (2000). The impact of the national curriculum on secondary school science teaching in England and Wales. *International Journal of Science Education*, *22*(3), 325–336. doi:10.1080/095006900289903

Jenkins, E. W., & Nelson, N. W. (2005). Important but not for me: Students' Attitudes towards Secondary School Science in England. *Research in Science & Technological Education*, *23*(1), 41–57. doi:10.1080/02635140500068435

Jensen, J. L., & Lawson, A. (2011). Effects of collaborative group composition and inquiry instruction on reasoning gains and achievement in undergraduate biology. *CBE Life Sciences Education*, *10*(1), 64–73. doi:10.1187/cbe.10-07-0089 PMID:21364101

Jewitt, C., Kress, G., Ogborn, J., & Tsatsarelis, C. (2001). Exploring Learning Through Visual, Actional and Linguistic Communication: The multimodal environment of a science classroom. *Educational Review*, *53*(1), 5–18. doi:10.1080/00131910123753

Jhuang, H., Serre, T., Wolf, L., & Poggio, T. (2007, October). A biologically inspired system for action recognition. In *Proceedings of Computer Vision*, (pp. 1-8). IEEE. doi:doi:10.1109/ICCV.2007.4408988 doi:10.1109/ICCV.2007.4408988

Jin, X., & Reynolds, R. G. (1999). Using knowledge-based evolutionary computation to solve nonlinear constraint optimization problems: a cultural algorithm approach. In *Proceedings of the 1999 Congress onEvolutionary Computation* (*Vol. 3*). IEEE.

Johnson, D. W., & Johnson, R. T. (2009). An educational psychology success story: Social interdependence theory and cooperative learning. *Educational Researcher*, *38*(5), 365–379. doi:10.3102/0013189X09339057

Johnson, D. W., Johnson, R. T., & Smith, K. A. (1998). *Active learning: Cooperation in the college classroom*. Edina, MN: Interaction Book Company.

Johnson, J. (2000). Learning communities and special efforts in the retention of university students: What works, what doesn't, and is the return worth the investment? *Journal of College Student Retention*, *2*(3), 219–238. doi:10.2190/V0PA-BL4B-1X2L-W5VT

Johnson, P., & Papageorgiou, G. (2010). Rethinking the introduction of particle theory: A substance-based framework. *Journal of Research in Science Teaching*, *47*(2), 130–150.

Johnson, P., & Tymms, P. (2011). The emergence of a learning progression in middle school chemistry. *Journal of Research in Science Teaching*, *48*(8), 849–877. doi:10.1002/tea.20433

Johnstone, A. H. (1982). Macro- and micro-chemsitry.[Notes and correspondence]. *The School Science Review*, *64*(227), 377–379.

Johnstone, A. H. (1991). Why is science difficult to learn? Things are seldom what they seem. *Journal of Computer Assisted Learning*, *7*(2), 75–83. doi:10.1111/j.1365-2729.1991.tb00230.x

Johnstone, A. H. (1997). … And Some Fell on Good Ground. *University Chemistry Education*, *1*, 8–13.

Johnstone, A. H., & El-Banna, H. (1986). Capacities, Demands and Processes: A Predictive Model for Science Education. *Education in Chemistry*, *23*(5), 80–84.

Johnstone, A. H., & Selepeng, D. (2001). A Language Problem Revisited. *Chemistry Education: Research and Practice in Europe*, *2*(1), 19–29.

Jolliffe, T., & Baron-Cohen, S. (1999). A test of central coherence theory: linguistic processing in high-functioning adults with autism or Asperger syndrome: is local coherence impaired? *Cognition*, *71*(2), 149–185. doi:10.1016/S0010-0277(99)00022-0 PMID:10444907

Jones, A., & Moreland, J. (2003). Developing classroom-focused research in technology education. *Canadian Journal of Science. Mathematics and Technology Education.*, *3*(1), 37–50. doi:10.1080/14926150309556551

Jonides, J. (1995). *Evaluation and dissemination of an undergraduate program to improve retention of at-risk students*. Ann Arbor, MI: University of Michigan College of Literature, Science, and Arts.

Ju, Y. (2005). The physics of having a swing. *Physics Education*, *40*(6), 534–536. doi:10.1088/0031-9120/40/6/004

Kaufman, A., Mennin, R. S., Waterman, S., Duban, C., Hansbarger, H., & Silverblatt, S. et al. (1989). The New Mexico experiment. *Academic Medicine*, *64*(6), 285–294. doi:10.1097/00001888-198906000-00001 PMID:2719785

Kavaloski, V. (Ed). (1997). Interdisciplinary education and humanistic aspiration: A critical reflection. In J. Kockelmans (Ed.), *Interdisciplinarity and Higher Education*. University Park, PA: The Pennsylvania State University Press.

Kay, K., & Greenhill, V. (2011). Twenty-first century students need 21st century skills. In G. Wan, & D. Gut (Eds.), *Bringing schools into the 21st century* (pp. 41–65). New York: Springer. doi:10.1007/978-94-007-0268-4_3

Kennedy, J. (2010). Particle swarm optimization. In Encyclopedia of machine learning (pp. 760-766). Springer US.

Keogh, B., & Naylor, S. (1999). Concept Cartoons, teaching and learning in science: An evaluation. *International Journal of Science Education*, *21*(4), 431–446. doi:10.1080/095006999290642

Key Stage 3 National Strategy. (2002). *Framework for teaching science: years 7, 8 and 9*. London: Department for Education and Skills.

Kilminster, S. (1997). *Vocational Education and Really Useful Knowledge*. Paper presented at the 27th Annual SCUTREA Conference. Crossing orders, Breaking Boundaries: Research in the Education of Adults Conference. London, UK.

Kim, M., & Hannafin, M. (2004). Designing Online Learning Environments to Support Scientific Inquiry. *Quarterly Review of Distance Education*, *5*(1), 1–10.

Kind, V., & Taber, K. S. (2005). Science: Teaching School Subjects 11-19. London: RoutledgeFalmer.

King, K., Shumow, L., & Lietz, S. (2001). Science education in an urban elementary school: Case studies of teacher beliefs and classroom practices. *Science Education*, *85*(2), 89–110. doi:10.1002/1098-237X(200103)85:2<89::AID-SCE10>3.0.CO;2-H

Kirschner, P. A., Sweller, J., & Clark, R. E. (2006). Why minimal guidance during instruction does not work: An analysis of the failure of constructivist, discovery, problem-based, experiential, and inquiry-based teaching. *Educational Psychologist*, *41*(2), 75–86. doi:10.1207/s15326985ep4102_1

Kleine-Staarman, J., & Mercer, N. (2010). The guided construction of knowledge: Talk between teachers and students. In K. Littleton, C. Wood & J. K. Kleine-Staarman (Eds.), International Handbook of Research of Psychology in Education (pp. 75-104). Bingley, UK: Emerald.

Knowles, M. S. (1990). *The Adult Learner: A Neglected Species* (4th ed.). London: Gulf Publishing Company.

Knox, A. (1977). *Adult Development and Learning*. San Francisco, CA: Jossey Bass.

Koballa, T. R. J., & Glynn, S. M. (2007). Attitudinal and motivational constructs in science learning. In S. K. Abell, & N. G. Lederman (Eds.), *Handbook of research on science education* (pp. 75–102). Mahwah, NJ: Lawrence Erlbaum Association Publishers.

Koffka, K. (1967). Principles of Gestalt Psychology. In J. A. Dyal (Ed.), *Readings in Psychology: Understanding human behavior* (2nd ed., pp. 9–13). New York: McGraw-Hill Book Company.

Kofoed, M. H. (2006). The Hiroshima and Nagasaki bombs: Role-play and students' interest in physics. *Physics Education*, *41*(6), 502–507. doi:10.1088/0031-9120/41/6/002

Kolb, D. A. (1984). *Experiential Learning: Experience as The Source of Learning and Development*. Upper Saddle River, NJ: Prentice Hall Inc.

Ktoridou, D. (2010, April). Applying an Inductive Method to a New, Multidiciplinary, Management of Innovation & Technology Course: Evidence from the University of Nicosia. In *Engineering Education Conference – The Future of Global Learning in Engineering Education* (pp. 452–460). IEEE. doi:10.1109/EDUCON.2010.5492422

Kuhn, T. S. (1996). *The Structure of Scientific Revolutions* (3rd ed.). Chicago: University of Chicago. doi:10.7208/chicago/9780226458106.001.0001

Kvale, S. (1996). *Interviews: An introduction to qualitative research interviewing*. Thousand Oaks, CA: Sage.

Lakatos, I. (1999). Lectures on Scientific Method. In M. Motterlini (Ed.), *For and Against Method* (pp. 19–109). Chicago: University of Chicago Press. doi:10.7208/chicago/9780226467030.001.0001

LAMP. (2013). *Land Administration and Management Project*. Retrieved from http://www.lamp.gov.me/?lang=en

Lange, C. (2003, March 16). *PBAT*. Retrieved from http://www.biostat.harvard.edu/clange/default.htm

Lave, J. (1993). Situated learning in communities of practice. In L. B. Resnick, J. M. Levine, & S. D. Teasley (Eds.), *Perspectives on Socially Shared Cognition* (pp. 63–82). Washington, DC: American Psychological Association.

Leach, J., & Scott, P. (2002). Designing and evaluating science teaching sequences: An approach drawing upon the concept of learning demand and a social constructivist perspective on learning. *Studies in Science Education*, *38*(1), 115–142. doi:10.1080/03057260208560189

Lee, K. W. L., Goh, N. K., Chia, L. S., & Wan,, Y. K. (2006). *Report on Creativity in Science Education*. National Institute of Education, Nanyang Technological University Q 183.4.555.

Lee, O., Eichinger, D. C., Anderson, C. W., Berkheimer, G. D., & Blakeslee, T. S. (1993). Changing middle school students' conceptions of matter and molecules. *Journal of Research in Science Teaching*, *30*(3), 249–270. doi:10.1002/tea.3660300304

Lehrer, R., & Schauble, L. (2000). Modeling in mathematics and science. In R. Glaser (Ed.), *Advances in instructional psychology* (Vol. 5, pp. 101–159). Mahwah, NJ: Lawrence Erlbaum Associates.

Lei, X. R. B. W. (2002). Artificial immune system: Principle, models, analysis and perspectives. *Chinese Journal of Computers*, *12*, 000.

Leibnitz, K., Wakamiya, N., & Murata, M. (2006). Biologically inspired self-adaptive multi-path routing in overlay networks. *Communications of the ACM*, *49*(3), 62–67. doi:10.1145/1118178.1118203

Liddell, G., & Macpherson, S. (2013). *SPICe Briefing: Post-16 Education (Scotland) Bill*. Edinburgh, UK: Scottish Parliament.

Light, R. (2001). *Making the most of college: StudentssSpeak their minds*. Cambridge, MA: Harvard University Press.

Lipton, L., & Wellman, B. (1998). *Patterns and practices in the learning-focused classroom. Guilford*. VT: Pathways Publishing.

Lister, T. (2006). *Cutting Edge Chemistry*. London: Royal Society of Chemistry.

Liu, X., & Lesniak, K. (2006). Progression in children's understanding of the matter concept from elementary to high school. *Journal of Research in Science Teaching*, *43*(3), 320–347. doi:10.1002/tea.20114

Liu, X., & Lesniak, K. M. (2005). Students' progression of understanding the matter concept from elementary to high school. *Science Education*, *89*(3), 433–450. doi:10.1002/sce.20056

Lopatto, D. (2004). Survey of undergraduate research experiences (SURE): First findings. *Cell Biology Education*, *3*(4), 270–277. doi:10.1187/cbe.04-07-0045 PMID:15592600

Lowry, M. (2008). Teaching universal gravitation with vector games. *The Physics Teacher*, *46*(9), 519–521. doi:10.1119/1.3023651

Lumpe, A. T., Haney, J. J., & Czerniak, C. M. (1998). Science teacher beliefs and intentions regarding the use of cooperative learning. *School Science and Mathematics*, *98*(3), 123–135. doi:10.1111/j.1949-8594.1998.tb17405.x

Lyman, F. (1987). Think-Pair-Share: An expanding teaching technique. *MAACIE Cooperative News*, *1*, 1–2.

Mahaffy, P. (2004). The Future Shape of Chemistry Education. *Chemistry Education: Research and Practice in Europe*, *5*(3), 229–245.

Mallow, J. (1982). *Science anxiety: Fear of science and how to overcome it*. New York, NY: Van Nostrand Reinhold.

Maltese, A. V., & Tai, R. H. (2010). Eyeballs in the fridge: Sources of early interest in science. *International Journal of Science Education*, *32*(5), 669–685. doi:10.1080/09500690902792385

Margel, H., Eylon, B. S., & Scherz, Z. (2008). A longitudinal study of junior high school students' conceptions of the structure of materials. *Journal of Research in Science Teaching*, *45*(1), 132–152. doi:10.1002/tea.20214

Marshall, J. C., Horton, R., Igo, B. L., & Switzer, D. M. (2009). K-12 science and mathematics teachers' beliefs about and use of inquiry in the classroom. *International Journal of Science and Mathematics Education*, *7*(3), 575–596. doi:10.1007/s10763-007-9122-7

*Martial Art Philosophy*. (n.d.). Retrieved March 29, 2014, from http://www.alljujitsu.com/martialartphilosophy.html

Martin, D. J. (2012). *Elementary science methods: A constructivist approach*. Belmont, CA: Wadsworth.

Matthews, M. R. (1994). *Science Teaching: The role of history and philosophy of science*. London: Routledge.

Mawn, M. V., Carrico, P., Charuk, K., Stote, K. S., & Lawrence, B. (2011). Hands-on and Online: Scientific Explorations through Distance Learning. *Open Learning*, *26*(2), 135–146.

McConnell, M. C. (1982). Teaching about Science, Technology and Society at the secondary school level in the United States. An educational dilemma for the 1980s. *Studies in Science Education*, *9*(1), 1–32. doi:10.1080/03057268208559893

McCormick, R. (1997). Conceptual and procedural knowledge. *International Journal of Technology and Design Education*, *7*(1-2), 141–159. doi:10.1023/A:1008819912213

McCullough, J., & McCullough, R. (2000). The Role of Toys in Teaching Physics. College Park, MD: American Association of Physics Teachers.

Mcgarvey, B., Marriott, S., Morgan, V., & Abbott, L. (1997). Planning for differentiation: The experience of teachers in Northern Ireland primary schools. *Curriculum Studies*, *29*(3), 351–364. doi:10.1080/002202797184080

McGaugh, J. L. (2003). *Memory and emotion: The making of lasting memories*. New York, NY: Columbia University Press.

McGervey, J. (1995). Hands-on physics for less than a dollar per hand. *The Physics Teacher*, *33*(4), 238–241. doi:10.1119/1.2344206

McHarg, J., Kay, E. J., & Coombes, L. R. (2012). Students' engagement with their group in a problem-based learning curriculum. *European Journal of Dental Education*, *16*(1), e106–e110. doi:10.1111/j.1600-0579.2011.00682.x PMID:22251332

McKeachie, W. J. (1986). *Teaching tips: strategies, research, and theory of college and university teachers*. Boston, MA: Houghton Mifflin.

McKinney, K., & Graham-Buxton, M. (1993). The use of collaborative learning groups in the large class: Is it possible? *Teaching Sociology*, *21*(4), 403–408. doi:10.2307/1319092

McNair, S., Parry, G., Brooks, R., & Cole, P. (2004). *Learning together: Age mixing in further education colleges*. Learning and Skills Research Centre.

McTighe, J., & Seif, E. (2003). *A summary of underlying theory and research base for Understanding by Design*. Retrieved from assets.pearsonschool.com/asset_mgr/ ubd_myworld_research.pdf

McTighe, J., & Thomas, R. (2003). Backward design for forward action. *Educational Leadership*, *60*(5), 52–55.

Mello, A. S., & Ruckes, M. E. (2006). Team composition. *The Journal of Business*, *79*(3), 1019–1039. doi:10.1086/500668

Merrill, B. (2001). Learning and Teaching in Universities: Perspectives from adult learners and lecturers. *Teaching in Higher Education*, *6*(1), 5–17. doi:10.1080/13562510020029563

Merritt, J. D., Krajcik, J., & Shwartz, Y. (2008, June). Development of a learning progression for the particle model of matter. In *Proceedings of the 8th International conference for the Learning Sciences* (Vol. 2, pp. 75-81). Utrecht, The Netherlands: International Society of the Learning Sciences.

Meyer, D. (2012). Designing Design Challenges: Getting the details right. Using engineering problems to enact inquiry learning. *Science Teacher (Normal, Ill.)*, *79*(2), 58–62.

Meyers, C., & Jones, T. B. (1993). *Promoting active learning: Strategies for the college classroom*. San Francisco, CA: Jossey-Bass.

Michaelsen, L. K. (2002). Team-based learning in large classes. In L. K. Michaelsen, A. B. Knight, & L. D. Fink (Eds.), *Team-based learning: A transformative use of small groups* (pp. 157–171). Westport, CT: Praeger Publishers.

Michel, M. C., Bischoff, A., & Jacobs, K. H. (2002). Comparison of problem- and lecture-based pharmacology teaching. *Trends in Pharmacological Sciences*, *23*, 168 – 170.

Michelsohn, A. M., & Hawkins, S. (1994). Current practice in science education of prospective elementary school teachers. In S. Raizen, & A. Michelsohn (Eds.), *The future of science in elementary schools: Educating prospective teachers* (pp. 151–161). San Francisco, CA: Jossey-Bass.

Millar, R. (2004, 3-4 June 2004). *The role of practical work in the teaching and learning of science*. Paper presented at the High School Science Laboratories: Role and Vision, National Academy of Sciences. Washington, DC.

Monk, D. (1994). Subject area preparation of secondary mathematics and science teachers and student achievement. *Economics of Education Review*, *13*(2), 125–145. doi:10.1016/0272-7757(94)90003-5

Mortimer, E. F., & Scott, P. H. (2003). *Meaning Making in Secondary Science Classrooms*. Maidenhead, UK: Open University Press.

Munck, M. (2007). Science pedagogy, teacher attitudes, and student success. *Journal of Elementary Science Education*, *19*(2), 13–24. doi:10.1007/BF03173660

Murdock, J. L., & Williams, A. M. (2011). Creating an Online Learning Community. *Innovations in Higher Education*, *36*(5), 305–315. doi:10.1007/s10755-011-9188-6

Murphy, C. (2012). Vygotsky and primary science. In B. J. Fraser, K. G. Tobin, & C. J. McRobbie (Eds.), *Second international handbook of science education* (pp. 177–187). Dordrecht, The Netherlands: Springer. doi:10.1007/978-1-4020-9041-7_14

Nagpal, R. (2002, July). Programmable self-assembly using biologically-inspired multiagent control. In *Proceedings of the First International Joint Conference on Autonomous Agents and Multiagent Systems: Part 1* (pp. 418-425). ACM. doi:doi:10.1145/544838.544839 doi:10.1145/544838.544839

Nakhleh, M. B., & Samarapungavan, A. (1999). Elementary school children's beliefs about matter. *Journal of Research in Science Teaching*, *36*(7), 777–805. doi:10.1002/(SICI)1098-2736(199909)36:7<777::AID-TEA4>3.0.CO;2-Z

Nakiboğlu, C., & Taber, K. S. (2013). The atom as a tiny solar system: Turkish high school students' understanding of the atom in relation to a common teaching analogy. In G. Tsaparlis & H. Sevian (Eds.), Concepts of Matter in Science Education (pp. 169–198). Dordrecht, The Netherlands: Springer. doi:doi:10.1007/978-94-007-5914-5_8 doi:10.1007/978-94-007-5914-5_8

Natarajan, M. M. (2006). Use of online technology for multimedia education. *Information Services & Use*, *26*(3), 249–256.

Nathans, D., & Smith, H. O. (1975). Restriction endonucleases in the analysis and restructuring of DNA molecules. *Annual Review of Biochemistry*, *44*(1), 273–293. doi:10.1146/annurev.bi.44.070175.001421 PMID:166604

National Commission on Teaching and America's Future. (1996). *What matters most: Teaching for America's future. Report of the National Commission on Teaching & America's Future*. New York, NY: Author.

National Research Council (NRC). (1996). *National science education standards*. Washington, DC: National Academy Press.

National Research Council (NRC). (2012). *A Framework for K-12 Science Education: Practices, Crosscutting Concepts, and Core Ideas*. Washington, DC: The National Academies Press. Retrieved on 15th September 2013 from http://www.nap.edu/openbook.php?record_id=13165&page=1

National Research Council (NRC). (2013). *Next Generation Science Standards: For States, By States*. Washington, DC: The National Academies Press. Retrieved on 15th September 2013 from http://www.nap.edu/NGSS/

National Research Council. (2000). *Inquiry and the National Science Education Standards: A guide for teaching and learning*. Washington, DC: National Academies Press.

National Science Board. (2013). *STEM Education Data and Trends*. Retrieved October 1, 2013, from http://www.nsf.gov/nsb/sei/edTool/timeline.html#5

National Science Foundation. (2006). *America's Pressing Challenge — Building a Stronger Foundation: Mathematics and Science Achievement is Critical*. Retrieved from http://www.nsf.gov/statistics/nsb0602/

National Science Teachers Association. (2004). *NSTA Position Statement: Science Teacher Preparation*. Retrieved from http://www.nsta.org/about/positions/preparation.aspx?print=true

Neville, A. J. (2009). Problem-Based Learning and Medical Education Forty Years on. *Medical Principles and Practice*, *18*(1), 1–9. doi:10.1159/000163038 PMID:19060483

Newell, W.H. (1990). Interdisciplinary curriculum development. *Issues in Integrative Studies*.

Newell, W. H. (1998). *Interdisciplinarity: Essays from the Literature* (8th ed.). New York: The College.

Newton, J. (2003). Implementing an institution-wide learning and teaching strategy: Lessons in managing change. *Studies in Higher Education*, *28*(4), 427–441. doi:10.1080/0307507032000122279

NGSS Lead States. (2013). *Next generation science standards: For states, by states*. Washington, DC: The National Academies Press.

NHS Education for Scotland. (2013). *Nursing and Midwifery: Supporting Scotland's nurses and midwives*. Retrieved 21 December 2013, from http://www.nes.scot.nhs.uk/education-and-training/by-discipline/nursing-and-midwifery.aspx

Niaz, M. (2005). How to Facilitate Students' Conceptual Understanding of Chemistry? A History and Philosophy of Science Perspective. *Chemical Education International*, *6*(1), 1–5.

Nickerson, R. S. (1998). Confirmation bias: A ubiquitous phenomenon in many guises. *Review of General Psychology*, *2*(2), 175–220. doi:10.1037/1089-2680.2.2.175

Nicolescu B.(1998). The transdisciplinary evolution of the university, condition for sustainable development. *Rencontres Transdisciplinaires,* 12.

Norman, E. (1993). Science for design. *Physics Education*, *28*(5), 301–306. doi:10.1088/0031-9120/28/5/010

Nunley, K. F. (2006). *Differentiating in the High School*. Thousand Oaks, CA: Corwin Press.

Nursing and Midwifery Council. (2013). *Our role*. Retrieved 21 December 2013, from http://www.nmc-uk.org/About-us/Our-role/

Nussbaum, J. (1998). History and philosophy of science and the preparation for constructivist teaching: The case of particle theory. In J. J. Mintzes, J. H. Wandersee, & J. D. Novak (Eds.), *Teaching science for understanding–A human constructivist view* (pp. 165–194). San Diego, CA: Academic Press.

Osborne, J. (2006). *Communicating Science: A BAI Roundtable Summary*. Retrieved 22 December 2013, from http://cils.exploratorium.edu/pdfs/CILS%20BAI_Osborne_Communicating%20Science.pdf

Osborne, J., Simon, S., & Collins, S. (2003). Attitudes towards science: A review of the literature and its implications. *International Journal of Science Education*, *25*(9), 1049–1079. doi:10.1080/0950069032000032199

Özmen, H. (2004). Some student misconceptions in chemistry: A literature review of chemical bonding. *Journal of Science Education and Technology*, *13*(2), 147–159. doi:10.1023/B:JOST.0000031255.92943.6d

Özmen, H., & Ayas, A. (2003). Students' difficulties in understanding of the conservation of matter in open and closed-system chemical reactions. *Chemistry Education: Research and Practice*, *4*(3), 279–290.

Pacifici, L. B., & Thomson, N. (2011). Undergraduate science research: A comparison of influences and experiences between premed and non–premed students. *CBE Life Sciences Education*, *10*(2), 199–208. doi:10.1187/cbe.11-01-0005 PMID:21633068

Palmer, D. H. (2006). Sources of self-efficacy in a science methods course for primary teacher education students. *Research in Science Education*, *36*(4), 337–353. doi:10.1007/s11165-005-9007-0

Parkin, A. J. (1993). *Memory: Phenomena, experiment and theory*. Hove, UK: Psychology Press.

Pashler, H., McDaniel, M., Rohrer, D., & Bjork, R. (2009). Learning styles: Concepts and evidence. *Psychological Science in the Public Interest*, *9*, 105–119.

Paterson, L. (2003). *Scottish Education in the Twentieth Century*. Edinburgh, UK: Edinburgh University Press.

Patrick, B. C., Hisley, J., & Kempler, T. (2000). "What's everybody so excited about?" The effects of teacher enthusiasm on student intrinsic motivation and vitality. *Journal of Experimental Education*, *68*(3), 217–236. doi:10.1080/00220970009600093

Patrick, H., Anderman, L. H., Ryan, R. M., Edelin, K. C., & Midgley, C. (2001). Teachers' communication of goal orientations in four fifth-grade classrooms. *The Elementary School Journal*, *102*(1), 35–58. doi:10.1086/499692

Patterson, D. W. (1998). *Artificial neural networks: theory and applications*. Prentice Hall PTR.

Pauk, W., & Owens, R. J. Q. (2005). *How to study in college* (11th ed.). Boston, MA: Houghton Mifflin.

Pearson Education. (2014). *eCollege* [Computer software]. Pearson Education, Inc. Retrieved from http://www.eCollege.com

Pellizzaro, A., Welker, G., Scott, D., Solomon, R., Cooper, J., & Farone, A. et al. (2012). Direct laser trapping for measuring the behavior of transfused erythrocytes in a sickle cell anemia patient. *Biomedical Optics Express*, *3*(9), 2190–2199. doi:10.1364/BOE.3.002190 PMID:23024913

Pell, T., & Jarvis, C. (2001). Development of attitude to science scales for use with children of ages from five to eleven years. *International Journal of Science Education*, *23*(8), 847–862. doi:10.1080/09500690010016111

Pendrill, A. M. (2005). Rollercoaster loop shapes. *Physics Education*, *40*(6), 517–521. doi:10.1088/0031-9120/40/6/001

Pendrill, A. M., & Williams, G. (2005). Swings and slides. *Physics Education*, *40*(6), 527–533. doi:10.1088/0031-9120/40/6/003

Perfilieva, I., & Močkoř, J. (1999). *Mathematical principles of fuzzy logic*. Springer.

Perović, S. (2013). The Levels of Integrality in Architecture and Urbanism Studies at the University of Montenegro. *Procedia: Social and Behavioral Sciences*, *93*(93), 654–658. doi:10.1016/j.sbspro.2013.09.256

Perović, S., & Popović, S. (2013). Reflections of Utopia and the Ideal City in the Development of Physical Structure of Nikšić Aspect of Visual Perception. *World Academy of Science, Engineering and Technology*, *79*, 1247–1255.

Perry, W. J. (1999). *Forms of Ethical and Intellectual Development in the College Years: A Scheme*. San Francisco: Jossey-Bass Inc.

Peterson, P. E., & Kaplan, P. (2013). Despite Common Core, States Still Lack Common Standards. *Education Next*, *13*(4), 44–49.

Pettersen, I. A. H., & Williams, G. (2004). Overhead projector doubles as a classroom 'rainbow machine'. *Physics Education*, *39*(6), 463. doi:10.1088/0031-9120/39/6/F02

Planinsic, G., Kos, M., & Jerman, R. (2004). Two-liquid Cartesian diver. *Physics Education*, *39*(1), 58–64. doi:10.1088/0031-9120/39/1/003

Polanyi, M. (1962). Personal knowledge: Towards a post-critical philosophy (Corrected version ed.). Chicago: University of Chicago Press.

Poli, R., Kennedy, J., & Blackwell, T. (2007). Particle swarm optimization. *Swarm Intelligence*, *1*(1), 33–57. doi:10.1007/s11721-007-0002-0

Prawat, R. S. (1992). Teachers' beliefs about teaching and learning: A constructivist perspective. *American Journal of Education*, *100*(3), 354–395. doi:10.1086/444021

Prensky, M. (2006). Listen to the natives. *Educational Leadership*, *63*(4), 8–13.

Project Learning Tree. (2013). *Pre K-8 environmental education activity guide*. Washington, DC: American Forest Foundation.

Project, W. E. T. (2011). *K-12 curriculum and activity guide*. Bozeman, MT: The Watercourse/Project WET International Foundation and the Council for Environmental Education.

Purcell, S., Neale, B., Todd-Brown, K., Thomas, L., Ferreira, M. A. R., & Bender, D. et al. (2007). PLINK: A toolset for whole-genome association and population-based linkage analysis. *American Journal of Human Genetics*, 599. PMID:17701901

Purser, R. (2010). *Problem-Based Learning*. Retrieved October 7, 2013 from: http://online.sfsu.edu/rpurser/revised/pages/problem.htm

QCA. (2000). *Key stage 3 schemes of work*. Qualification and Curriculum Authority.

QCA. (2005). *Science: 2004/5 annual report on curriculum and assessment*. London: Qualifications and Curriculum Authority.

QCA. (2007). *Science: Programme of study for key stage 3 and attainment targets*. London: Qualifications and Curriculum Authority.

Qin, Z., Johnson, D. W., & Johnson, R. T. (1995). Cooperative versus competitive efforts and problem solving. *Review of Educational Research*, *65*(2), 129–143. doi:10.3102/00346543065002129

Raizen, S. A., & Michelsohn, A. M. (Eds.). (1994). *The future of science in elementary schools: Educating prospective teachers*. San Francisco, CA: Jossey-Bass.

Ramadas, J. (2009). Visual and spatial modes in science learning. *International Journal of Science Education*, *31*(3), 301–318. doi:10.1080/09500690802595763

Raviv, D. (2003). Do We Teach Them How to Think? In *Proceedings of the 2002 American Society for Engineering Education Annual Conference and Exposition*. American Society for Engineering Education Session.

Reid, N., & Skryabina, E. A. (2003). Gender and Physics. *International Journal of Science Education*, *25*(4), 509–536. doi:10.1080/0950069022000017270

Reiss, M. R. (2004). Students' attitudes towards science: A long-term perspective. *Mathematics and Technology Education*, *4*(1), 97–109. doi:10.1080/14926150409556599

Renström, L., Andersson, B., & Marton, F. (1990). Students' conceptions of matter. *Journal of Educational Psychology*, *82*(3), 555–569. doi:10.1037/0022-0663.82.3.555

Repko, A. F. (2009). *Assessing Interdisciplinary Learning Outcomes* (Working Paper). School of Urban and Public Affairs, University of Texas at Arlington.

Report of the ASCD Working Group on Humanistic Education. (1978). *Humanistic Education: Objectives and Assessment*. Washington, DC: Association of Supervision and Curriculum Development.

Report of the project on liberal education and the sciences. (1990). *The liberal art of science: Agenda for action*. Washington, DC: American Association for the Advancement of Science.

Resnick, M., Berg, R., & Eisenberg, M. (2000). Beyond Black Boxes: Bringing Transparency and Aesthetics Back to Scientific Investigation. *Journal of the Learning Sciences*, *9*(1), 7–30. doi:10.1207/s15327809jls0901_3

Reynolds, R. G. (1994, February). An introduction to cultural algorithms. In *Proceedings of the third annual conference on evolutionary programming* (pp. 131-139). World Scientific.

Reynolds, R. G., & Sverdlik, W. (1994, June). Problem solving using cultural algorithms. In *Proceedings of the First IEEE Conference onEvolutionary Computation* (pp. 645-650). IEEE.

Rideout, V. J., Foehr, U. G., & Roberts, D. F. (2010). *Generation M 2: Media in the lives of 8-to-18-year-olds*. Menlo Park, CA: The Henry J. Kaiser Family Foundation. Last retrieved July 12, 2011 from http://www.kff.org/entmedia/mh012010pkg.cfm

Rieser, C. J. (2004). *Biologically inspired cognitive radio engine model utilizing distributed genetic algorithms for secure and robust wireless communications and networking*. (Doctoral dissertation). Virginia Polytechnic Institute and State University, Blacksburg, VA.

Riggs, I. M., & Enochs, L. G. (1990). Toward the development of an elementary teacher's science teaching efficacy belief instrument. *Journal of Science Education*, *74*(6), 625–637. doi:10.1002/sce.3730740605

Roberts, G. G. (2002). *SET for success. The supply of people with science, technology, engineering and mathematics skills. The report of Sir Gareth Roberts' Review*. London: HM Treasury.

Robinson, K. (2006). *Do schools kill creativity?* Paper presented at the Technology, Entertainment and Design (TED) Conference. Retrieved September 1, 2013, from http://www.youtube.com/watch?v=iG9CE55wbtY

Rogers, M. (2007). An Inquiry-based Course Using "Physics?" in Cartoons and Movies. *The Physics Teacher*, *45*(1), 38–41. doi:10.1119/1.2409508

Ro, H., & Choi, Y. (2011). Student Team Project: Gender Differences in Team Project Experience and Attitudes Toward Team-Based Work. *Journal of Teaching in Travel & Tourism*, *11*(2), 149–163. doi:10.1080/15313220.2011.575022

Ross, J. A. (1992). Teacher efficacy and the effect of coaching on student achievement. *Canadian Journal of Education*, *17*(1), 51–65. doi:10.2307/1495395

Roveri, N., Falini, G., Sidoti, M. C., Tampieri, A., Landi, E., Sandri, M., & Parma, B. (2003). Biologically inspired growth of hydroxyapatite nanocrystals inside self-assembled collagen fibers. *Materials Science and Engineering C*, *23*(3), 441–446. doi:10.1016/S0928-4931(02)00318-1

Rowe, M. P., Pugh, E. N. Jr, Tyo, J. S., & Engheta, N. (1995). Polarization-difference imaging: A biologically inspired technique for observation through scattering media. *Optics Letters*, *20*(6), 608–610. doi:10.1364/OL.20.000608 PMID:19859271

Rowett, T. (2010). *Paper Clip Top*. Grand Illusions Videos. Retrieved September 1, 2013, from http://www.youtube.com/watch?v=qgSiMHyJcxc

Royal College of Nursing. (2013). *About us*. Retrieved 21 December 2013, from http://www.rcn.org.uk/aboutus

Ruiz, M. J. (2006). Lenz's Law magic trick. *The Physics Teacher*, *44*(2), 96–98. doi:10.1119/1.2165439

Ruthven, K., Howe, C., Mercer, N., Taber, K. S., Luthman, S., Hofmann, R., & Riga, F. (2010). Effecting Principled Improvement in STEM Education: Research-based pedagogical development for student engagement and learning in early secondary-school physical science and mathematics. In M. Joubert & P. Andrews (Eds.), *British Congress of Mathematics Education* (pp. 191-198). British Society for Research into Learning Mathematics.

Ruthven, K., Laborde, C., Leach, J., & Tiberghien, A. (2009). Design Tools in Didactical Research: Instrumenting the Epistemological and Cognitive Aspects of the Design of Teaching Sequences. *Educational Researcher*, *38*(5), 329–342. doi:10.3102/0013189X09338513

Ryhammar, L., & Brolin, C. (1999). Creativity research: Historical considerations and main lines of development. *Scandinavian Journal of Educational Research*, *43*(3), 259–273. doi:10.1080/0031383990430303

Saleh, S. M., Asi, Y. M., & Hamed, K. M. (2013). Effectiveness of integrating case studies in online and face-to-face instruction of pathophysiology: A comparative study. *Advances in Physiology Education*, *37*(2), 201–206. doi:10.1152/advan.00169.2012 PMID:23728138

Sandars, J. (2013). Technology-enhanced learning. *Education for Primary Care*, *24*(4), 300–301. PMID:23906176

Sarle, W. S. (1994). *Neural networks and statistical models*. Academic Press.

Sarquis, J., Hogue, L., Sarquis, M., & Woodward, L. (1997). *Investigating Solids Liquids and Gases with Toys*. Middletown: McGraw-Hill. Miami University.

Sarquis, J., Sarquis, M., & Williams, J. P. (1995). *Teaching chemistry with TOYS: activities for grades K-9. McGraw-Hill*. Learning Triangle Press.

Sarquis, M. (1997). *Exploring Matter with TOYS - Using and understanding the senses*. McGraw-Hill, Inc.

Schauble, L. (1996). The development of scientific reasoning in knowledge-rich contexts. *Developmental Psychology*, *32*(1), 102–119. doi:10.1037/0012-1649.32.1.102

Schlichting, H. J., & Suhr, W. (2010). The buzzer—a novel physical perspective on a classical toy. *European Journal of Physics, 31*(3), 501–510. doi:10.1088/0143-0807/31/3/007

Schmidt, H. G. (1993). Foundations of problem-based learning: Some explanatory notes. *Medical Education, 27*(5), 422–432. doi:10.1111/j.1365-2923.1993.tb00296.x PMID:8208146

Schneider, B., Wallace, J., Blikstein, P., & Pea, R. (2013). Preparing for future learning with a tangible user interface. The case of neuroscience. *IEEE Transactions of Learning Technologies, 6*(2), 117–129. doi:10.1109/TLT.2013.15

Schwartz, M. S., Sadler, P. M., Sonnert, G., & Tai, R. H. (2009). Depth versus breadth: How content coverage in high school science courses relates to later success in college science coursework. *Science Education, 93*(5), 798–826. doi:10.1002/sce.20328

Schwarz, C. V., Reiser, B. J., Davis, E. A., Kenyon, L., Achér, A., & Fortus, D. et al. (2009). Developing a learning progression for scientific modeling: Making scientific modeling accessible and meaningful for learners. *Journal of Research in Science Teaching, 46*(6), 632–654. doi:10.1002/tea.20311

Science and Technology Committee. (2006). *10th Report of Session 2005/2006*. London: Authority of the House of Lords.

Scottish Executive. (2001). *A Science Strategy for Scotland*. Glasgow, UK: Scottish Executive.

Scottish Executive. (2003). *Life Through Learning, Learning Through Life: The Lifelong Learning Strategy for Scotland*. Glasgow, UK: Scottish Executive.

Scottish Funding Council. (2012). *Facts & Figures: The 2012 at a glance guide to the Scottish Funding Council*. Edinburgh, UK: Scottish Funding Council.

Scottish Funding Council. (2013a). *Colleges we fund*. Edinburgh, UK: Scottish Funding Council.

Scottish Funding Council. (2013b). *Infact Database*. Retrieved 21 December 2013, from https://stats.sfc.ac.uk/infact/index.htm

Scottish Funding Council. (2013c). *Learning for All*. Retrieved 21 December 2013, from http://www.sfc.ac.uk/funding/FundingOutcomes/Access/learningforall/LearningforAll.aspx

Scottish Funding Council. (2013d). *SFC Corporate Publication: Annual Report and Accounts 2012-13*. Edinburgh, UK: Scottish Funding Council.

Scottish Government & Scottish Funding Council. (2011). *College Regionalisation: Proposals for implementing Putting Learners at the Centre*. Edinburgh, UK: Scottish Government & Scottish Funding Council.

Scottish Government. (2007). *Skills for Scotland: A Lifelong Skills Strategy*. Edinburgh, UK: Scottish Government.

Scottish Government. (2008). *Science for Scotland*. Edinburgh, UK: Scottish Government.

Scottish Government. (2010). *Science and Engineering 21 - Action Plan for Education for the 21st Century*. Edinburgh, UK: Scottish Government.

Scottish Government. (2011). *Putting Learners at the Centre: Delivering our Ambitions for Post-16 Education*. Edinburgh, UK: Scottish Government.

Scottish Government. (2012). *Supporting Scotland's STEM Education and Culture. Science and Engineering Education Advisory Group (SEEAG). Second Report: January 2012*. Edinburgh, UK: Scottish Government.

Scottish Qualifications Authority. (1993). *National Certificate Module: Unit Specification General Information - Introduction to Chemistry*. Retrieved 22 December 2013, from http://www.sqa.org.uk/files/nu/3181213.pdf

Scottish Qualifications Authority. (2006). *Annual Statistical Report 2005*. Glasgow, UK: Scottish Qualifications Authority.

Scottish Qualifications Authority. (2013a). *About the Scottish Qualifications Authority (SQA)*. Retrieved 21 December 2013, from http://www.sqa.org.uk/sqa/5656.html

Scottish Qualifications Authority. (2013b). *Scottish Credit and Qualifications Framework*. Retrieved 21 December 2013, from http://www.sqa.org.uk/sqa/5659.html

Scottish Qualifications Authority. (2013c). *What We Do*. Retrieved 21 December 2013, from http://www.sqa.org.uk/sqa/5659.html

Seymour, E., & Hewitt, N. M. (1997). *Talking about leaving: Why undergraduates leave the sciences*. Boulder, CO: Westview Press.

Shapiro, B. L. (1996). A case study of change in elementary student teacher thinking during an independent investigation in science: Learning about the "Face of science that does not yet know.". *Science Education*, *80*(5), 535–560. doi:10.1002/(SICI)1098-237X(199609)80:5<535::AID-SCE3>3.0.CO;2-C

Shipstone‡, D. M., Rhöneck, C., Jung, W., Kärrqvist, C., Dupin, J.-J., Johsua, S., & Licht, P. (1988, July). A study of students' understanding of electricity in five European countries. *International Journal of Science Education*, *10*(3), 303–316. doi:10.1080/0950069880100306

Shulman, L. S. (1986). Those who understand: Knowledge growth in teaching. *Educational Researcher*, *15*(2), 4–14. doi:10.3102/0013189X015002004

Shymansky, J. A., Kyle, W. C., & Alport, J. M. (1983). The effects of new science curricula on student performance. *Journal of Research in Science Teaching*, *20*(5), 387–404. doi:10.1002/tea.3660200504

Siegler, R. S. (2006). Microgenetic analyses of learning. In D. Kuhn & R. S. Siegler (Eds.), Handbook of child psychology: Volume 2: Cognition, perception, and language (6th ed., pp. 464-510). Hoboken, NJ: Wiley.

Siegler, R. S. (1996). *Emerging minds: The process of change in children's thinking*. New York: Oxford University Press.

Siegler, R. S. (1998). *Children's thinking* (3rd ed.). Upper Saddle River, NJ: Prentice Hall.

Siegler, R. S. (2007). Cognitive variability. *Developmental Science*, *10*(1), 104–109. doi:10.1111/j.1467-7687.2007.00571.x PMID:17181707

Silk, E., Higashi, R., Shoop, R., & Schunn, C. (2010). Designing technology activities that teach mathematics. *Technology Teacher*, *69*(4), 21–27.

Sills, T. W. (1999). *Science Fun with Toys*. Chicago: Dearborn Resources.

Sirhan, G., & Reid, N. (2001). Preparing the Mind of the Learner - Part 2. *University Chemistry Education, 5*, 52–58.

Slater, T. F. (1996). Portfolio assessment strategies for grading first-year university physics students in the USA. *Physics Education, 31*(5), 329–333. doi:10.1088/0031-9120/31/5/024

Smith, C. L., Solomon, G. E., & Carey, S. (2005). Never getting to zero: Elementary school students' understanding of the infinite divisibility of number and matter. *Cognitive Psychology, 51*(2), 101–140. doi:10.1016/j.cogpsych.2005.03.001 PMID:16081058

Solomon, J. (1992). *Getting to Know about Energy - in School and Society*. London: Falmer Press.

Sousa, D. A. (2001). *How the Brain Learns* (2nd ed.). Thousand Oaks, CA: Corwin Press.

Springer, L., Stanne, M. E., & Donovan, S. S. (1999). Effects of small-group learning on undergraduates in science, mathematics, engineering, and technology: A meta-analysis. *Review of Educational Research, 69*(1), 21–51. doi:10.3102/00346543069001021

Stables, K. (1997). Critical issues to consider when introducing technology education into the curriculum of young learners. *Journal of Technology Education, 8*(2), 50–65.

Starting Point Teaching and Learning Economics. (2010). *Interdisciplinary Approaches in Learning*. Retrieved Sept 20 2013, from: http://serc.carleton.edu/econ/interdisciplinary/why.html

Stead, K. (1985). An Exploration, Using Ajzen and Fishbein's Theory of Reasoned Action, of Students' Intentions to Study or not to Study Science. *Research in Science Education, 15*(1), 76–85. doi:10.1007/BF02356528

Stein, M., Larrabee, T. G., & Barman, C. R. (2008). A study of common beliefs and misconceptions in physical science. *Journal of Elementary Science Education, 20*(2), 1–11. doi:10.1007/BF03173666

STEM Smart Brief. (n.d.). *Preparing and support STEM educators*. Retrieved from http://successfulstemeducation.org/sites/successfulstemeducation.org/files/Preparing%20Supporting%20STEM%20Educators_FINAL.pdf

Stemler, S., & Tsai, J. (2008). Best practices in interrelater reliability: Three common approaches. In J. Osborne (Ed.), *Best practices in quantitative methods* (pp. 29–49). Thousand Oaks, CA: Sage Publications. doi:10.4135/9781412995627.d5

Stevens, R., & Casillas, A. (2006). *Artificial neural networks. Automated Scoring of Complex Tasks in Computer Based Testing: An Introduction* (pp. 259–312). Mahwah, NJ: Lawrence Erlbaum.

Stevens, S. Y., Delgado, C., & Krajcik, J. S. (2010). Developing a hypothetical multidimensional learning progression for the nature of matter. *Journal of Research in Science Teaching, 47*(6), 687–715. doi:10.1002/tea.20324

Subramaniam, R., & Riley, J. P. II. (2008). Physics trick gets students interested. *Physics Education, 43*(4), 355–356. doi:10.1088/0031-9120/43/4/F06

Subramaniam, R., & Tiang, N. H. (2004). Pendulums swing into resonance. *Physics Education, 39*(5), 395. doi:10.1088/0031-9120/39/5/F11

Sumners, C. (1997). *Toys in Space: Exploring Science With The Astronauts*. McGraw-Hill, Blue Ridge Summit.

Sweet, M., & Pelton-Sweet, L. M. (2008). The social foundation of team-based learning: Students accountable to students. *New Directions for Teaching and Learning*, *2008*(116), 29–40. doi:10.1002/tl.331

Taber, K. S. (2013, May). The right medicine for educational research? *Education in Chemistry*.

Taber, K. S. (2001). When the analogy breaks down: Modelling the atom on the solar system. *Physics Education*, *36*(3), 222–226. doi:10.1088/0031-9120/36/3/308

Taber, K. S. (2003). Lost without trace or not brought to mind? - a case study of remembering and forgetting of college science. *Chemistry Education: Research and Practice*, *4*(3), 249–277.

Taber, K. S. (2007). *Enriching School Science for the Gifted Learner*. London: Gatsby Science Enhancement Programme.

Taber, K. S. (2008). Towards a curricular model of the nature of science. *Science & Education*, *17*(2-3), 179–218. doi:10.1007/s11191-006-9056-4

Taber, K. S. (2010). Challenging gifted learners: General principles for science educators; and exemplification in the context of teaching chemistry. *Science Education International*, *21*(1), 5–30.

Taber, K. S. (2011). Constructivism as Educational Theory: Contingency in learning, and optimally guided instruction. In J. Hassaskhah (Ed.), *Educational Theory* (pp. 39–61). New York: Nova.

Taber, K. S. (2013). Revisiting the chemistry triplet: Drawing upon the nature of chemical knowledge and the psychology of learning to inform chemistry education. *Chemistry Education Research and Practice*, *14*(2), 156–168. doi:10.1039/c3rp00012e

Taber, K. S. (Ed.). (2009). *Progressing Science Education: Constructing the scientific research programme into the contingent nature of learning science*. Dordrecht: Springer. doi:10.1007/978-90-481-2431-2

Taber, K. S. (Forthcoming). Student Thinking and Learning in Science: Perspectives on the Nature and Development of Learners'. *Ideas: Routledge*.

Taber, K. S., de Trafford, T., & Quail, T. (2006). Conceptual resources for constructing the concepts of electricity: The role of models, analogies and imagination. *Physics Education*, *41*(155-160).

Tai, R. H., Liu, C. Q., Maltese, A. V., & Fan, X. T. (2006). Planning early for careers in science. *Science*, *312*(5777), 1143–1144. doi:10.1126/science.1128690 PMID:16728620

Tanner, K. D. (2009). Talking to learn: Why biology students should be talking in classrooms and how to make it happen. *CBE Life Sciences Education*, *8*(2), 89–94. doi:10.1187/cbe.09-03-0021 PMID:19487494

Tao, P. K. (1994). Comprehension of Non-Technical Words in Science: The case of students using a 'foreign' language as the medium of instruction. *Research in Science Education*, *24*(1), 322–330. doi:10.1007/BF02356359

Tasker, R., & Dalton, R. (2006). Research into practice: Visualisation of the molecular world using animations. *Chemistry Education Research and Practice*, *7*(2), 141–159. doi:10.1039/b5rp90020d

Taylor, B. A. P., Poth, J., & Portman, D. J. (1995). *Teaching physics with toys: activities for grades K-9*. Learning Triangle Press.

The Association for Science Teacher Education. (2014). *ASTE Position Statement on Technology in Science Teacher Education*. Retrieved from http://theaste.org/about/aste-position-statement-on-technology-in-science-teacher-education/

The Independent. (2000). "Report." Report Produced in Association with UCAS. *University Chemistry Education.*, *3*(1), 1–7.

The National Strategies Secondary. (2008). *Explaining how electric circuits work: Science teaching unit*. Department for Children, Schools and Families.

The Royal Society. (2011). *Preparing for the transfer from school and college science and mathematics education to UK STEM higher education: A 'state of the nation' report*. London: The Royal Society.

*The Simpsons* wiki (2013). Retrieved on 14th June 2013 from http://simpsons.wikia.com/wiki/index.php?search=atom&fulltext=Search

Thiry, H., Laursen, S. L., & Hunter, A. (2011). What experiences help students become scientists? A comparative study of research and other sources of personal and professional gains for STEM undergraduates. *The Journal of Higher Education*, *82*(4), 357–388. doi:10.1353/jhe.2011.0023

Thompson, G., & Mathieson, D. (2001). The Mirror Box. *The Physics Teacher*, *39*(8), 508–509. doi:10.1119/1.1424606

Timmis, J., & Neal, M. (2001). A resource limited artificial immune system for data analysis. *Knowledge-Based Systems*, *14*(3-4), 121–130. doi:10.1016/S0950-7051(01)00088-0

Timmis, J., Neal, M., & Hunt, J. (2000). An artificial immune system for data analysis. *Bio Systems*, *55*(1-3), 143–150. doi:10.1016/S0303-2647(99)00092-1 PMID:10745118

Tinto, V. (1987). *Leaving college: Rethinking the causes and cures of student attrition*. Chicago, IL: University of Chicago Press.

Tobias, S. (1992). *Revitalizing undergraduate science: Why some things work and most don't*. Tucson, AZ: Research Corporation.

Tobin, K., Tippins, D. J., & Gallard, A. J. (1994). Research on instructional strategies for science teaching. In D. L. Gabel (Ed.), *Handbook of research on science teaching and learning* (pp. 45–93). New York, NY: Macmillan Publishing Company.

Tomei, A., Dawson, E., & Dillon, J. (2013). *A study of science, technology, engineering and mathematics education in the United Kingdom, Consultant report: securing Australia's future: STEM: country comparisons*. Melbourne: Australian Council of Learned Academies.

Torrance, E. P. (1979). *The search for satori and creativity*. Buffalo, NY: Creative Education Foundation and Creative Synergetic Associates.

Torrance, E. P. (1990). *Torrance Tests of Creative Thinking*. Beakonville, IL: Scholastic Testing Services.

Torrance, H., & Coultas, J. (2004). *Do summative assessment and testing have a positive or negative effect on post-16 learners' motivation in the learning and skills sector? A review of the research literature on assessment in post-compulsory education in the UK*. London: Learning and Skills Research Centre.

Tosun, T. (2000). The beliefs of pre-service elementary teachers toward science and science teaching. *School Science and Mathematics*, *100*(7), 374–379. doi:10.1111/j.1949-8594.2000.tb18179.x

Treagust, D. F., Chittleborough, G., & Mamiala, T. L. (2002). Students' understanding of the role of scientific models in learning science. *International Journal of Science Education*, *24*(4), 357–368. doi:10.1080/09500690110066485

Trumbo, J. (2006). Making science visible: Visual literacy in science communication. In L. Pauwels (Ed.), *Visual culture of science: Re-thinking representational practices in knowledge building and science communication Hanover* (pp. 266–283). NH: Dartmouth College Press, University Press of New England.

Tschannen-Moran, M., Hoy, A. W., & Hoy, W. K. (1998). Teacher efficacy: Its meaning and measure. *Review of Educational Research*, *68*(2), 202–248. doi:10.3102/00346543068002202

Tsui, C.-Y., & Treagust, D. (2010). Evaluating Secondary Students' Scientific Reasoning in Genetics Using a Two-Tier Diagnostic Instrument. *International Journal of Science Education*, *32*(8), 1073–1098. doi:10.1080/09500690902951429

Tuan, H.-L., Chin, C.-C., & Shieh, S.-H. (2005, January). The development of a questionnaire to measure students' motivation towards science learning. *International Journal of Science Education*, *27*(6), 639–654. doi:10.1080/0950069042000323737

Turner, R. C. (1983). Toys in physics teaching: Cartesian diver. *American Journal of Physics*, *51*(5), 475–476. doi:10.1119/1.13482

*TurnItIn*. (2014). Retrieved from http://www.turnitin.com

Tytler, R., & Osborne, J. (2012). Student attitudes and aspirations towards science. In B. J. Fraser, K. G. Tobin, & C. J. McRobbie (Eds.), *Second international handbook of science education* (pp. 597–625). Dordrecht, The Netherlands: Springer. doi:10.1007/978-1-4020-9041-7_41

Tytler, R., & Peterson, S. (2000). Deconstructing learning in science—Young children's responses to a classroom sequence on evaporation. *Research in Science Education*, *30*(4), 339–355. doi:10.1007/BF02461555

Tytler, R., Prain, V., Hubber, P., & Waldrip, B. G. (Eds.). (2013). *Constructing Representations to Learn in Science*. Rotterdam: Sense Publishers. doi:10.1007/978-94-6209-203-7

Tytler, R., Symington, D., & Smith, C. (2009). A curriculum innovation framework for science, technology and mathematics education. *Research in Science Education*, 1–20.

U.S. Department of Labor, Bureau of Labor Statistics. (2014, December 8). *Occupational outlook handbook, 2014-15 Edition: Kindergarten and Elementary School Teachers*. Retrieved from http://www.bls.gov/ooh/education-training-and-library/kindergarten-and-elementary-school-teachers.htm

Ucke, C. (2002). Professor Sakai's paper-clip tops. *Physics Education (India)*, *19*(2), 97–100.

UK Commission for Employment and Skills. (2013). *How can education and training providers use NOS?* Retrieved 21 December 2013, from http://nos.ukces.org.uk/about-nos/Pages/Education-and-Training-Providers.aspx

UNESCO. (1998). *Transdisciplinarity stimulating synergies, integrating knowledge*. UNESCO. Retrieved from http://unesdoc.unesco.org/images/0011/001146/114694eo.pdf

Venter, G., & Sobieszczanski-Sobieski, J. (2003). Particle swarm optimization. *AIAA Journal*, *41*(8), 1583–1589. doi:10.2514/2.2111

Venville, G., & Donovan, J. (2005, Spring). Searching for clarity to teach the complexity of the gene concept. *Teaching Science*, *51*(3), 20–24.

Venville, G., & Donovan, J. (2007). Developing Year 2 students' theory of biology with the concepts of gene and DNA. *International Journal of Science Education*, *29*(9), 1111–1131. doi:10.1080/09500690600931079

Venville, G., & Donovan, J. (2008). How pupils use a model for abstract concepts in genetics. *Journal of Biological Education*, *43*(1), 6–14. doi:10.1080/00219266.2008.9656143

Verner, C. (1962). *Conceptual scheme for the identification and classification of processes of adult education association*. Washington, DC: Adult Education Association.

Vernon, P. E. (1989). The nature-nurture problem in creativity. In J.A. Glover, R.R. Ronning, & C.R. Reynolds (Eds.), Handbook of creativity: perspectives on individual differences. New York: Plenum Press.

Vertes, R. P. (2004). Memory Consolidation in Sleep. *Neuron*, *44*(1), 135–148. doi:10.1016/j.neuron.2004.08.034 PMID:15450166

Vose, M. D. (1999). *The simple genetic algorithm: foundations and theory* (Vol. 12). The MIT Press.

Vosniadou, S. (2012). Reframing the classical approach to conceptual change: Preconceptions, misconceptions and synthetic models. Second International Handbook of Science Education, 1, 119-130.

Vosniadou, S. (Ed.). (2008). *International Handbook of Research on Conceptual Change*. London: Routledge.

Vygotsky, L. (1978). *Mind in Society*. Cambridge, MA: Harvard University Press.

Vygotsky, L. S., & Cole, M. (1978). *Mind in society: The development of higher psychological processes*. Cambridge, MA: Harvard University Press.

Wang, C.-H., Shannon, D. M., & Ross, M. E. (2013). Students' characteristics, self-regulated learning, technology, self-efficacy, and course outcomes in online learning. *Distance Education*, *34*(3), 302–323. doi:10.1080/01587919.2013.835779

Wang, D., & Allen, M. (2003). Understanding by Design Meets Integrated Science. *Science Teacher (Normal, Ill.)*, *70*(7), 37–41.

Wang, L., & Fu, K. (2008). *Artificial neural networks*. John Wiley & Sons, Inc.

Watts, M. (1983). Some alternative views of energy. *Physics Education*, *18*(5), 213–217. doi:10.1088/0031-9120/18/5/307

Wayne, A. J., & Youngs, P. (2003). Teacher characteristics and student achievement gains: A review. *Review of Educational Research*, *73*(1), 89–122. doi:10.3102/00346543073001089

Weisenthal, J. (2013). Here's the New Ranking of Top Countries in Reading, Science, and Math. *Business Insider*. Accessed February 16, 2014 at http://www.businessinsider.com/pisa-rankings-2013-12

Wenger, E. (1998). Communities of practice: Learning as a social system. *Systems Thinker*, *9*(5), 2–3.

Wenglinsky, H. (2002). The link between teacher classroom practices and student academic performance. *Education Policy Analysis Archives*, *10*(12).

Wenner, G. (1993). Relationship between science knowledge levels and beliefs toward science instruction held by pre-service elementary teachers. *Journal of Science Education and Technology*, *2*(3), 461–468. doi:10.1007/BF00694428

Westerback, M. E. (1982). Studies on attitude toward teaching science and anxiety about teaching science in pre-service elementary teachers. *Journal of Research in Science Teaching*, *19*(7), 603–616. doi:10.1002/tea.3660190710

Westerback, M. E., & Long, M. J. (1990). Science knowledge and the reduction of anxiety about teaching earth science in exemplary teachers as measured by the science teaching state-trait anxiety inventory. *School Science and Mathematics*, *90*(5), 361–374. doi:10.1111/j.1949-8594.1990.tb17225.x

West, T. (1997). *In the Mind's Eye: Visual Thinkers, Gifted People with Dyslexia and Other Learning Difficulties, Computer Images and the Ironies of Creativity*. Amherst, NY: Prometheus Books.

Whitehouse, M. (2002). Supporting Physics Teaching (11-14). *Physics Education*, *37*(5), 363.

White, R. T., & Gunstone, R. F. (1992). *Probing Understanding*. London: Falmer Press.

Whitley, D. (1994). A genetic algorithm tutorial. *Statistics and Computing*, *4*(2), 65–85. doi:10.1007/BF00175354

Wiebe, E. N., Clark, A. C., & Hasse, E. V. (2001). Scientific Visualization: Linking science and technology education through graphic communications. *The Journal of Design and Technology Education*, *6*(1), 40–47.

Wiggins, G., & McTighe, J. (1998). *Understanding by Design*. Alexandria, VA: ASCD.

Wilensky, U., & Resnick, M. (1999). Thinking in Levels: A Dynamic Systems Approach to Making Sense of the World. *Journal of Science Education and Technology*, *8*(1), 3–19. doi:10.1023/A:1009421303064

Wilkinson, E.T., Cheifetz, S., De Grandis S.A. (1995). Development of competitive PCR and the QPCR system 5000 as a transcription-based screen. *PCR Methods Applications*, (6), 363-7.

Williams, K. (2014). Research Institutions: Research-Based Teaching through Technology. In Cases on Research Based Teaching Methods in Science Education. Academic Press.

Willingham, D. T. (2008). What Is Developmentally Appropriate Practice? *American Educator*, *32*(2), 34.

Wiser, M., & Smith, C. L. (2008). Learning and teaching about matter in grades K-8: When should the atomic-molecular theory be introduced. In International handbook of research on conceptual change, (pp. 205–239). Academic Press.

Wolf, T. (2010). Google Apps for Education Users Grow to 10 Million. *TMCnet Education Technology*. Retrieved October 1 2011, from: http://education.tmcnet.com/topics/education/articles/109042-google-apps-education-users grow-10-million.htm

Wolfe, P. (2010). *Brain matters: Translating research into classroom practice*. Alexandria, VA: ASCD.

Woolnough, B. E. (1993). Teachers' perception of reasons students choose for, or against, science and engineering. *The School Science Review*, *75*(270), 112–117.

Wright, N. S., & Drewery, G. P. (2006). Forming cohesion in culturally heterogeneous teams: Differences in Japanese, Pacific Islander and Anglo experiences. *Cross Cultural Management: An International Journal*, *13*(1), 43–53. doi:10.1108/13527600610643475

Yair, Y., & Yair, Y. (2004). "Everything comes to an end": An intuitive rule in physics and mathematics. *Science Education*, *88*(4), 594–609. doi:10.1002/sce.10142

Yang, X., & Zheng, J. (2009). Artificial neural networks. In Handbook of Research on Geoinformatics. Academic Press.

Young, A. (2006). Why young people are turned off science. *The Herald*. Retrieved 20 December 2013, from http://www.heraldscotland.com/sport/spl/aberdeen/why-young-people-are-turned-off-by-science-1.21353

Young, A., & Norgard, C. (2006). Assessing the quality of online courses from the students' perspective. *The Internet and Higher Education*, *9*(2), 107–115. doi:10.1016/j.iheduc.2006.03.001

Yuruk, N. (2011). The predictors of pre-service elementary teachers' anxiety about teaching science. *Journal of Baltic Science Education*, *10*(1), 17–26.

Zadeh, L. A. (1965). Fuzzy sets. *Information and Control*, *8*(3), 338–353. doi:10.1016/S0019-9958(65)90241-X

Zike, D. (2004). Dinah Zike's big book of science for elementary K-6. San Antonio, TX: Dinah-Might Adventures, LP.

Ziman, J. (2002). *Real science: What it is and what it means*. Cambridge, UK: Cambridge University Press.

# About the Contributors

**Eugene de Silva** has been an educator for over 25 years. As a professor of physics and chemistry, he developed the first online college physics programs under the auspices of Walters State Community College and the Board of Regents for the State of Tennessee. He established the National Accrediting Commission for Martial Arts (NACMA)- a registered charity- and Virginia Research Institute (VRI) – a non-profit organization in the United States in 2004. These foundations were set in place to spark the younger generation's interest in furthering their education and research. He has been the President of the Tennessee Science Department Chairs Association since 2008. He is also the Chair of Institute of Physics UK/USA Branch (South Eastern USA). He is an Executive Member of the Tennessee Academy of Sciences. De Silva has written several books including textbooks in physics and chemistry and is an internationally recognized educator with his name entered in the Who's Who in the World, Who's Who in American Education, and Who's Who in America. He developed the first Martial Arts educational degree in the world in 1993 when he was in the UK. He holds the highest Dan grade in martial arts and is the founder of an international charity known as The Society of Martial Arts, UK. He, as a practitioner of martial arts, introduced a syllabus of teaching physics through martial arts in 2007. He introduced Physics Day in the USA, which has been in place since 2005 for high school students where mechanics section of physics is taught through martial arts. He has also won the "Innovative Teaching Award," "Above and Beyond Award," and has received recognition awards from the Tennessee Academy of Sciences, USA and the Institute of Chemistry, Sri Lanka. He is a chartered chemist, chartered scientist, and chartered physicist; he was also elected as a fellow of the Royal Society for the encouragement of Arts, Manufactures, and Commerce in recognition of his outstanding work in the field of education. The World Head of Family Sokeship Council also inducted him to the Hall of Fame in 2001 in Florida, USA. His novel teaching model, " START," is now being introduced in the teaching of science through the Virginia Research Institute. He also holds two world records in breaking cinder blocks on different parts of his body.

* * *

**Judy Alhamisi** is Assistant Professor and Coordinator of Elementary Education at Marygrove College in Detroit, Michigan. She is the coordinator of elementary education at Marygrove. Her primary research focus is teacher preparation, teacher effectiveness, and student achievement. Her recent publications have appeared in the Journal of the National Association for Alternative Certification, Tennessee Educational Leadership Journal, and Educational Facility Planner. In addition, Dr. Alhamisi has presented her research in national, state, and local educational conferences.

**Nazir Amir** is a Lead Teacher in Greenview Secondary School in Singapore. He recently completed his PhD at Nanyang Technological University, and his research explores the hybridization of Physics with Design and Technology. His publications include four peer-reviewed papers published in international journals and a chapter in the book of an international publisher. He was recently conferred the President's Award for Teachers in Singapore.

**Heather L. Barker** has spent nearly a decade teaching at the elementary, middle, and high school levels. Her teaching experience includes single-grade and multi-grade classrooms in the United States as well as American schools located in Taiwan. She has a passion for environmental education and has taught in a variety of informal educational settings such as nature centers and outdoor schools. Heather holds a degree in elementary education, a masters degree in outdoor education, and is currently pursuing a doctorate in mathematics and science education at Middle Tennessee State University. A member of numerous professional associations, Heather has presented at regional and national academic conferences. She plans to be a professor of education and has a passion for mentoring pre-service teachers.

**Tianxing Cai** is a reseracher in the Dan. F Smith Department of Chemical Engineering at Lamar University. Tianxing specialized in the research fields of modeling, simulation and optimization for the industrial operation, process safety, and environment protection. His major research is the development of optimization models(Linear Programming, Quadratic Constraint Programming, Nonlinear Programming, Mixed Integer Programming, Relaxed Mixed Integer Programming, Mixed Integer Quadratic Constraint Programming, Mixed Integer Nonlinear Programming, and Relaxed Mixed Integer Quadratic Constraint Programming) to realize the synthesis of energy and water systems, manufacturing planning and scheduling and plant wide optimization. Besides that, he also involves the software application of

Aspen, HYSYS, ProII, MATLAB, and gPROMS to conduct simulation and optimization for the process design, environment impact reduction, and safety assessment.

**Thomas Cheatham** earned his Bachelor's of Science from Campbellsville College in mathematics and his Masters of Science and PhD in mathematics from the University of Kentucky. Currently, he directs the Tennessee STEM Education Center at Middle Tennessee State University. Prior to that he was Dean of MTSU's College of Basic and Applied Sciences and former Chair of the MTSU Computer Science Department. He has been recognized over the years for his excellent teaching in mathematics and computer science. His current research focuses on mathematics and science education.

**Eugenie de Silva** holds a Bachelors degree and Masters degree in Intelligence Studies with a concentration in Intelligence Analysis from the American Military University. She is a current student at the Harvard University Division of Continuing Education in the final stages of completing her Masters in Liberal Arts with a concentration in Legal Studies. She is also a current PhD student at the University of Leicester in England. She has given eight oral presentations at academic conferences in fields, such as teaching physics and chemistry, online software programs, biometrics, intelligence studies, and Denial and Deception (D&D). Her research is mainly multidisciplinary in nature. She also holds the world record for being the youngest person to graduate with a Bachelors degree in Intelligence Analysis at the age of 14.

**Jennifer Donovan** is a lecturer in Science Education at the University of Southern Queensland, Australia. From a BSc (Botany), she "fell" into teaching bringing skills acquired from teaching guitar for several years into the science classroom. The intended science career never eventuated, as she realised her love of engaging students with science concepts, processes, and skills. Her 14 years of school teaching was followed by 10 years of teaching anything from mosses to muscles and stars to cells at Edith Cowan and Curtin Universities. Following this were three years at the Curriculum Council as the Science Officer for Western Australia during the implementation of a new outcomes-focused curriculum framework. After that, another "fall" into research created a second passion. This culminated in a belated PhD studying the influence of the mass media on primary students' understandings of genes and DNA. Recent research explores teaching atomic theory to primary students.

**Nancy El-Farargy** is an educationalist with scientific and healthcare underpinnings, and has interests in the delivery of science education for organizations and learners. She has delivered a number of presentations at a variety of conferences, and has published a range of articles on learning, teaching and development. She works within the UK National Health Service, and previously, she worked as a lecturer in science and healthcare. She obtained her PhD from the University of Glasgow, and has a range of interests in educational and workforce development.

**Grant Gardner** received a BS in Biological Sciences from Vanderbilt University and an MS in Zoology from North Carolina State University. As a masters student, he developed a love for teaching while serving as a graduate TA and became a community college biology instructor for several years. He became fascinated by research-based instructional strategies during his teaching tenure which led to completion of a PhD in Science Education from North Carolina State University. He currently is an Assistant Professor of Biology at Middle Tennessee State University, serving as faculty in the Math/Science Education PhD program. His research interests include understanding faculty implementation of research-based instructional strategies, student small group negotiation of motivation in classroom contexts, and interdisciplinary themes related to nanotechnology education.

**Blanche Jackson Glimps** is a Professor in the Department of Teaching and Learning in the College of Education at Tennessee State University. She has served as Chairperson of Education Departments and as a Vice President for Academic Affairs at different colleges and universities in the United States. She is involved in the preparation of teachers and teaches primarily special education courses. Among her research and writing interests include the following: spirituality and teaching, culturally responsive teaching, diversity issues, and the overrepresentation of specific ethnic and gender groups in special education classrooms in the United States.

**Carole Haeusler** is a lecturer in Science Education at the University of Southern Queensland, Australia. She began her career in the sciences, gaining a PhD in theoretical chemistry, but decided against a career in pure science, preferring rather to engage with students and teachers in science education. She has extensive teaching experience across a range of education sectors including lecturing in chemistry and science education, teaching secondary school chemistry and science, and consulting with schools in science curriculum and assessment. Her research interests are in the areas of science teachers' education and science curriculum and assessment. Her current research project is investigating the capacity of elementary school children to understand atomic-molecular theory. This work has gained public recognition in Australia through an Australian university and federal government sponsored in an on-line forum called *The Conversation* and was selected to be presented at a symposium in the national capital, Canberra.

**Riikka Hofmann** is a Research and Teaching Associate in the Faculty of Education, University of Cambridge. Her research interests focus on classroom interaction, learning through talk, innovations in teaching, student engagement, and professional learning. She teaches widely on educational and social science research methods and coordinates the Faculty's post-graduate research methods courses. Prior to coming to Cambridge, Riikka studied and worked at the University of Helsinki and LMU Munich. Having completed her PhD in the Faculty in 2008, for the last 5 years Riika has been involved with the ESRC-funded *epiSTEMe* project led by Kenneth Ruthven, Christine Howe, Neil Mercer, and Keith Taber.

**Jeffrey Horner** has been the Dean and Professor of Natural Sciences at Walters State Community College since 1995. He received his doctorate in Educational Leadership and Policy Analysis from East Tennessee State University in 2005. As the Dean of Natural Sciences at Walters State, he also received several academic awards in recognition of his contributions to the field, such as the WSCC Faculty of the Year and the "WOW" award for outstanding work in higher education. He has also presented many papers at domestic and international conferences in which he has focused on the necessity and importance of mobilization in education, in addition to discussing the challenge-based pedagogy. Since 2013, Dr. Horner has also been the consultant for the Tennessee Board of Regents (TBR) Office of Emerging Technologies and Mobilization. He was also the WSCC Fellow of TBR Teacher Preparation Partnership Project funded by the NSF for a three-year grant of about 1.25 million. He is currently an academic auditor for TBR and a member of the Chancellor's Executive Council on Mobilization, TBR Mobile Executive Council, Regents Academic Leadership Institution, Leadership WSCC program, WSCC Executive Council on Mobilization, and WSCC Mobilization Team. Additionally, Dr. Horner also acts as a Merlot Biology Reviewer.

**Christine Howe** is a Professor of Education at the University of Cambridge, and Fellow of Lucy Cavendish College. She is a psychologist whose main research interests are children's reasoning in science and mathematics, peer interaction and conceptual growth, and communication and social relations amongst children. She has published seven books and over 150 journal articles and book chapters. Christine has edited three academic journals and served on many editorial boards. She has been elected to the Academy of Social Sciences, and holds an Associate Fellowship of the British Psychological Society, whose Developmental Section she has chaired. She has contributed to research administration and policy making at the local, national and international levels.

**Latoya N. Johnson** received her PhD from the Interdisciplinary Program in Molecular and Cellular Biology from Tulane University in August 2006. She completed her postdoctoral research appointment at Emory University School of Medicine's Alcohol and Lung Biology Program. She worked as a Science Writer/Editor with the National Home Office of the American Cancer Society, as a Program Coordinator in the Dean's Office of the School of Medicine, and as an Adjunct Faculty Member for Grand Canyon University and Georgia Gwinnett College. Dr. Johnson presently works in the College of Health Sciences at Walden University teaching Public Health Biology and in the General Studies department at Beulah Heights University teaching Principles of Science.

**Pamela L. Knox** is Associate Vice Chancellor for Academic Affairs for the Tennessee Board of Regents. As the Associate Vice Chancellor, her system-wide responsibilities include leadership of initiatives in academics, research, and internationalization. Prior to joining the Board of Regents, she held faculty and administrative responsibilities at Middle Tennessee State University, Tennessee State University, Idaho State University, Oklahoma State University, University of Kansas, and Penn State University. Dr. Knox received her PhD in Psychology (Counseling Psychology) in 1984 from Virginia Commonwealth University after completing her internship at The University of Texas –Austin. She has published numerous professional journal articles and book chapters presenting extensively nationally and internationally. She is a tenured Professor at Tennessee State University and a Licensed Psychologist (Health Services Provider) in Tennessee.

**Despo Ktoridou** is currently an Associate Professor and Head of Management and the MIS Department at the University of Nicosia in Cyprus. Dr. Ktoridou holds a BSc (1991) and an MSc (1993) In Computer Engineering and a PhD (2000) in the field of Expert Systems from Saint Petersburg State Electrotechnical University in Russia. Dr. Ktoridou has worked as a Senior Computer Engineer for different organizations in Cyprus (1992 – 1999), from 2000 – 2007, as an Assistant Professor of Educational Technology and currently as an Associate Professor of MIS at the University of Nicosia. Dr. Ktoridou's research focuses on areas of ICT-Information Communication Technologies application in education, Innovation, and Technology Management in Education and Business and Innovative Teaching Learning Pedagogies in Higher Education. Dr. Ktoridou has presented papers in numerous refereed international conferences and has published several papers in refereed journals. Dr. Ktoridou participated in EU and local funded programs and has been invited by foreign universities as a guest lecturer.

**Stefanie Luthman** holds an MSc and PhD in psychology from the University of Kiel, Germany. She has extensive experience in conducting survey studies and experiments in health psychology, media psychology, and education psychology. In the epiSTEMe project at the Faculty of Education in Cambridge, Stefanie investigated the effect of a teaching intervention on students' attainment and attitudes towards math and science. Prior to her doctoral studies, Stefanie worked as a researcher in a project at the University of Lübeck, Germany, where she examined the effect of a public-health campaign on general practitioners' attitudes towards screening and minimizing patients' alcohol consumption. In her current position at Quantify Research in Stockholm, Stefanie has developed expertise in the field of breast cancer prognosis, economic evaluations of Alzheimer's Disease treatments and cost-effectiveness studies of the prevention of domestic violence.

**Neil Mercer** is a Professor of Education at the University of Cambridge, where he is also the Chair of the Psychology and Education Group and Vice-President of the college Hughes Hall. Previously, he was Professor of Language and Communications at Open University. He is a psychologist with a particular interest in the development of children's language and reasoning and in the effective use of talk for learning in school. His research with colleagues has generated the *Thinking Together* practical approach to classroom pedagogy, and he has worked extensively and internationally with teachers, researchers, and educational policy makers on improving talk for learning in schools. His most recent books are *Exploring Talk in School* (with Steve Hodgkinson), *Dialogue and the Development of Children's Thinking* and *Interthinking: putting talk to work* (both with Karen Littleton).

**Donald Nelson** is a Professor of Mathematics and Chair of the Department of Mathematical Sciences at Middle Tennessee State University. He earned his undergraduate degree in mathematics from Mississippi College and a Masters of Arts and PhD from Vanderbilt University. His current research is in graph theory.

**Chukwunyere E. Okezie** is an Associate Professor of Education at Marygrove College in Detroit, Michigan. He received his doctorate from the University of Pittsburgh, with a concentration in social and comparative analysis in higher education. He is the former Chair of the Education Department at Marygrove College. He is also Coordinator of Secondary Education and the Griot Graduate Programs. His current research centers on culturally responsive teaching, Understanding by Design, African American males, and career changers.

**Svetlana Perovic**, born in Niksic, Montenegro,1978. Completed her elementary and high school in Niksic with great success. She graduated from the University of Novi Sad, Faculty of Technical Sciences, Department of Architecture and Civil Engineering, with an average grade of 9.00. Her graduation thesis was graded with a 10 (ten). At the same University, she completed an Academic Masters studies program and was awarded the title Master of Science (M.Sc.) in Architecture. Currently, she is a PhD student at the University of Novi Sad, in the field of "Modern Architecture and Urbanism." Her Phd thesis is titled: Transdisciplinary research paradigm in sustainable development of the physical structure of the city. During 2007, was employed as a designer at the company "Reality" d.o.o. (Association for design, construction, and investment) in Podgorica. She has been a Teaching Assistant at the Faculty of Architecture in Podgorica since 2005. She has about 30 major original projects realized in the field of architectural design and urbanism. She is actively engaged in scientific research in the field of Architecture and Urbanism.

**Fran Riga** has taught science and mathematics in secondary schools in South Africa, Greece, and England. She is currently completing doctoral research in student learning in astronomy, and has been a research assistant in the Faculty of Education at the University of Cambridge where she has worked on projects such as: ASCEND (Able Students Collectively Exploring New Demands), Assessment in Secondary Schools, epiSTEMe, Pathway, and CamTalk. She has also contributed to academic publications associated with these projects. Her particular areas of interest are conceptual development and inquiry-based science education.

**Ginger Holmes Rowell** is Professor of Mathematics at Middle Tennessee State University where she primarily teaches graduate and undergraduate statistics courses. She earned a Bachelors of Science Degree in Mathematics with a concentration in Secondary Education from Birmingham Southern college. Her Masters of Arts in Mathematics and her PhD in Applied Mathematics are both from the University of Alabama in Huntsville. Her current research interest are in statistics education.

**Kenneth Ruthven** after teaching in schools in Scotland and England, Ken(neth) Ruthven joined the Faculty of Education at the University of Cambridge, where he is now Professor of Education. His research focuses on curriculum, pedagogy and assessment, especially in school mathematics and science, and particularly in the light of technology change. Information about his research and other professional activities can be found at http://ww.educ.cam.ac.uk/people/staff/ruthven/.

**Brittany Smith** is a graduate student in Middle Tennessee State University's Mathematics and Science Education PhD program, with a concentration in Mathematics Education. She earned a Masters of Science and a Bachelors of Science both in mathematics from MTSU. Ms. Smith is a graduate research assistant studying factors that affect retention of STEM majors with a weak math background as part of MTSU's National Science Foundation STEM Talent Expansion Project.

**Cindi Smith-Walters** is a biology professor at Middle Tennessee State University and co-directs their Center for Environmental Studies. Holding degrees in biology, curriculum and instruction, and environmental science, she has worked with thousands of young people, teachers, pre-service teachers and non-formal educators. A prolific grant-writer, author, and presenter, she is active in state and national committees and is a past president and Fellow of the Tennessee Academy of Science. Cindi has served on National Science Teachers Association committees including the Childrens' Book Council and the *Journal of College Science Teaching*. Her latest book chapters include 'Using Field Guides with your Child,' and 'Diversity of Life.' Her awards include the MTSU Foundation Award for Excellence in Teaching and the Excellence in Public Service Award (twice), the Presidents' Environmental Challenge Award received in a White House Ceremony, the Tennessee Environmental Education Association's Distinguished Service Award, and as a National Project Learning Tree Outstanding Educator.

**D. Christopher Stephens** is a Professor of Mathematics at Middle Tennessee State University. He earned his Bachelor of Science degree in mathematics from Cumberland College in Kentucky. His Masters of Science and PhD were earned at Vanderbilt University. His research and publications are in graph theory.

**R. Subramaniam** is an Associate Professor at the National Institute of Education in Nanyang Technological University. He has several research interests, with the principal ones being science education and science communication. He has over 100 publications, and these include papers in peer-reviewed international journals, refereed chapters in books of international publishers, and books by international publishers.

**Keith S. Taber** is a Reader in Science Education at the University of Cambridge, where he is currently chair of the Science, Technology, and Mathematics Education Academic Group. After graduating in chemistry, Keith Taber trained as a teacher of chemistry and physics. Whilst teaching in English comprehensive schools and a further education college, he studied part-time and obtained an MSc (with a dissertation project on girls' underrepresentation in physics classes) and PhD (with a thesis project on students' conceptual understanding in chemistry). When he first moved to the Faculty of Education at Cambridge he was heavily involved in initial teacher education, but currently teaches educational research methods to graduate students. Dr Taber was seconded to the Royal Society of Chemistry for a year (2000-2001) as its Teacher Fellow and is the editor of the journal *Chemistry Education Research and Practice*.

**Elaine Bouldin Tenpenny** is an Associate Professor of Mathematics at Middle Tennessee State University (MTSU). She earned an undergraduate degree in Mathematics from David Lipscomb University and a Masters of Science in Mathematics from MTSU. She was recognized for her excellent teaching as the 1996 MTSU Teacher of the Year.

**Dana-Marie Thomas** earned a PhD in Public Policy and Administration with an emphasis in Epidemiology and Community Health from Virginia Commonwealth University. Dr. Thomas completed a postdoctoral fellowship in Nutrition Sciences at the University of Alabama at Birmingham (UAB) where her research focused on bio-behavioral factors that influence obesity-related traits, nutrigenomics, and cancer prevention and control. While at UAB, Dana-Marie was also a Scholar in the Health Disparities Training Program funded by the Morehouse School of Medicine and at UAB, by the Comprehensive Cancer Center and Minority Health and Research Center. During this training, Dr. Thomas examined the genetic contribution of taste as a potential marker of food preferences and micronutrient intake in women at high risk for cervical cancer. Dr. Thomas was also a postdoctoral trainee in the Department of Preventive Medicine/Vascular Biology and Hypertension Program supported by the NIH NRSA Institutional Training Grant. Dr. Thomas has expertise in healthcare policy analysis, genetic and biobehavioral markers of dietary compliance, biopsycho-sociocultural factors affecting weight related behaviors and cognitions, women's health, and body image and quality of life. She holds a Masters in Tourism from Temple University with an emphasis in global health and a BS in Marketing from Morgan State University. Dr. Thomas has over 15 years of experience in human resource management and health policy. Previously, she served as Chief of Applied Research and Health Policy at the Transformational Development Consortium, LLC. Dr. Thomas has published in peer-reviewed academic journals specializing in public health. She has taught courses in public administration and health sciences at the graduate level and health administration at the undergraduate level. Currently, Dr. Thomas is an administrator in the School of Public Policy and Administration at Walden University. She is responsible for academic governance and the maintenance and enhancement of quality assurance and standards across academic programs.

**Kristi L. Walters** is currently a Masters Candidate in Biology at East Carolina University in Greenville, NC. She expects to graduate in May 2014. Her thesis examines the significance of undergraduate students' motivation to learn biology when designing collaborative learning groups and the impact on the students' views and attitudes of science, perceptions of biology and biologists, and knowledge of biology. She has a Bachelors degree in Marine Biology from Roger Williams University in Bristol, RI and has taught natural science at aquariums, environmental education camps, and nature centers along the Atlantic coast for almost two decades. After she completes her degree, she hopes to teach biology at either the high school or college level. She resides in North Carolina with her extremely patient husband, two wonderful children, two spoiled dogs, a naughty cat, and a rather boring goldfish.

**K. Y. Williams** received his BS in Chemistry from Dillard University in his hometown of New Orleans, and one of his MS degrees in Biochemistry and Biophysics from Rensselaer Polytechnic Institute. His PhD research focused on Computational Models of Phytoestrogens, Mycoestrogens, and Diethylstilbestrol Derivates as Estrogens. His research earned him his Doctorate degree from Tulane University, also located in his hometown. He joined the University of Alabama at Birmingham Section on Statistical Genetics in June 2009. Where his research interests focused on genetic inheritance of various disorders, protein structure prediction, and drug development from environmental compounds. Dr. Williams joined the US Food and Drug Administration (US FDA) as an ORISE Fellow in September 2011 at the same time he became a faculty member at Walden University. His work within the US FDA focused on policy and genomics. In January 2013, Dr. Williams also became faculty at Kaplan University before heading to the Department of Defense – Defense Threat Reduction Agency.

**Jennifer Yantz** earned a Bachelors of Science and a Masters of Science in Mathematics from Middle Tennessee State University (MTSU). She also earned a PhD in Mathematics Education from MTSU. She was the 2013 Outstanding Doctoral Student in Mathematics and Science Education at MTSU. Her research is about student perceptions of the connections between arithmetic and algebra. She will be an Assistant Professor of Mathematics at Austin Peay State University in 2014.

# Index

## U



## V



## W

www.ingramcontent.com/pod-product-compliance
Lightning Source LLC
LaVergne TN
LVHW080610300926
849162LV00005B/1102
*9781466663756*